GW01607206

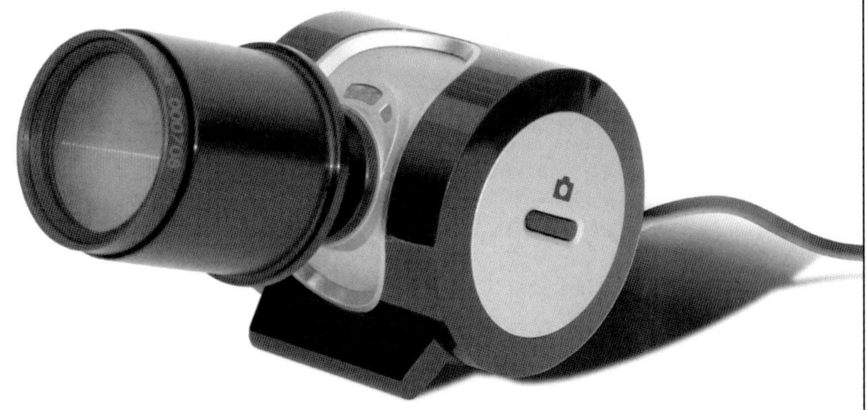

INTRODUCTION TO WEBCAM ASTROPHOTOGRAPHY

Imaging the Universe with the amazing, affordable webcam

Robert Reeves

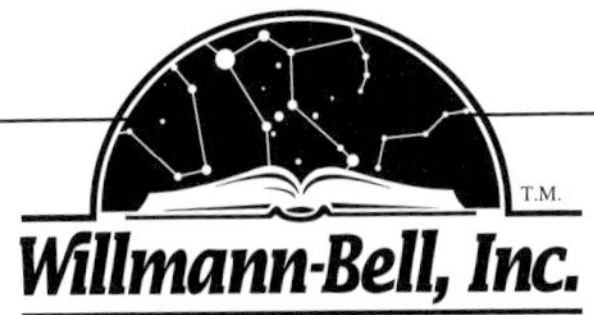

P.O. Box 35025 • Richmond, VA 23235 • TOLL FREE 1 (800) 825-7827 • (804) 320-7016 • FAX (804) 272-5920
www.willbell.com

Published by Willmann-Bell, Inc., P.O. Box 35025, Richmond, Virginia 23235

Printed in the United States of America

Library of Congress Cataloging in Publication Data

Reeves, Robert, 1946-
Introduction to webcam astrophotography : imaging the universe with the amazing affordable webcam / Robert Reeves.
p. cm.
Includes bibliographical references and index.
ISBN: 0943396867 (alk. paper)
1. Webcams. 2. Astronomical photography--Equipment and supplies.
3. Imaging systems in astronomy. I. Title

QB121.5 .R44 2006
522'.63--dc22 2006042805

06 07 08 09 1011 9 8 7 6 5 4 3 2 1

Table of Contents

Acknowledgements

In a field as technically vast as astrophotography, no single person can be considered an expert in all areas. Although I have now written three books on this subject, *Wide-Field Astrophotography*, *Introduction to Digital Astrophotography*, and now *Introduction to Webcam Astrophotography*, I still do not consider myself an expert in celestial imaging—I'm merely the messenger that spreads an accumulated wealth of knowledge that is increased piece by piece by many individuals from around the world.

I have fun shooting images of the sky, and I often produce results I am proud of, but on many occasions I've had to seek the advice of others. Fortunately, a global astronomical "brotherhood" now exists where amateurs from all over the world are in constant contact via Internet and email. Spread throughout the world, many mail lists, and Internet portals this "brotherhood" numbers thousands of individuals with a common interest in astronomy or celestial imaging. It has been my experience that within this worldwide gathering of astronomers no sincere question goes unanswered for long.

In the writing of this book I gratefully received help from many members of the astronomical community who resided in four separate continents, yet we all spoke the common language of astronomy, In alphabetical order by first name, I want to acknowledge the following people:

Adrian New for use of his LPI camera and help with Meade and JMI electric focusers. Art Kunkle of Philips Electronics for providing me with an advance version of the Philips SPC900NC webcam. Cor Berrevoets for help with the wonderful RegiStax video-processing program. Craig Stark for help with SAC and Orion cameras. Doug Anderson for use of his Shoestring Astronomy webcam autoguider interface. George Cushing for help with older webcams. Jim Ferreira for help with the Atik Instruments filter wheel. John Cordiale at Adirondak Video Astronomy for use of an Atik ATK-2HS, focal reducer and filters, and help with the PADC. Jordan Blessing at Scopetronix for help with the ToUcam 840, telescope adapter, and IR filter. Martin Burri for help with the WcCtrl camera control program. Mary Reeves, my wife, for her blessing and support while writing this book. Matt Taylor for help with the Outback Cooler used on the Meade

DSI cameras. Matthias Meijer for early help "dissecting" the Philips SPC900NC camera. Pedro Pereira at BrightStar Telescopes in Portugal for use of an SPC900NC adapter. Perry Remaklus at Willmann-Bell, Inc. for "pushing" me into this area of astrophotography. Peter Katreniak for help with the K3CCD Tools camera control program. Rick Sage at Lumicon for help with Lumicon Multiple Filter Selector. Robin Leadbeater for help with the Star Analyzer spectrograph. Steven Mogg for help with camera adapters, focal reducers, and camera lens adapters.

The collective wisdom of two Yahoo Groups mail lists was invaluable in the preparation of this book. The ToUcam list and the QuickCam and Unconventional Imaging Astronomy Group (known by the acronym QCUIAG) list have a combined membership of over 8000 astronomers worldwide. Beginners and experts alike find quick answers to questions and fast help for problems from the friendly and dedicated list members. Anyone entering webcam astrophotography is encouraged to join these mail lists.[1]

Photo contributors to this book are Steve Barkes, Martin Burri, Russell Croman, Jim Ferreira, Steve Foster, Frank Hinson, Robin Leadbeater, Paul Maxson, Matthias Meijer, Damian Peach, Zac Pujic, Ashley Roecklein, Glenn Schaeffer, Craig Stark, Dave Street, Tim Tasto, Matt Taylor, Jim Thommes, and Jason Ware.

In conclusion, I would like to make a special note of my two highly talented editors, Jeff Kanipe and John D. Koester, and a useful website hosted by Jan Timmermans.[2] Although none of Jan's photographs appear in this book, his influence permeates it. Timmerman's website was an invaluable resource filled with guidance, answers, and inspiration for me when I began my webcam astrophotography adventures. I therefore recommend that anyone entering this field spend some time browsing his extensive list of tutorials, examples, and demonstrations of "real world" webcam astrophotography.

[1] http://groups.yahoo.com/group/ToUcam/ and http://groups.yahoo.com/group/QCUIAG/
[2] http://www.madpc.net/~firmament/

Preface

In a span of less than five years, webcam astrophotography has exploded onto the astronomy scene. It has evolved from short exposure six-bit black-and-white imagery into long-exposure full-color 16-bit per channel imagery of such quality that it rivals "conventional" means of astrophotography. Indeed, webcams have become the method of choice for planetary imaging. This is the equivalent of compressing the entire 150-year history of film astrophotography, from the daguerreotype to hypered Technical Pan film, into a span of five years. Such incredible progress will surely continue as new developments in webcam-derived astrophotography appear constantly. The learning process is never-ending and it is amazing what can be done with such an affordable device.

My objective here is to provide a comprehensive introduction to the subject and not a step-by-step manual that will instantly become out of date. While I make frequent reference to specific cameras and software products I do so within the framework of how they work. I firmly believe that anybody who reads this book today or five years from now will smoothly move into the technology of that instant no matter how it has evolved—which it most certainly will. The tools for imaging have changed, but not the principles. I say this with the conviction of a "geezer" who has been at astronomical imaging for over 45 years and has seen and participated in the evolution of film to digital imaging. This concept was the same over three quarters of a century ago when master photographer Ansel Adams produced his classic photographs. Not once did he mark and publish the location of his tripod holes. Rather, he dealt with principles, techniques, and the need for preparation. His advice had nothing to do with equipment, thus 75 years later his words still guide us.

The message of this book is that you too can participate in the revolution without spending very much money. You do not need to invest $10,000 in a CCD camera, telescope and software. A basic webcam costs about the same as a "so-so" eyepiece. Software that will get you going is free. If you have the telescope (practically any telescope that will track) and a computer you are ready. This book is therefore a portal to show the newcomer what has been accomplished so far and to guide them toward sources of recent information and techniques which appear so quickly that nobody can capture the latest within the pages of a book.

There have been hundreds of different brands and styles of webcams on the market over the past half-decade. A majority of them use a complementary metal oxide semiconductor (CMOS) imaging sensor, a device that has proven inferior for webcam astrophotography; but others use charge-coupled device (CCD) sensors that have proven their worth in this application. One of these is the Philips brand of CCD-based webcams. My experience is limited to this brand and I discuss its models at length in this text, but good results are also achieved with Logitech CCD-based webcams.

The use of webcams for lunar and planetary imaging actually precedes the recent explosion in the use of "conventional" digital cameras for celestial photography. Thought of as insensitive, low-resolution imaging devices by most people, webcams have become a surprisingly powerful tool in astrophotography. Creative modifications to the electronics of inexpensive webcams and the development of camera control and image processing software that utilizes the streaming video output of these cameras have made them a force to be reckoned with. Indeed, today's webcams produce lunar and planetary images that are rivaled only by the magnificent images returned by NASA space probes.

While webcams have proven to be capable performers for bright solar system objects, they traditionally lacked the long-exposure capability needed for deep-sky imaging. This changed in 1999 when Dave Allman discovered that by clipping a single wire in the Connectrix (now Logitech) black-and-white webcam, it could be used for long exposures beyond the limits of its normal software. Steve Chambers followed the innovation path in 2001 by developing modifications to the low-cost CMOS Trust SpaceC@m and Philips CCD-based Vesta webcams. From these humble beginnings fostered by the tinkering of amateur astronomers, an industry has sprung up that has produced the capable SAC, Atik, and Artemis cameras as well as the Celestron, Meade, and Orion webcam-style planetary and deep-sky imaging devices. Legions of do-it-yourself electronics buffs—in the tradition of the *Cookbook Camera* of the early 1990s—have modified their own Philips Vesta, ToUcam, and SPC900NC cameras as well as the Logitech 4000 series of cameras. The introduction of Luminera high-performance low-noise USB 2.0 cameras has opened even more opportunities for the celestial photographer. In the blink of an eye the field of webcam imaging has blossomed from non-existence to the latest hot item in celestial imaging.

The success of adapting low-cost webcams to various forms of astrophotography has opened celestial imaging to many who otherwise could

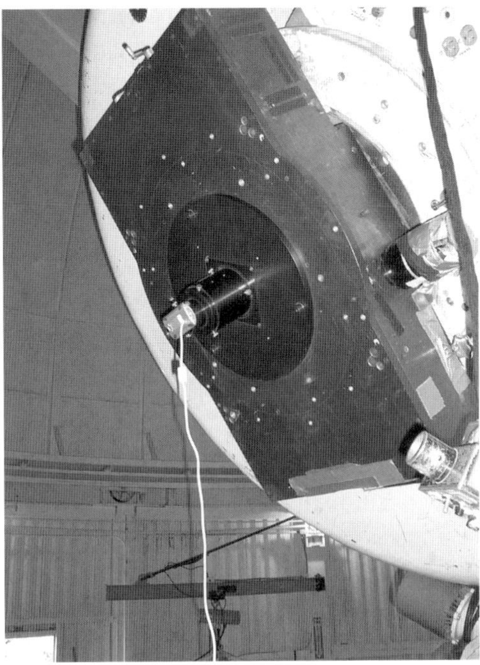

Inspired by the webcam planetary work performed by amateurs, successful experiments in planetary imaging using a ToUcam on the Lunar and Planetary Laboratory's 61-inch telescope have led to the application of webcam-style cameras for planetary research. Photo by Rick Hill.

not afford it or had telescopic equipment not suited for use with heavier conventional film or digital cameras. Just within the past several years there has been a complete paradigm shift in the nature of how high-resolution lunar and planetary images are achieved. Traditional approaches using film cameras to capture the Moon and planets were rendered obsolete. The essentially free method of electronic image capture (after initial equipment purchase) has liberated the photographer from the cost and time burdens of photographic film and developing. This allows more extensive and frequent imaging leading to a noticeable amateur upsurge in high-quality lunar and planetary observing.

Paralleling the developments in webcam long-exposure capabilities are software packages that allow control of modified cameras and processing of their output into amazingly detailed images. Programs like K3CCDTools authored by Peter Katreniak allow easier and more precise astronomical control of the webcam than possible with the programs bundled by the manufacturer. Astrovideo by COAA and Vega by Colin Bowness allow integration of large numbers of short exposures into high-depth 32-bit Fits files simulating long exposures. RegiStax, created by Cor Berrevoets, and Astro-Stack, authored by Robert J. Stekelenburg, allow the stacking of hundreds of lunar and planetary video frames into a single high-resolution image that out-resolves any single film or digital camera exposure. Camera control programs like K3CCDTools and commercial packages available with the SAC, Atik, Artemis, Meade, Celestron, and Orion cameras also allow modified cameras to take unlimited-length exposures.

Performing real astronomical science is now a possibility using both standard and long-exposure modified webcams. The technicalities of such activity are beyond the scope of this book, but anybody who wishes to apply their camera to such activity should know that modified webcams are very capable of such observations. The procedures for testing a CCD to calibrate it for science acquisition and the techniques of performing spectroscopy (analyzing the spectrum of a celestial object), astrometery (measuring the exact position of an astronomical object), and photometery (measuring the brightness of an object) are well documented in *The Handbook of Astronomical Image Processing with AIP4Win Software*, an excellent book and software package by Richard Berry and James Burnell. Readers interested these subjects will find Chapters 6 through 11 in the second edition (2005) of HAIP to be of interest. A plus is the included *AIP4Win* software that does both scientific and "pretty picture" image processing without compromise.

Although modified webcam deep-sky imaging is not as sensitive as dedicated CCD astrocameras, the low cost and simple operation of the webcam is drawing enthusiastic legions to this new form of astrophotography. For minimal investment, you can image the heavens with a webcam and share your results over the Internet with thousands of like-minded astronomers.

Regardless of how you apply a webcam to astrophotography, you will derive a number of benefits. Working with them has been accurately described as interesting, challenging, and fulfilling. They are capable of producing beautiful astrophotos that create a lasting record of your astronomical experience. Webcam astrophotography is an area where any enthusiast can, with relatively little investment, join an ever-increasing army of highly capable imagers in exploring an exciting new means of capturing the beauties of the sky and preserving them for all to enjoy. The pages that follow will guide the reader into this fascinating topic and allow them to become a participant in this latest wave of astrophotography progress.

Master Webcam Imagers

I began my telescopic astrophotography adventures in 1961 and over the past four and a half decades I have pursued the ultimate lunar and planetary image. I have many that I am proud of, but all pale in comparison to those captured by the "masters" of this art. In the preparation of this book I was privileged to be in contact with some of the finest amateur lunar and planetary astrophotographers. In my opinion, the work of Damian Peach from the United Kingdom and Zac Pujic from Brisbane, Australia, stands out from the rest. Pujic and Peach both use the same webcam equipment available to everyone and their telescopes are modest in size, ranging from 9.25-inch to 14-inch aperture. By careful application of webcam astrophotography techniques and seeking the most stable seeing conditions, they routinely produce images of solar system objects that are not only beautiful, but are of such detail that they have genuine scientific value. The image gallery displayed here presents the work of both these astrophotographers as an inspiration to all of us who enjoy celestial photography.

Damian and Zac capture incredible images of both lunar and planetary targets, but in this gallery I present Damian's planetary work and Zac's lunar images. Zac uses a homemade 12-inch reflector featuring a mirror crafted by Australian Mark Suchting. The mirror rests in a Novak cell inside a Parks fiberglass tube and is mounted on a Meade DS-16 equatorial head. To achieve the high *f*-ratio needed for high-resolution solar system imaging, Zac handcrafted a projection adapter to house a large 9-mm Nagler eyepiece (see Figure Gallery 8) and a Philips ToUcam 840. For his astrophotography adventures, Damian now chooses between 9.25-, 11-, or 14-inch Celestron Schmidt-Cassegrain telescopes. The secret to Damian's incredibly detailed planetary images is his quest for the best seeing conditions. He achieves this through frequent trips from his home in the village of Loudwater in Buckinghamshire, England, to the Canary Islands which enjoy some of the best astronomical seeing in the world. Recently, Peach has switched from a ToUcam 840 to a high-speed black-and-white Luminera LU075M USB 2.0 camera.

As you browse Damian and Zac's websites,[1] you will notice that in

[1] At http://www.damianpeach.com and http://users.bigpond.net.au/metaplace/home.html, respectively.

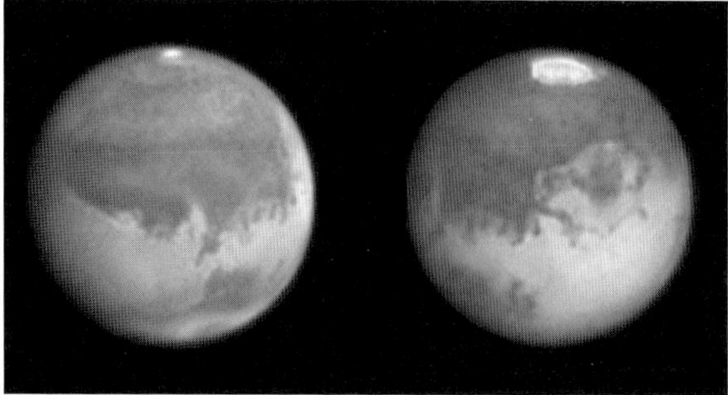

Figure Gallery 1 *A Luminera LU075M camera and a Celestron-14 operating at f/39 were used to capture the left image of Mars on October 16, 2005.The image is centered on Mare Erythraeum. On the right, Solis Lacus, the "Eye of Mars," is the prominent feature captured on August 22, 2003, with a 10-inch Schmidt-Cassegrain telescope at f/55 and a ToUcam 840. Photos by Damian Peach.*

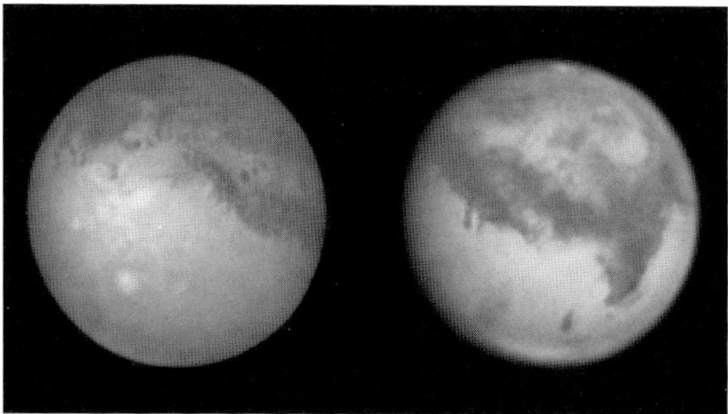

Figure Gallery 2 *(Left) The extinct volcano Olympus Mons is visible as a bright spot recorded on November 6, 2005. (Right) Taken on October 26, 2005, the image displays the prominent Syrtis Major region. Celestron-14 at f/40 using a Luminera LU075M camera. Photos by Damian Peach.*

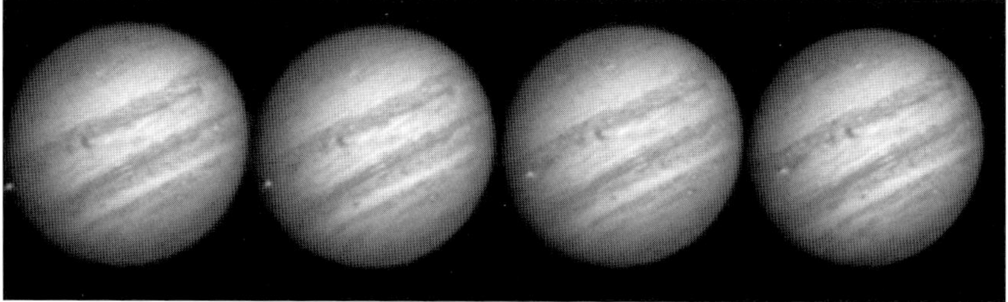

Figure Gallery 3 *The May 16, 2004 transit of Io across the face of Jupiter. Celestron-11 Schmidt-Cassegrain telescope at f/31 using an Atik 1HS camera. Photos by Damian Peach.*

addition to the "landmark" lunar features displayed on these pages, there are numerous obscure and previously ignored features and craters. This is because the tremendous detail-gathering power of webcam astrophotography now allows imagers to record the smaller lunar features that have been out of reach when using conventional imaging methods.

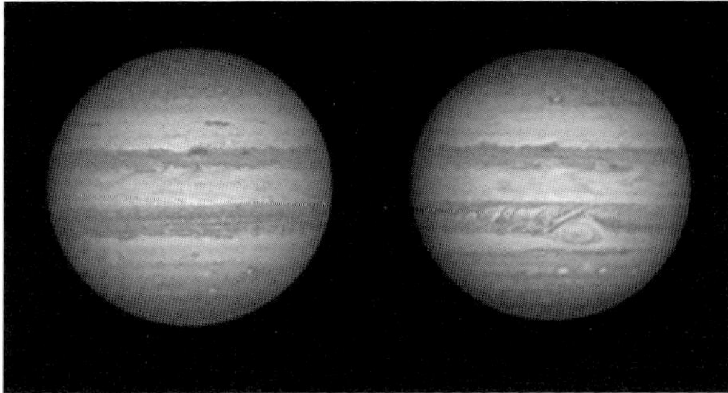

Figure Gallery 4 *The "dull" side of Jupiter (left) was imaged on February 23, 2004, through an 11-inch SCT at f/22 using a ToUcam 840. The same telescope and camera, operating at f/31, imaged the Great Red Spot on February 19, 2004. Photos by Damian Peach.*

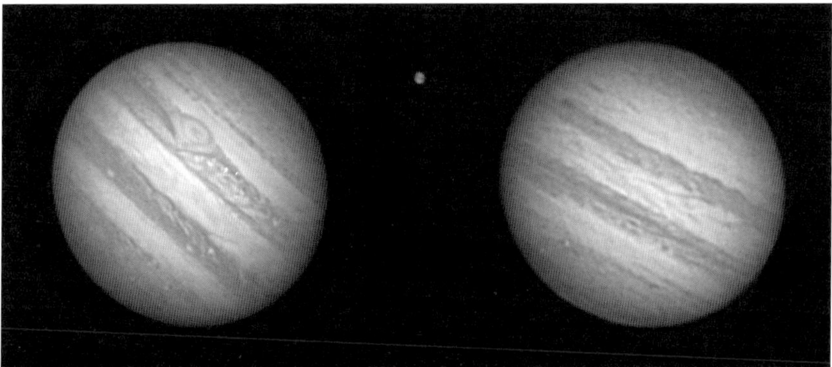

Figure Gallery 5 *A Celestron-9.25 SCT and Luminera LU075M camera were used to take the left image of the Great Red Spot on April 30, 2005. On the right, a Celestron-11 operating at f/22 and a ToUcam 840 captured detail on both Jupiter and its moon Ganymede on February 20, 2003. Photos by Damian Peach.*

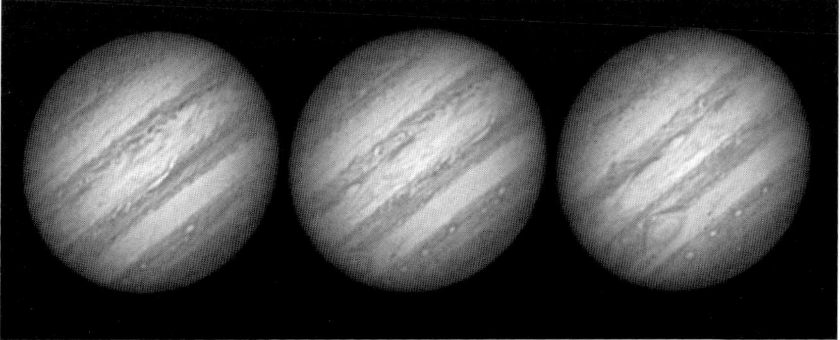

Figure Gallery 6 *A Celestron-11 operating at f/31 and an Atik 1HS captured this series of images of the Great Red Spot rotating into view on March 1, 2004. Notice the white rift cutting into the South Equatorial Belt on the left and center images. Photos by Damian Peach.*

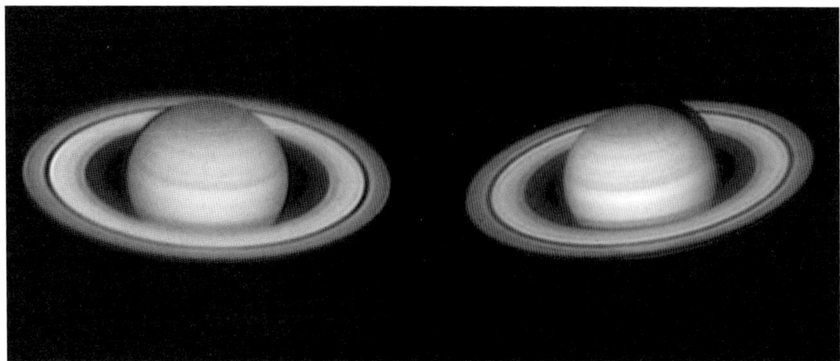

Figure Gallery 7 *The atmospheric features of Saturn remain relatively static for long periods of time; however, the tilt of the ring system and the aspect of Saturn's shadow on it constantly change. These two images were taken a little over a year apart and show noticeable changes. The left one was taken on December 16, 2003, through a Celestron-11 Schmidt-Cassegrain telescope at f/31 using an Atik 1HS camera. The right image was taken on April 25, 2005, through a Celestron-9.25 Schmidt-Cassegrain telescope using a Luminera LU075M camera. Photos by Damian Peach.*

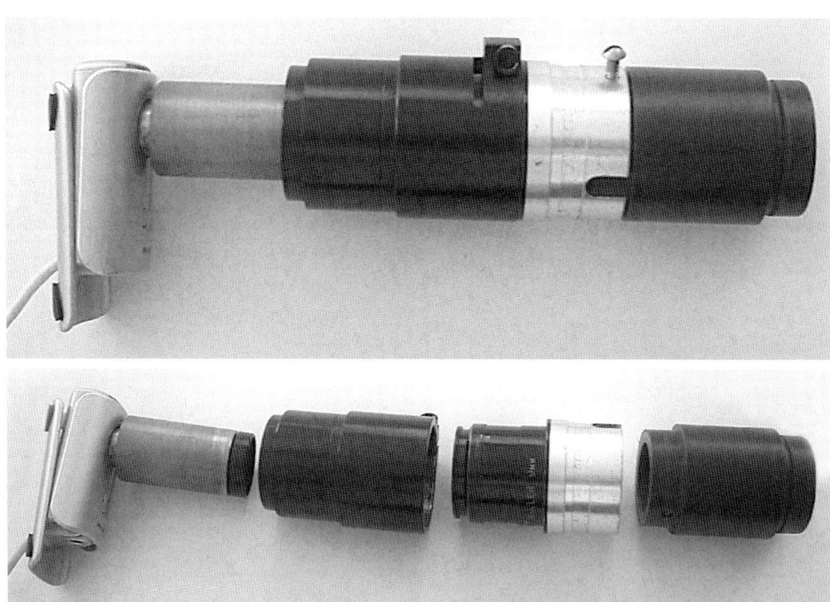

Figure Gallery 8 *The custom-made eyepiece projection system Zac Pujic uses with a 12-inch Newtonian telescope to acquire his amazingly detailed lunar images is shown here. A brass extension tube attaches to the camera and inserts into an adapter tube that slides over a 9mm Nagler eyepiece. Pujic removes the Barlow lens that is built into the Nagler eyepiece in order to achieve the desired focal ratio. A custom-made barrel locks into the eyepiece and inserts into the telescope focuser. Photos by Zac Pujic.*

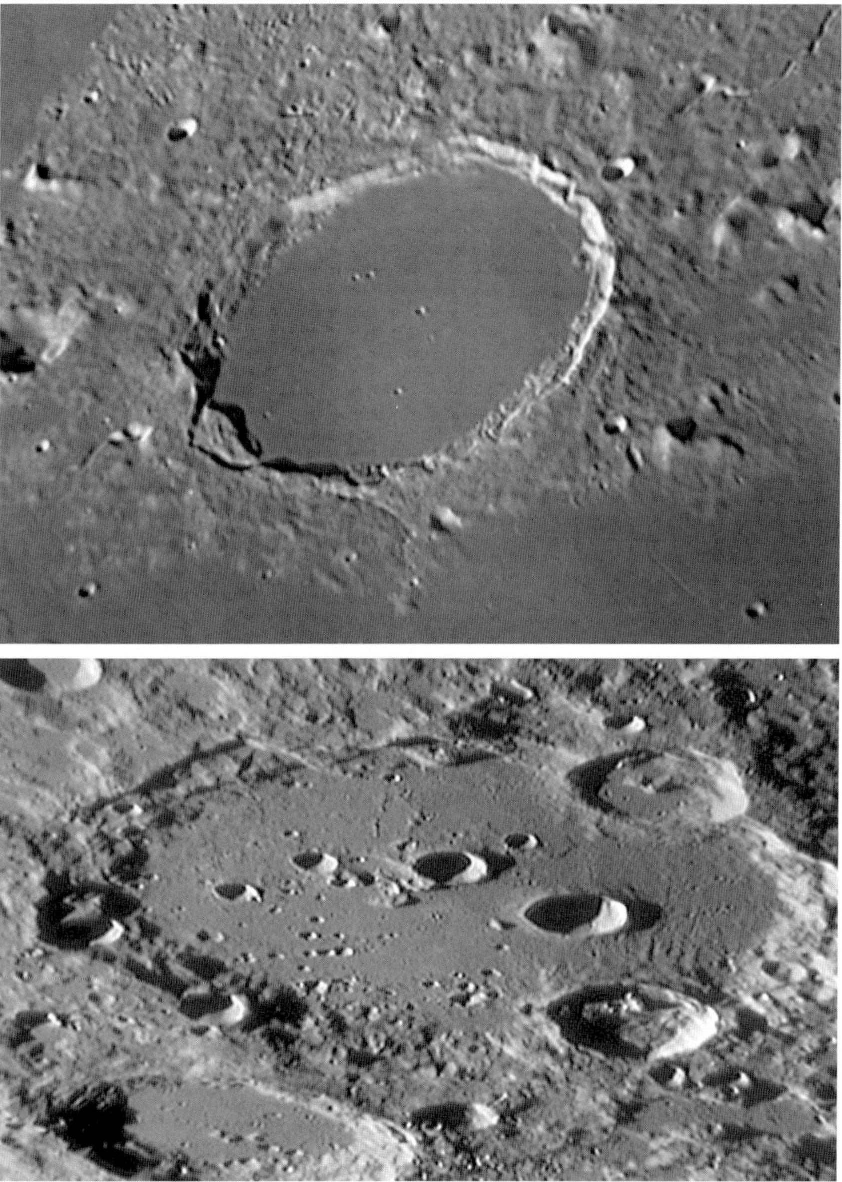

Figure Gallery 9 *Many craterlets are visible on the smooth floor of the crater Plato (top). A myriad of tiny craterlets are visible within the giant crater Clavius (bottom). The younger craters Porter and Rutherfurd overlay the upper and lower crater walls of Clavius. Photos by Zac Pujic.*

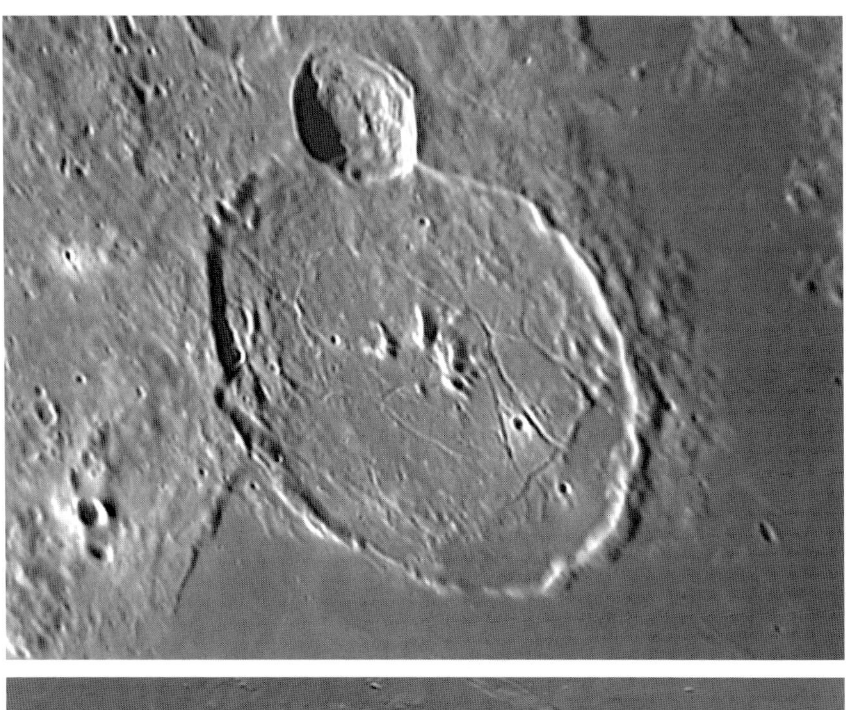

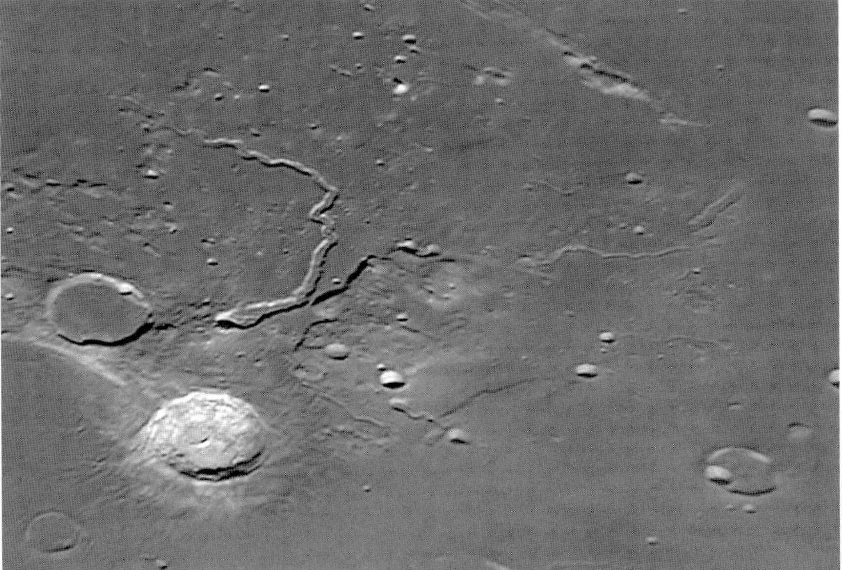

Figure Gallery 10 *The fractured lava-filled crater Gassendi lies just north of Mare Humorum (top). Schroeter's Valley cuts through the flat plain of Oceanus Procellarum to the west of the bright crater Aristarchus (bottom). Photos by Zac Pujic.*

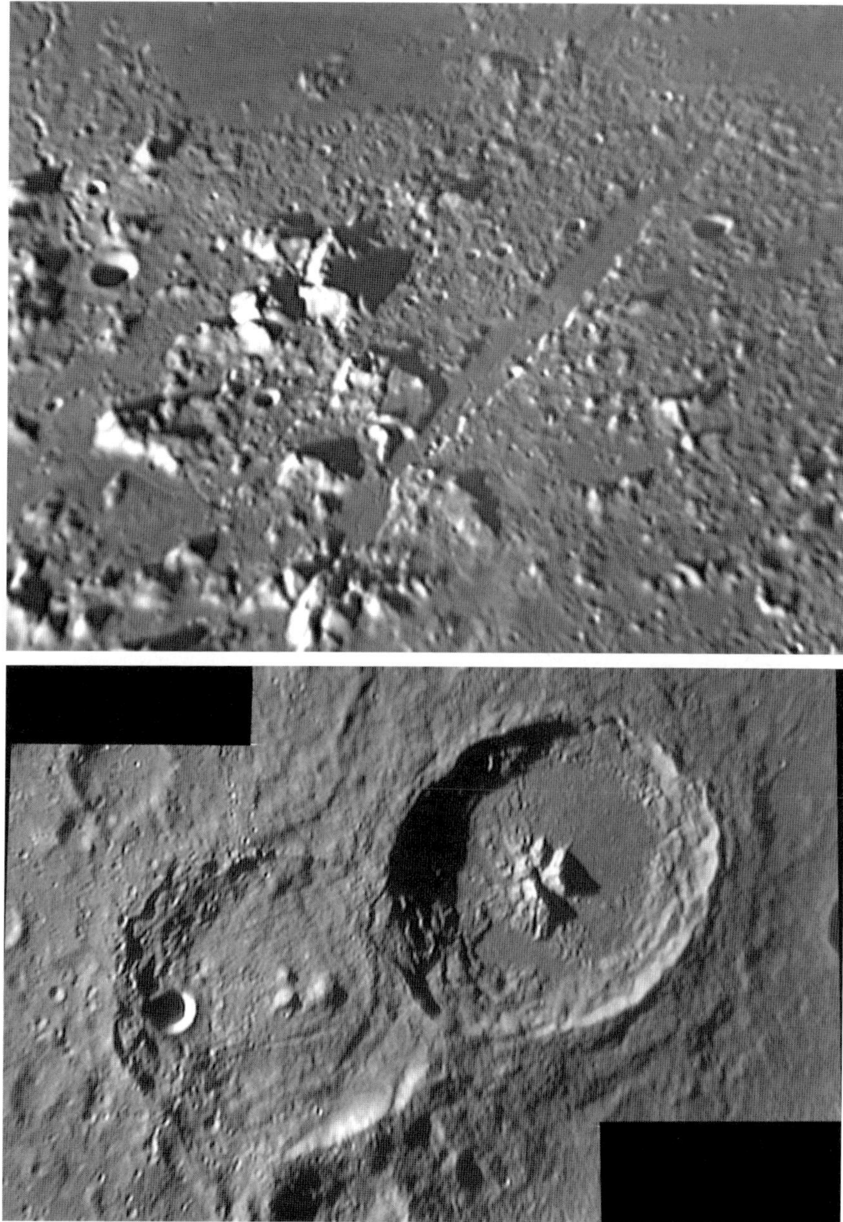

Figure Gallery 11 *The elusive rille running through the center of the Alpine Valley is visible (top). The twin craters Cyrillus (left) and Theophilus (right) lie to the south of Mare Tranquillitatis (bottom). Photos by Zac Pujic.*

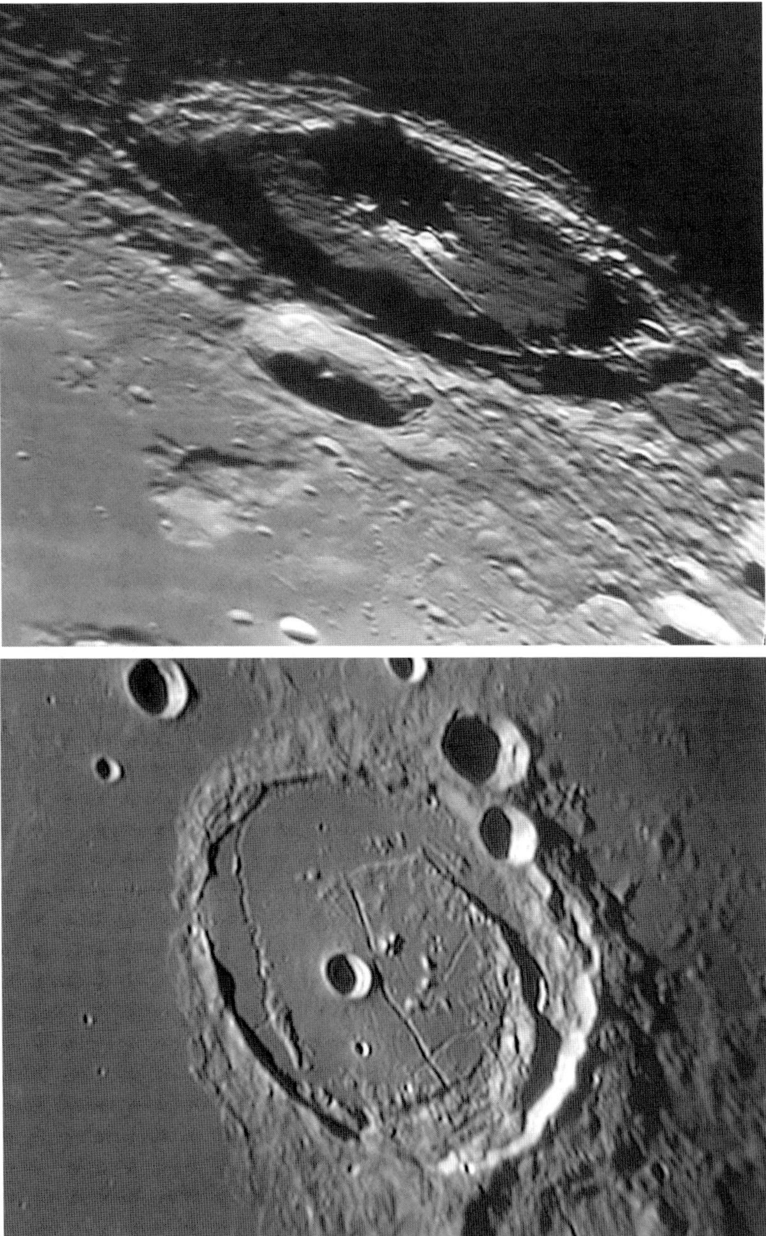

Figure Gallery 12 *The unusual fractured floor of Petavius near the eastern limb of the Moon is highlighted as the Sun sets on the crater (top). Another fractured crater, Posidonius, is visible along the eastern edge of Mare Serenitatis (bottom). Photos by Zac Pujic.*

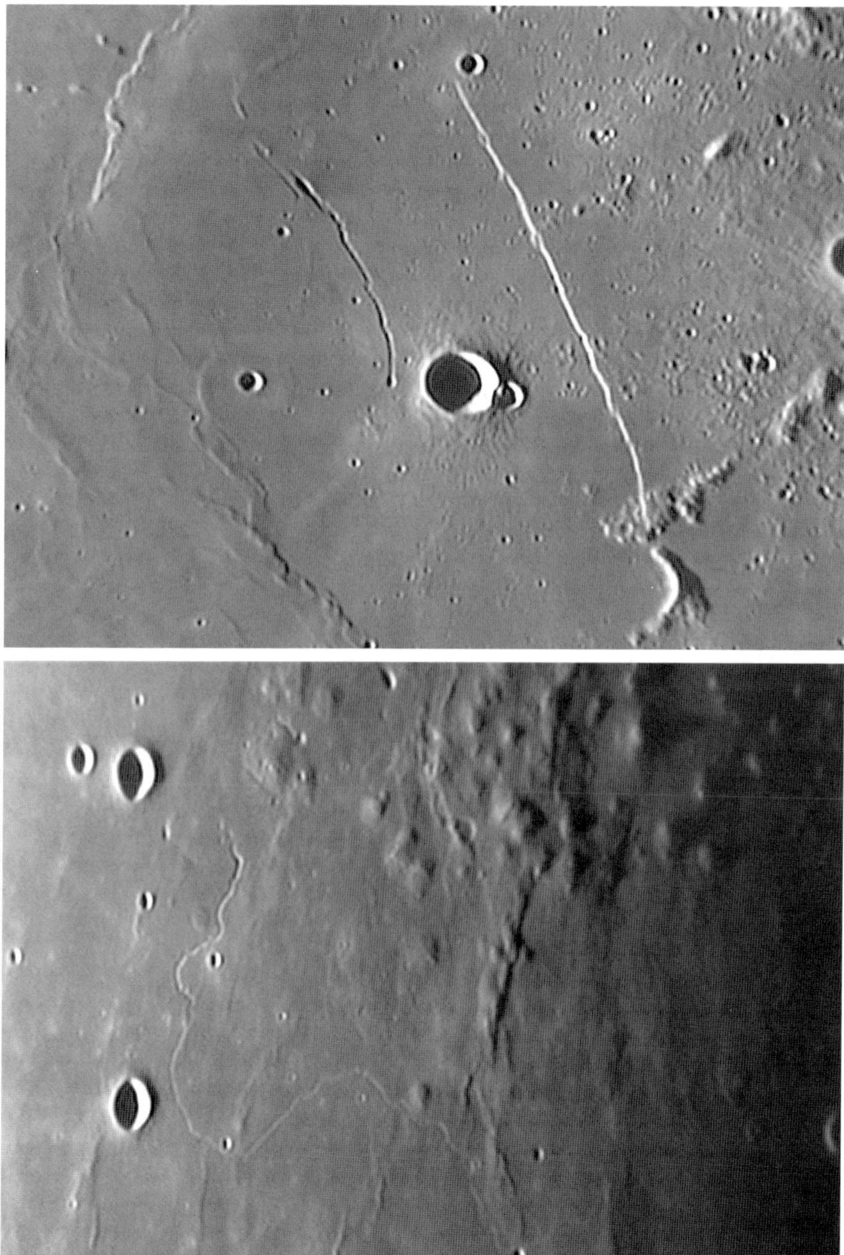

Figure Gallery 13 *Rupes Recta, also known as the Straight Wall, is visible on the plains of Mare Nubium (top). The Marius Rille snakes past the Marius Hills to the north of the crater of the same name (bottom). Photos by Zac Pujic.*

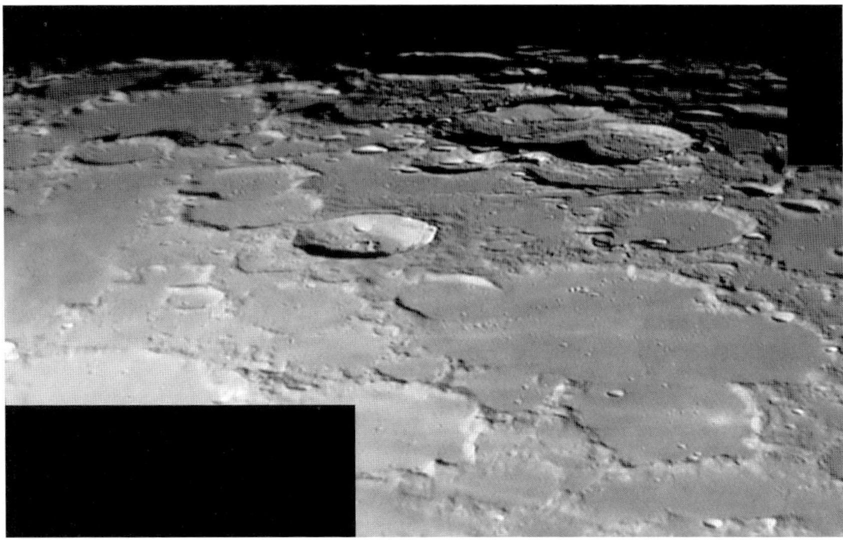

Figure Gallery 14 *At the center of this two-image mosaic of the Moon's northern limb the sharp-rimmed crater Scoresby lies to the upper left of the broad flat Meton crater complex. Photo by Zac Pujic.*

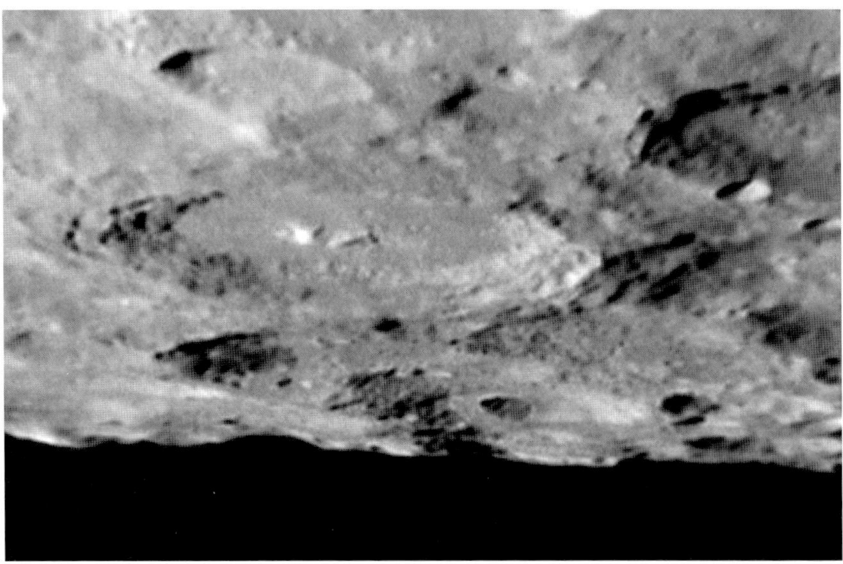

Figure Gallery 15 *At the center of this two-image mosaic (top) of the Moons' northern limb the sharp-rimmed crater Scoresby lies to the upper left of the broad flat Meton crater complex. Along the Moon's southern limb (bottom), the crater Moretus features terraced walls surrounding a flat floor and central peaks. Photos by Zac Pujic.*

Chapter 1
Introduction to Webcam Astrophotography

1.1 Some Background

The idea that charge-coupled devices, or CCDs, have completely revolutionized the field of professional and amateur astrophotography over the past several decades is no longer up for debate. It is a fact. Starting in the late 1980s, commercially available CCD cameras began making significant inroads into an area of astronomy that had been dominated by film photography for over a century. Early CCD cameras possessed relatively small imaging sensors that translated into small, pixilated fields of view. This keyhole perspective on the universe, coupled with high cost and new and complex operating procedures that required the use of a computer, made the transition to digital imaging an expensive and challenging option.

By the 1990s, computers were a common item in most homes and could be adapted to amateur astronomy as quickly as a new eyepiece design. The price of electronic gadgets also declined and the price of entry-level astronomical CCD cameras soon became competitive with mainstream film cameras. As the 21st century arrived, digital astrophotography with CCD cameras had taken an equal footing with the older technique of using conventional film cameras to image the universe. At the same time, however, the advent of CCD imaging polarized the astrophotography community into "purists," who remained loyal to film, and "tinkerers," who saw new possibilities with digital imaging.

But the true challenge to the dominance of film astrophotography came from another direction. Consumer digital snapshot cameras have been around since the mid-1990s, but early versions were clearly inferior to film cameras for anything except lunar and planetary imaging. Then, fittingly, it seemed, with the arrival of the new millennium, Canon introduced the SLR-style EOS 30D, followed soon by the 60D, and astrophotographers around the world took note. When the long exposure-

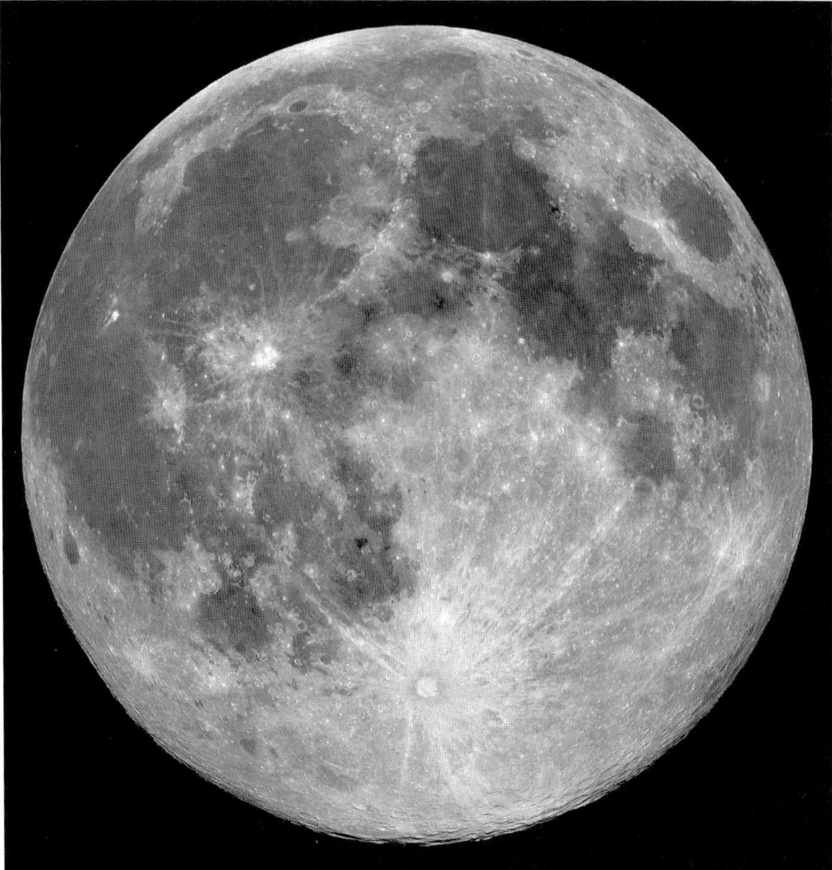

Fig. 1.1 *This mosaic of 49 webcam images showing the January 24, 2005 full Moon is actually the inspiration that led the author to write this book. The mosaic (details on mosaic assembly are in Appendix D) was taken with a Phillips ToUcam Pro II 840K on a Celestron-8 telescope at f/10 and assembled using freeware software programs and simple webcam imaging techniques. The completed image shows greater detail than the author has been able to achieve in over 40 years of lunar imaging with conventional film and digital cameras. Photo by Robert Reeves.*

capable 6-megapixel EOS 10D and 300D cameras were released in 2003, digital imaging entered mainstream astrophotography.

Although it is fashionable to say that digital astrophotography became mainstream in the early 21st century, the basic truth is that another form of digital astrophotography had already quietly become popular with a large number of astro-imagers. Before consumer digital cameras evolved to the point of surpassing film cameras in popularity for sky-shooting, webcams had been enthusiastically applied to imaging the Moon and planets. At the same time, a number of pioneering individuals discovered that with some modifications, certain webcam models were capable of taking long expo-

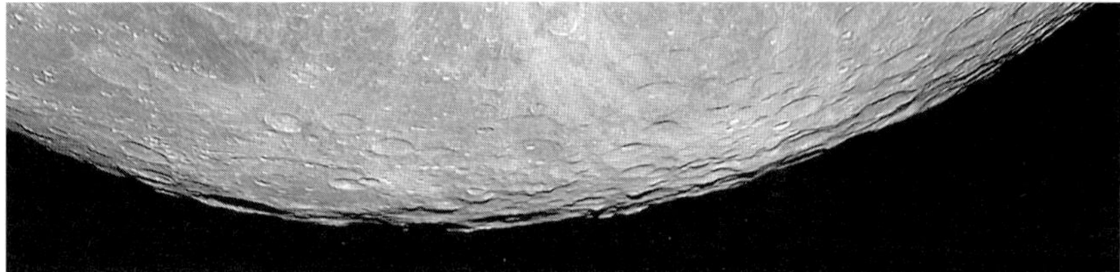

Fig. 1.2 *The cropped and enlarged southern edge of this January 24, 2005 lunar mosaic shows the amazing detail that can be captured using webcam imaging techniques. In this view benefiting from an unusually high southern lunar libration angle, the viewer is actually peering over the peaks of lunar mountains and crater rims. Photo by Robert Reeves.*

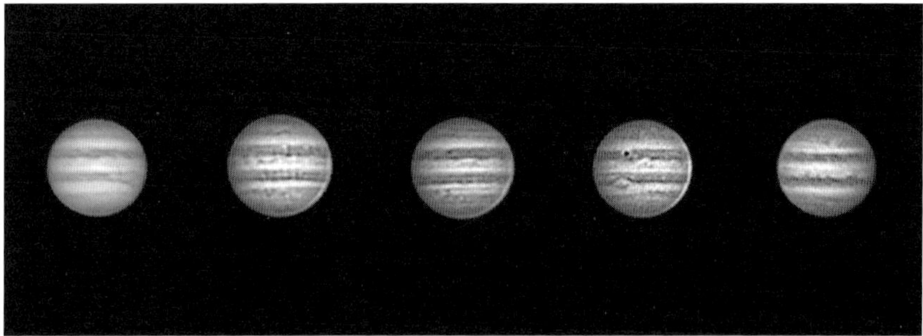

Fig. 1.3 *Jupiter should never be considered a "one-shot" target. The planet is a dynamic subject that presents a constantly changing appearance as cloud structures evolve and change in the planet's atmosphere. These images were taken over a period of months using a ToUcam 840 and a Celestron-8 telescope operating at f/10. Photos by Robert Reeves.*

sures of deep-sky objects. The results achieved with these simple little cameras were amazing. At the time, most people regarded webcams as a means of getting started in digital astrophotography without having to spend a lot of money, but they soon proved themselves to be not only simple to operate, but capable of achieving high-resolution images of the Moon and planets. Amateurs using nothing more than an 8-inch telescope and a webcam produced very NASA-like, high-resolution, research-grade images of solar system objects from their backyard!

Today, digital astrophotography has come full circle. With the adaptation of webcams to celestial imaging, we are now back to the point of using imaging devices that provide a small keyhole perspective of the universe and require the use of a computer at the telescope. But in contrast to the expensive early astronomical CCD cameras, these devices are cheap and widely available at electronics and computer outlets. Exposure times can range from $1/10,000$- to $1/25$-second, and in some cases as long as $1/5$-second—more than sufficient to image the brighter planets and the Moon. The resulting images can be easily processed using specialized freeware pro-

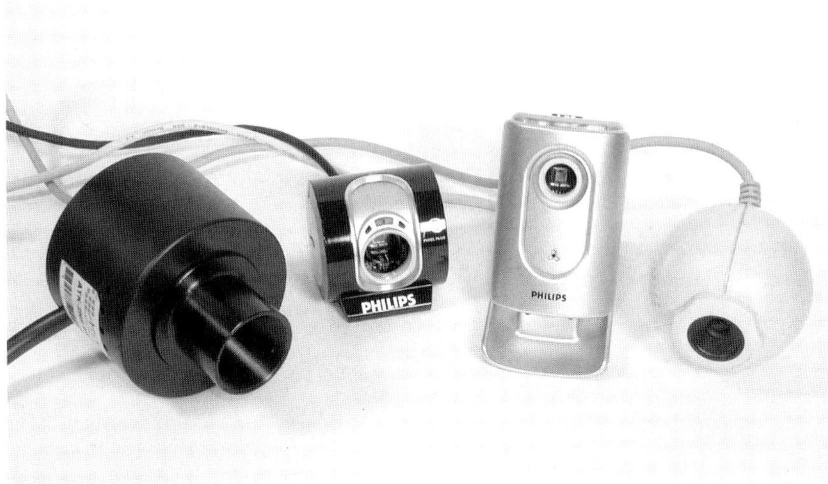

Fig. 1.4 *The astronomical webcam "family tree" is shown above. At the right is the original spherical Connectix black-and-white Quickcam that started the long-exposure webcam astrophotography revolution. The Philips ToUcam Pro II that matured webcam astrophotography is next followed by the new generation Philips SPC900NC. At the left is the Atik ATK-2HS, a commercially-modified air-cooled long exposure-capable version of the ToUcam that has allowed the novice to achieve CCD-quality electronic astrophotography. Photo by Robert Reeves.*

grams designed for webcam astrophotography and software tools widely available for standard digital photography. While their 640 x 480 pixel format is small compared to modern 8-megapixel and higher digital still cameras, when properly processed, webcam output is very high quality. This makes a webcam a great starting point for someone who wants to enter the field of astrophotography.

1.2 A New Imaging Era

Over the past several decades, a majority of consumer electronic goods have branched off a single technological breakthrough—the ability to convert analog data into digital binary data. This breakthrough has allowed the development of still and video digital cameras that have redefined the ways that we handle and process video and audio information. CDs, DVDs, HDTV, and image-processing programs like Photoshop® are some of the ways we digitally manipulate images. The webcam is a direct offshoot of these breakthroughs in electronic photography.

The name "digital" refers to the binary representation of data as bits and bytes. A bit is the smallest unit of information that a computer can store and process. It represents an electrical state of either "off" or "on" with a value of either zero or one. The zero and one are used to create binary numbers, usually with eight bits associated together to form what is called a byte, representing a number between 0 and 255.

The digital photography term "pixel" is a contraction of PICture ELement. It is the smallest point of a bitmapped image that can be assigned color and intensity. Explaining further, a bitmap is the digital representation of a picture where all the pixels are arranged in a grid and assigned a specific color and intensity.

When they first became available to consumers, digital video cameras were costly. As they became popular and production increased, the price dropped, sometimes considerably. Webcams evolved from this technology as a low-cost digital video camera for computer video conferencing. Though small enough to hide in the palm of your hand, webcams are technologically complex devices, essentially little computers filled with imaging, memory, and processing chips as well as the optical components needed to create the image. The camera software has to handle megabytes of data every second. Additionally, the camera has to be able to network with computers in order to download the video images for display and processing.

If you want to start a good scrap between fellow astrophotographers, just ask which is best for astronomical imaging: film, digital, CCD, or webcams. In spite of the pontification of each medium's proponents, the basic answer is, "all of the above." No single medium can do all things in celestial imaging; each has its strengths and weaknesses. Film can still outperform digital media in some areas while digital-based imaging will outperform film in others. Indeed, if you want to silence the critics and supporters of each medium in the field of lunar and planetary imaging, just show them the amazing work being done on these subjects with simple video webcams. The icing on the cake is the fact that planetary images from the latest "Googlepixel" camera processed in "Pricemax" software can be outresolved by a simple webcam costing less than a good eyepiece while using free software!

The transition to all forms of digital imaging quickly overtook the astronomy community. The chemical photographic darkroom is a thing of the past. Following the publication of my first book, *Wide-Field Astrophotography*, I was invited to give many astrophotography presentations. A question I would ask the audience is how many process their images on a computer and how many still use a darkroom. In all the presentations I was privileged to give, I found a grand total of three practitioners of the darkroom arts.

While film is still cost-effective compared to the price of advanced astronomical CCD cameras, webcams are so inexpensive their use in

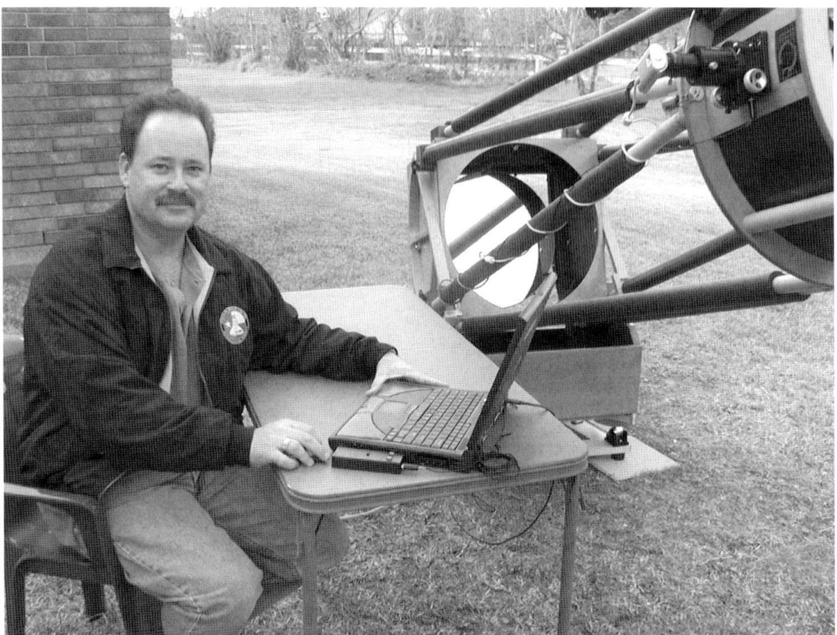

Fig. 1.5 *Glenn Schaeffer from Houston, Texas, uses a typical webcam planetary imaging setup. He couples a Philips ToUcam Pro II 840K to his 20-inch Obsession Dobsonian mounted on an Osypowski dual-axis equatorial platform to image the Moon and planets from his backyard. Photo by Glenn Schaeffer.*

astronomy cannot be ignored. Even those who never previously owned a film camera are attracted to the ease and instant results achieved with a webcam. Moreover, webcam imaging allows us to do some things that were not possible with film. The constant feedback from being able to view results instantly allows us to correct mistakes on the fly.

1.3 Minimum Requirements for Imaging

Now that we have seen that webcams are more than simple video conferencing devices and can be used for astronomy imaging, what is the minimum equipment you need to begin applying a webcam to lunar and planetary digital photography? Experience has shown that with good seeing and focus, and good telescope optics, a webcam astrophotographer will be able to record lunar and planetary detail that exceeds the Dawes Limit, the usual benchmark for visual capability through a given telescope. Indeed, it is not unusual to see far more detail on Jupiter with processed webcam images than can be seen visually through the same telescope under equal viewing conditions.

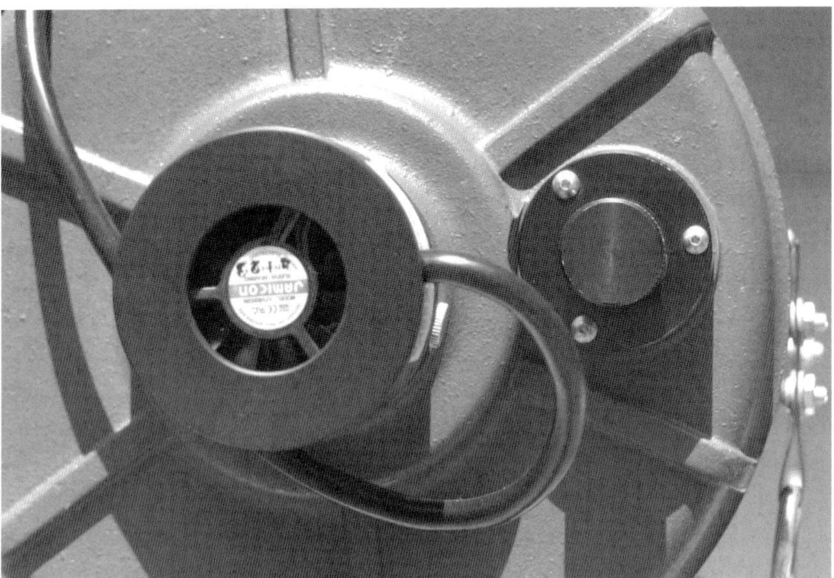

Fig. 1.6 *It is easier to view images when the camera's display shows north at the top. When installing the camera, orient it so the display is properly aligned to begin with. With an ATK-2HS on a Celestron-8, aligning the camera cable with the telescope focus knob aligns north at the top of the field. Photo by Robert Reeves.*

The required basics to achieve good images of solar system objects are surprisingly simple, but as with anything in astronomy, there is room to grow and expand capabilities as one's interest develops.

The requisites are:

- Any telescope with a removable 1¼-inch eyepiece.
- A good quality webcam with a CCD sensor instead of CMOS and a device to adapt it to the telescope focuser.
- A telescope mount capable of tracking a celestial target.
- A Barlow lens for additional magnification (but this is not an immediate requirement).
- A computer, preferably a laptop for easier portability.

With the exception of the adapter to allow mounting a webcam at the focus of the telescope, most amateur astronomers already possess the minimum equipment to begin webcam imaging. Now that webcam astrophotography has entered mainstream astronomy, an entire industry has evolved to supply the needed imaging adapters through the popular sources that supply our standard astronomy needs.

1.4 Advantages and Disadvantages

Webcams can be very attractive, not only for entry-level astrophotography, but for some forms of advanced imaging as well. But each form of astronomical imaging, whether film, CCD, digital, or webcam, has a set of inherent advantages and disadvantages over other types. Since this book deals with webcam imaging, we will deal primarily with these issues as related to webcams.

Let's start with the advantages:

- Webcams are far cheaper than conventional cameras or astronomical CCDs.
- Webcams use USB plug-and-play technology, meaning they are easy to install on a computer and have fast image download times.
- Webcams provide real-time feedback. Focus and exposure are adjusted on the fly to insure best results.
- Each webcam video imaging session produces a single unique image.
- There are no film costs.
- There is no need to wait for a full roll of images to be taken before developing them.
- Webcams can create animated images and movies of changing events such as transits of Jovian moons or lunar occultations of bright stars and the planets.
- Webcam images are digital-friendly, meaning they can be printed, posted on the Internet, or emailed without the need for developing or scanning.
- Except with special black-and-white cameras, webcam images are in color. There is no need for combining tri-color images when doing lunar and planetary work.
- With the use of proper software, some of which is free, webcams can be used as autoguiders for other forms of astrophotography.
- A webcam shutter, the device that controls the length of an exposure, is entirely electronic. This gives webcams a great advantage over standard cameras in that there is no mechanical shutter to wear out or malfunction. A webcam can easily take more individual exposures in a single evening of lunar photography than someone will take in a lifetime of snapshot photography. For instance, the full Moon mosaic on page 2 is the summation of 28,800 individual video frames. This is the equivalent of snapping 800 rolls of 36-exposure 35mm film.

As good as all this sounds, these devices also have inherent disadvantages that must be recognized:

- Standard webcams are not as sensitive as conventional cameras or CCDs.

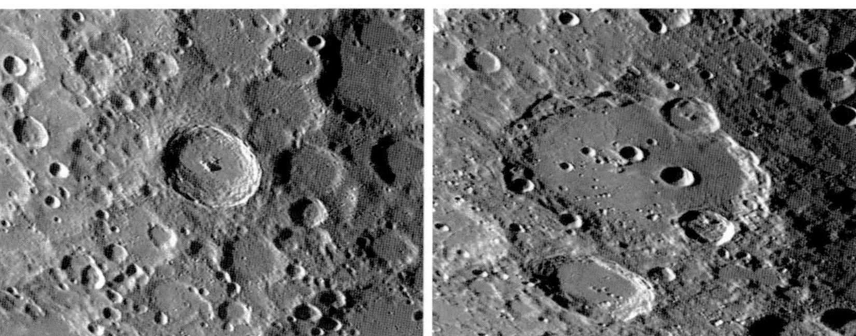

Fig. 1.7 *Webcam astrophotography techniques have allowed the users of moderate size telescopes to achieve lunar and planetary imaging results that are equal to that of professional observatories. The craters Tycho (left) and Clavius (right) were imaged from the author's urban backyard using an ATK-2HS camera and a Celestron-8 telescope operating at f/20. Photos by Robert Reeves.*

They have an inherent long-exposure limitation of about 1⁄25 of a second (as we shall later see, the actual exposure is longer, but still less than a second). This limits their practical use to bright solar system objects.

- Webcams have small imaging sensors that typically produce 640 x 480 VGA resolution. This is of little consequence with planetary imaging as a planet will be smaller than the field of view, but it limits the amount of lunar or solar territory that can be imaged through a telescope on a single frame.
- Webcams are not designed for astronomical use and must be modified to allow the long exposures needed for deep-sky imaging. (For those inclined to tinker with interesting astronomical gadgets, perhaps this could be considered an advantage.)
- Although the results can be rewarding, it is tedious to sort through hundreds of video frames to select the best ones for stacking. Software does a good job of automating the process, but best results are still achieved by "manually" sorting the best images.

1.5 Comparison of Imaging Techniques

Several things are evident about using a webcam for astrophotography and about the people who use them in this way. First, these devices are addictive! They are fun, inexpensive, and provide surprisingly good results, especially with solar system targets. Astrophotography with a webcam is as much about tinkering with electronics and software as it is about doing astronomy. The satisfaction derived from the "tinker factor" is very high. Second, people who delve into this activity become passionate about it. Their enthusiasm spreads to others, like-minds attract each other, and user

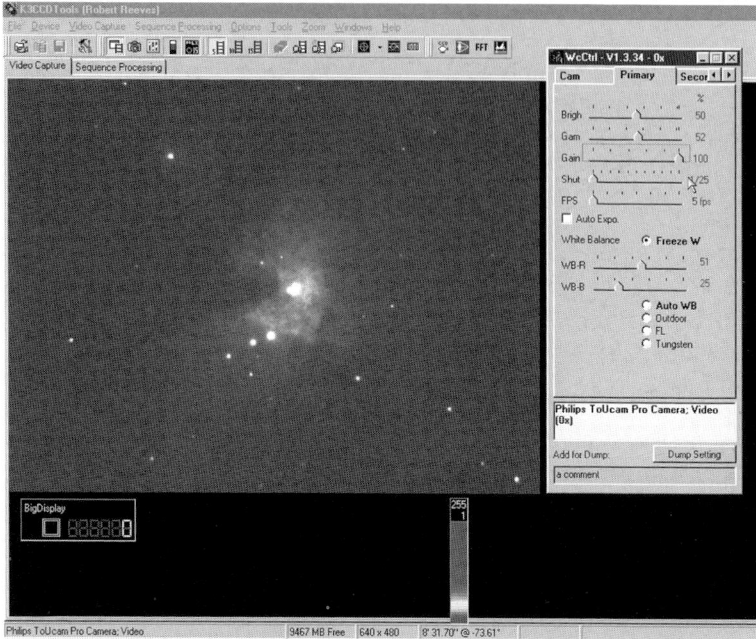

Fig. 1.8 *Standard unmodified webcams such as the ToUcam 840 and SPC900NC have a maximum exposure time of only a fraction of a second. These cameras are superior for lunar and planetary imaging but their maximum exposure time is too brief to properly image even the brightest deep-sky targets such as M42. Photo by Robert Reeves.*

groups appear on the Internet. From these gatherings of enthusiasts, new techniques and technologies to improve a webcam's celestial performance are constantly advanced.

The choice of any imaging device in astrophotography is driven by a set of compromises. It does not matter if the choice is a top-of-the-line dedicated astronomical CCD camera or a bargain webcam, there are still compromises. Let's look at some of the factors involved in making such a choice in Table 1.1. From this table we can draw some conclusions that rate webcams against other types of astronomical imaging devices.

1.5.1 Cost

First, webcams are cheaper than their alternatives often by orders of magnitude. A dedicated astronomical CCD such as an SBIG ST-10, plus supporting hardware and software, can cost nearly $10,000. For best results, it should be used on a high-end telescope and mount. Used advanced film cameras are becoming inexpensive as these models are phased out in favor of digital imaging. However, factoring in their support equipment like film-

Table 1.1 Comparison of Various Astronomical Imaging Options

	SBIG ST-10	Film SLR	Digital DSLR	Webcam
Cost	Very high	moderate	moderate	low
Flexibility	low	high	high	low
Complexity	high	moderate	moderate	low
"Frustration" factor	high	moderate	moderate	low
Ease of use	low	moderate	moderate	high
"Tinker" factor	low	moderate	moderate	high
"Satisfaction" factor	high	high	high	high
Bang for buck	low	moderate	moderate	high

developing apparatus, film scanners, and software, they can still cost over $1,000. True SLR-style digital cameras that are versatile enough for all types of astrophotography are still in the $1,000-to-$1,500 range. But the digital single lens reflex has the advantage of not needing film-developing apparatus or scanners, which would add to the cost, so the initial out-of-pocket expenses for film and digital are about even. By contrast, even the best webcams, along with their supporting hardware to attach to the telescope, cost only several hundred dollars and the important software is free.

1.5.2 Flexibility

The vast majority of imagers look at a piece of imaging equipment from the broad viewpoint of how versatile is it? Can it do both astrophotography and family pictures? In this case, film and digital SLRs are clearly superior as they can take pictures of the family vacation by day and image the universe later that night on the same roll of film or memory card. It is extremely impractical to use an ST-10 or a webcam to take pictures of your child's birthday party!

1.5.3 Complexity

Top-of-the-line CCD cameras used on a high-end telescope can produce research-quality astrophotos. But these images are acquired at the expense of having to use complex equipment and software. Although film and digital consumer SLR cameras have been designed for ease of use by the general public, their application in celestial imaging makes them more complex. Webcams, on the other hand, represent a virtual plug-and-play imaging system where results are viewed in real time and corrections applied on the fly.

1.5.4 Low "Frustration Factor"

For the electronic novice, modifying a webcam for long exposures can be

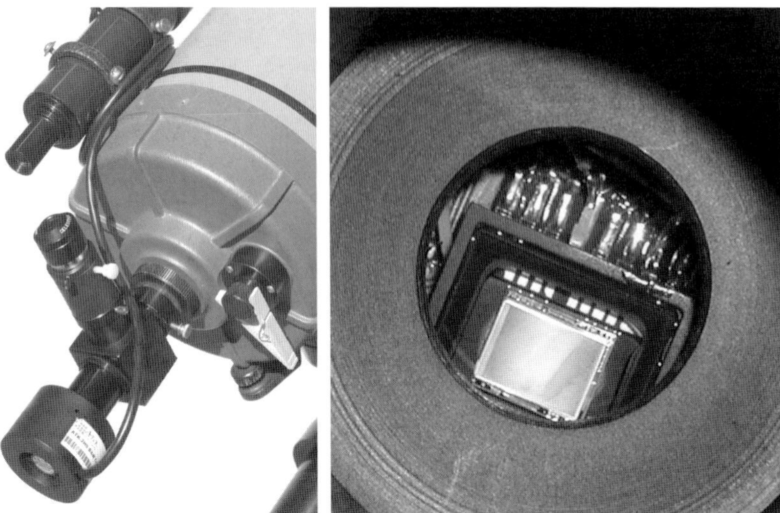

Fig. 1.9 *Always loop the camera cable around a solid telescope part, such as the finder bracket, (left) as a safety to prevent the camera from falling. As telescope parts cool at night, adapter set-screws can loosen and the camera can fall out. The author learned this could happen the hard way (right), resulting in a dislodged CCD in an ATK-2HS. The camera still functioned after landing on the concrete, but the sensor was loose and tilted, preventing accurate focus. Photos by Robert Reeves.*

frustrating. The astrophotographer on a budget will also find a great deal of frustration in the price tag attached to a capable dedicated CCD camera. But the application of webcam imaging of the Moon and planets is a simple pleasure. Take it from someone who spent almost five decades imaging the Moon with many different cameras and telescopes. High-resolution lunar astrophotography is hard. The demands on equipment and technique are high. Adding to the mix is the need for excellent seeing and just plain good luck. Webcam lunar imaging techniques, on the other hand, allow you to find the correct exposure, framing, and tracking parameters on the fly, and to some degree work around less than perfect seeing. In a nutshell, I find webcams offer both the lowest cost and the easiest lunar imaging techniques coupled with the highest success ratio.

Any new endeavor requires a learning process to master new devices and the skills needed to operate them. Experienced film and digital astrophotographers will find their introduction to advanced CCD cameras easier to master than a complete novice because their knowledge of camera basics will ease the transition into celestial photography. But virtually no photographic knowledge is needed to apply a webcam to lunar and planetary imaging beyond being able to properly operate the telescope used.

1.5.5 Ease of Use

This factor is closely tied to the issue of complexity. It is easier to use a device that can be quickly set up and has few peripheral support activities. Advanced CCD cameras are basically not that hard to use. However, the astonishingly beautiful results they are capable of producing is dependent on a high user skill level. The astronomical imaging skill level required to get results from these devices that are consistent with their dollar cost is only acquired through lengthy experience. The astrophoto techniques applied to film and digital SLR cameras also represent an application of techniques not found in ordinary snapshot photography. While not hard to learn, they do present challenges to the novice. Webcams, on the other hand, attach to the telescope and operate in the same manner for which they were designed in recording short video clips. The application of specialized software to process the captured video into a photograph is the only additional step required.

1.5.6 The "Tinker Factor"

To some people, one of the appealing factors about astrophotography is the opportunity it affords to tinker with devices and equipment. The modification of a non-astronomical item in order to make it applicable to astrophotography is as enjoyable as the astrophotography itself. In this capacity, dedicated CCD cameras offer little room for the average user to tinker because they are designed for a specific task and there is little left to do to optimize them for astrophotography. Film and digital cameras allow some room for innovation to make them perform better for astrophotography, but generally these devices are designed for long-exposure applications to begin with. Webcams on the other hand are designed for delivering streaming video over a computer. To get images of deep-sky objects or use a webcam as an autoguider requires some degree of tinkering, either with actual modifications to the camera itself or through the application of specialized software. Thus, the "tinker factor" is high and a great deal of satisfaction can be achieved through modifying an inherently insensitive imaging device to achieve stunning images of galaxies and star clusters.

1.5.7 Satisfaction Factor

Looking at all the above factors—price, ease of use, modification potential, low frustration factor, and the ability to achieve quick results, it is apparent that the satisfaction factor with astronomical webcams can be high. The bottom line is—webcams offer the biggest astrophoto bang for the buck.

1.6 How Digital Imaging Works

A webcam is basically a small video camera that relies on a standard home computer to supply some of the components needed to produce video files. The camera body houses the imaging sensor, electronics for reading and amplifying the video signal, and a lens for focusing the image onto the sensor. The camera is electronically connected to a computer via a USB cable. The computer supplies the electrical power for the camera and the operating system, or driver, that controls the camera, and the computer's hard drive acts as a recording medium for the resulting video files.

The heart of a webcam is the electronic imaging sensor. Various webcams use either a Charge-Coupled Device (CCD) or Complementary Metal Oxide Semiconductor (CMOS) as an imaging sensor. The CCD sensor uses a special manufacturing process to create the means of transporting electrical charges across the sensor without introducing distortion in the charge. CMOS sensors, on the other hand, rely on standard manufacturing techniques used to make computer microprocessor chips and can be fabricated on any silicon production line. CMOS sensors have been manufactured for a longer time, represent a more mature technology, and use up to 100 times less power than CCD sensors. The ease of manufacture and lower power requirements of a CMOS sensor would seem to make them attractive for webcam applications, but in general the small inexpensive CMOS sensors used in webcams are less sensitive to low light than CCD sensors, thus astrophotographers value CCD-equipped webcams.

Because the CCD is best suited to astronomical photography our discussion will concentrate on that style of sensor. This CCD was introduced to astronomy in 1976 and has become the imaging detector of choice in the decades since. CCDs have the advantage of high quantum efficiency, which means that they are very good at capturing what little starlight falls upon them, and they have a broad range of spectral sensitivity. The physics of how CCDs are constructed and how photons interact with the silicon structure of the CCD dictate how they react to photons of various energy levels. Between the 500-nanometer visible wavelengths and the near-infrared at 900 nanometers, high-quality scientific, and expensive, CCDs can achieve quantum efficiencies of between 40 and 90 percent compared to only one percent for photographic film-based imaging. At the same time, these physical construction characteristics also place an 1100-nanometer upper limit on spectral sensitivity while quantum efficiency at the shorter blue wavelengths drops off considerably, often to as low as 20 percent. The loss of blue sensitivity is of little concern for normal webcam astrophotography, but practitioners of advanced tri-color imaging using a commercially modified black-and-white webcam will find the need for longer blue exposures to compensate.

CCDs are composed of an array of tiny metal-oxide semiconductor capacitors arranged in a rectangular grid on a silicon substrate. Each small capacitor is known as a photosite and represents a pixel in the image created by the CCD. Light is captured by the individual photodiodes within the imaging sensor. The photons that strike the sensor are converted into electrons that arc proportional in number to the intensity of the light, and which are then stored at the photodiode location on the sensor. This is somewhat analogous to the chemical film process, in which photons strike the individual film grains causing a chemical reaction that becomes visible when the film emulsion is developed in a chemical bath.

The inexpensive and lower quality 640 x 480 CCD detectors used in standard webcams are admittedly not as sensitive as dedicated astronomical CCDs because webcams are not designed to be scientific instruments. Indeed, scientific grade CCDs such as those used in dedicated astronomical cameras often read the data off the CCD at the rate of only 50,000 photosites per second in order to preserve data with exacting precision, while the more than 300,000 photosites on a webcam are read as fast as 30, or even 60, times per second on USB 2 cameras. Some image quality sacrifices must be made in order to read data off the sensor that fast. But remember, webcams are imaging devices whose design is driven foremost by the concept of keeping them affordable. The fact that they are adaptable to something as exacting as astrophotography speaks highly of the operating characteristics of even simple entry-level CCD devices.

The major operational difference between the CCD and CMOS sensor is that the CCD reads the electrical charge from a particular photodiode by transporting the charge across each row of pixels on the sensor and then reading it at the edge of the sensor before converting it into a digital value. A CMOS sensor will directly read the charge at a particular photodiode without first moving it. Transistors present at each photodiode amplify and read the charge without having to move the charge across the sensor before amplifying it.

The maximum number of photo-generated electrons that can be stored at each photosite of a CCD is called its "electron well depth." The intensity of brightness recorded at any single photosite is proportional to the ratio of photo-generated electrons versus the well depth capacity of the photosite. The well depth for a particular imaging sensor is a finite number. If enough light strikes the sensor to create the maximum number of electrons that a photodiode's well depth can hold, the image pixel at that location will be saturated to maximum brightness. If the well depth capacity of

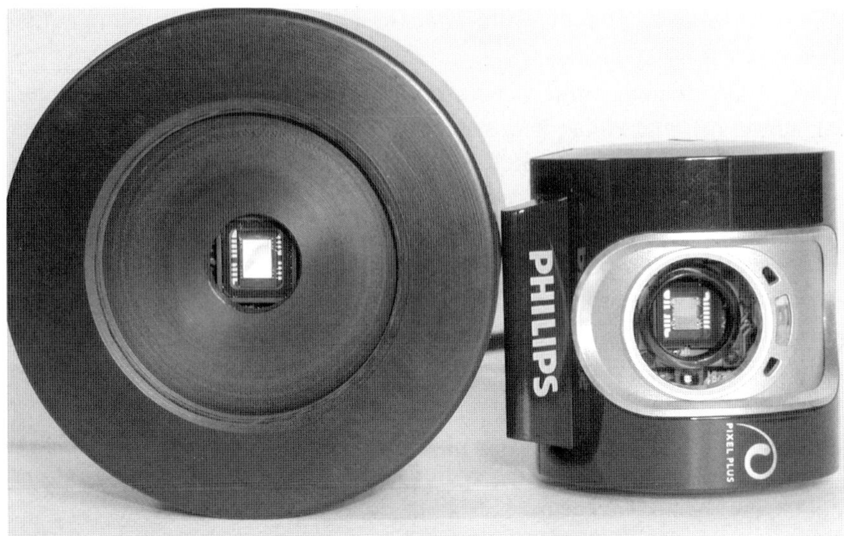

Fig. 1.10 *Modified re-chipped cameras such as the ATK-2HS have a larger CCD than a standard webcam. The Sony ICX424AQ CCD with 7.4-micron pixels is over 50% larger than the 5.6-micron pixel Sony ICX098BQ CCD in the Philips SPC900NC. Photo by Robert Reeves.*

a particular photosite is reached during an exposure, no amount of additional exposure will increase the number of electrons stored at that photosite.

Another term used when describing imaging sensors is "fill factor." This refers to the size of the actual light-sensitive portion of each photosite on the CCD sensor. The larger the light-sensitive area of the photosite, the more incoming light it can capture, resulting in a higher fill factor. An image sensor with a large fill factor is desirable in astrophotography because they are more efficient at capturing incoming light.

An easy way to understand how a digital imaging sensor works is to compare its array of individual photodiodes to a large array of thousands of solar cells. As the optics focus varying light intensities upon the different cells across the array, they produce varying amounts of electricity. The number of electrons produced by each solar cell is proportional to the brightness of the image of its particular location on the array. We can use the electricity from each solar cell to illuminate an equal sized array of light bulbs to reproduce the image. The brightness of each bulb on the "display" array will be proportional to the electrical output of the corresponding solar cell on the camera array. A CCD sensor in a webcam performs a similar task, but the electrons produced by each photodiode on the sensor are amplified by electronics in the camera, then sent to a computer where the camera's operating program processes the signals, displays the image on screen, and stores the images on the computer's hard drive.

Once the electrical charge created by the interaction of the photons striking the photosite is amplified, it is converted into a binary number representing a digital value by an analog-to-digital (AD) converter. The AD converter is the brain of a digital camera's imaging system. It converts the signal from the image sensor into the digital data that create the final image. The CCD sensor outputs a voltage, or analog signal, for each of the pixels its array. This voltage is proportional to the amount of light the sensor receives. The voltage is amplified and processed by the camera's AD converter, which changes the voltage to a binary number. The binary number, in turn, is stored in a file containing an array of other binary numbers representing the other pixels on the image sensor. Early black-and-white webcams used 6-bit AD converters resulting in a limited dynamic range of only 64 levels of gray scale. Today, consumer cameras use an 8-bit AD converter. This means that the differing voltages produced by the photodiode can have up to 256 distinct values of brightness ranging from completely black to completely white. A webcam is capable of displaying 24-bit color with 256 ranges of brightness in each of the red, green, and blue colors in the Bayer color mosaic.

Once the data have been digitized, the software interpolates colors from the Bayer color filter mosaic (a process that will be discussed later in this chapter). Then the image data are transferred to a computer's hard drive.

While the CCD detectors used in a variety of imaging applications, including webcams, produce excellent images, they are not without fault. Image artifacts are created by a number of factors and the longer the exposure the more these factors must be compensated for in order to limit their effect. Short-exposure images of the Moon or planets with unmodified webcams do not suffer from artifacts as much as longer exposures with modified ones do, so these factors can be ignored for the most part. Users who "graduate" to long-exposure capable cameras will begin to encounter the image defects inherent with CCDs. We need to be aware that in addition to the image signal produced by the CCD, it can also produce undesirable forms of electrical signal output capable of degrading the image. I will explain how to deal with these defects in later chapters.

These other forms of signal are:

1. A bias voltage that is constant, and present even if there is no exposure.
2. Systematic variations in the bias voltage.
3. Random variations in the bias voltage.
4. "Dark current" that increases with elevated temperature.

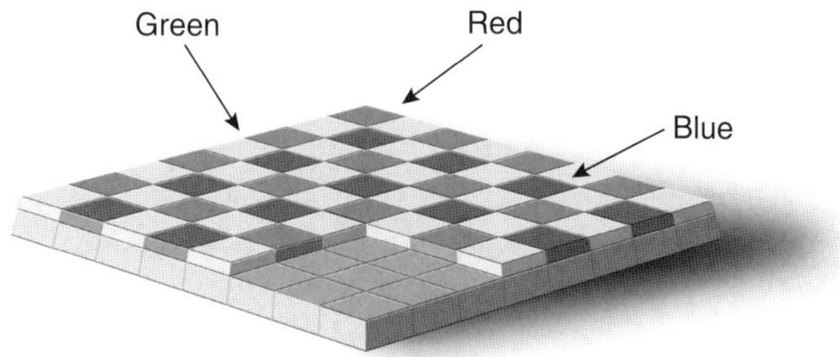

Fig. 1.11 *To create a color image using a black-and-white sensor, a mosaic of red, green, and blue color filters, called a Bayer Pattern array, is placed over the sensor so each photosite is covered by one individual color filter. The camera's software creates a color image by sampling the signal from other colors surrounding each single-color photosite.*

5. Systematic variations in the dark current.
6. Random variations in the dark current.
7. Random variations in the signal amplifier output known as readout noise.
8. Variations in the quantum efficiency of different photosites on the CCD array.

The good thing is that one does not have to know a lot about computers to use a webcam. The devices are geared toward ease of use. When applied to celestial photography, the user will have to apply advanced image-processing software to process the resulting videos into photographs, but the programs are not hard to master and require no more actual computer skill than that is required to operate a good word processing program.

1.7 Color Arrays and Image Sensors

The image sensor in a webcam is actually a monochrome device, that is, it only registers light intensity, not colors. To create a color image, a color filter array, called the Bayer pattern filter, is laid over the sensor. This filter is a grid of multicolored filters that looks much like the pattern of a checkerboard. A majority of cameras uses what is called an RGB filter that passes red, green, and blue wavelengths. As a result, each photosite only records the brightness information for one of the three colors.

The Bayer pattern filter is the most commonly used and attempts to mimic the human eye by passing more of the middle wavelengths in the green portion of the visible spectrum. For example, on a 640 x 480 pixel

RBG sensor, the pattern would be:

320 x 240	RED	76,800 photodiodes
320 x 240	BLUE	76,800 photodiodes
320 x 480	GREEN	153,600 photodiodes

The actual color of the image at each photosite is estimated by the camera's software using the different intensity information from the adjacent red, blue, and green sensitive photosites. This information is combined to create one pixel, or spot, of color.

The downside of the Bayer pattern filter is that it limits the amount of light reaching the sensor by filtering each photodiode to capture just one color; thus mosaic filter sensors capture only 25 percent of the red and blue light, and just 50 percent of the green.

The terms "photosites" and "pixels" are often used interchangeably in reference to a digital imaging sensor, but in reality they are two different items. A pixel represents the brightness data gathered by each photosite on a black-and-white CCD, or the combination of RGB data gathered by four adjoining photosites on a color CCD. As Figure 1.11 illustrates, this does not mean there are four times fewer pixels than photosites, but it does take data from four photosites to create the color information in one pixel. In fact, there are an equal number of photosites and resulting pixels in a given sensor. Technically, the resulting pixels that form the image are offset between photosites because each pixel shares data with four photosites.

Digital sensors react differently to light than film emulsion (see Figure 1.12). If film is exposed to light that is too weak, it has no effect until the light reaches a certain intensity limit called the threshold. Whether film has been exposed or not, it has a certain density when developed because of the natural density of the film material and a small amount of chemical-induced emulsion fog. Together these are known as "film base plus fog." Above the threshold level, exposure starts to rise very gradually, then more steeply. A graphical plot of a film's response to light intensity is called its characteristic curve. The initial slow rise in this curve is called its toe. The middle of the curve where it rises steadily is called the straight-line section, and at its top, where the film has recorded its maximum exposure, the curve starts to level off again in an area called the shoulder. The contrast range of the film image depends on the slope of the curve; the steeper the slope, the higher the contrast. Film photographs tend to have low contrast in the shadows, which normally reside in the shallow toe of the curve, and often have little detail in highlights if they are in the shoulder.

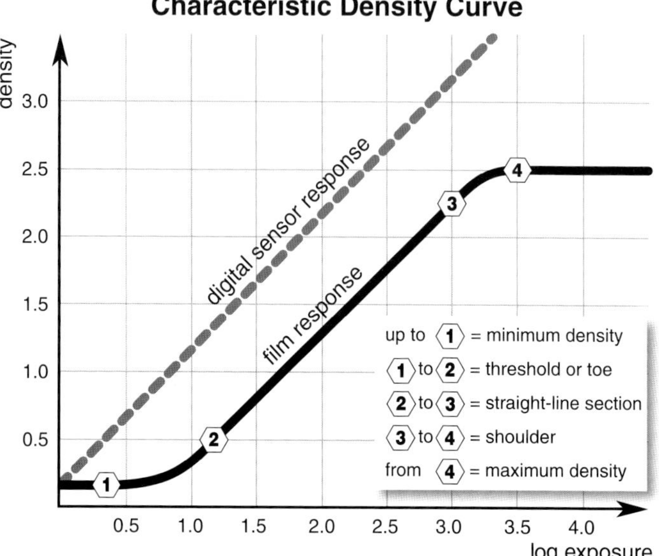

Fig. 1.12 *A comparison of the typical characteristic density curve for film compared to the response of a digital image sensor.*

Digital sensors also have a certain threshold below which they do not react to light. Above the threshold, sensors react to light increases in a directly proportional way. This means there is no toe to the response curve; it begins a straight climb up the slope until the sensor becomes saturated with photons and abruptly reaches a point where additional light has no effect at all. Thus, the characteristic curve for a digital sensor is essentially a straight line without either a toe or shoulder.

With film, underexposing or overexposing the negative results in thin shadow detail or dense highlights that are hard to print. The consequence of digital's straight-line response to light is that with underexposure there is no image at all and with overexposure the detail in the saturated pixels is also gone. There is nothing to work with. Above the threshold point however, we can achieve better shadow detail with digital because of its more linear response.

Unlike the single-size format used with 35mm film cameras, digital image sensors can have many sizes. These differing sizes can be confusing to a photographic enthusiast who has grown up with only the single 35mm format. Fortunately, webcams and their commercially modified astrocamera cousins only use several size designations. These are described by a fraction such as ¼ or ⅓. This designation dates back to the 1950s when television vidicon tubes were measured using this system. The designation does not actually refer to the size of the imaging surface but to the outer diameter of the round glass envelope of the tube, expressed as a fraction of an inch. Television engineers found that the usable area of the imaging tube was approximately ⅔ of the diameter of the tube, and a similar designation has been carried over into the digital imaging field. There is no specific mathematical model that determines the relationship between the diameter of the imaging circle and today's digital imaging sensor size, but it is always roughly two-thirds. Most users feel this size description system is outdated for digital imaging sensors, but it is still used by the electronics industry that supplies them.

Webcams offer the option of reduced resolution settings such as 320 x 240 instead of 640 x 480 pixels. But unlike astronomical CCDs that "bin" pixels together to achieve higher sensitivity at the expense of resolution, webcams do not bin pixels. Instead, they merely discard every other scan line and every other pixel in the scan lines that are used.

The aspect ratio of a CCD detector is characteristic of its physical shape, or the relation of the resulting image width and height in pixels. With conventional 35mm film photography, we are used to the 24mm x 36mm frame size where the image is half again as wide as it is high, giving us an aspect described as either 3:2 or 1.5. Webcams traditionally give us images that are 640 x 480 pixels in size, with an aspect ratio of 4:3 or 1.33.

Aspect ratio can also refer to the size of the individual pixels. Some astronomical CCDs have photosites that are rectangular. Fortunately, the CCDs in webcams, both those with 5.6- and 7.4-micron pixels, have square photosites that produce square pixels making it easier to perform geometric computations on the resulting images. Since webcams use lower quality CCDs than science instruments, some degree of post-exposure processing is almost always needed, but square pixels ease the computational requirements for such processing.

1.8 Interpolated Resolution

The ToUcam and SPC900NC are both capable of producing images with pixel dimensions of 1280 x 960. Astronomically, however, we do not use this option. The camera's native resolution is 640 x 480 and thus achieves the higher pixel count through interpolation, a process that doubles each existing pixel. Terrestrially, images with smooth tones look good at the higher resolution, but highly detailed scenes, such as the lunar surface, show no additional detail while the resulting image file size is much larger.

1.9 A Look at Two Popular Cameras

The Philips ToUcam Pro II 840K and the newer SPC900NC are very popular webcams for lunar and planetary imaging. Each weighs only about 100 grams, lighter than many telescope eyepieces, and they do not affect telescope balance. Philips has ceased production of the ToUcam 840, but many are still available on the used equipment market. Fortunately, the SPC900NC retains all the exposure functions that make the Philips cameras very desirable for astrophotography. Internal examination of the former's electronics reveals that the newer camera is essentially a repackage of the older ToUcam, but there is a definite improvement in image quality produced by the SPC900NC even though it uses the same CCD sensor as the ToUcam. Another good feature is that the SPC900NC USB cable is two feet longer than the one found on the ToUcam which often allows the computer to be more conveniently positioned.

The ToUcam Pro II comes with a removable 6mm *f*/2 lens that gives the device a normal vertical field of view of 33°. The SPC900NC uses a similar *f*/2.2 lens. These tiny lenses are designed for the cameras' normal terrestrial applications and are of little astronomical use. For the latter, the lens is removed by unscrewing it and substituting a threaded telescope adapter. The lens cover on the SPC900NC must be popped off the camera body before the lens can be removed. The adapter threads into the camera body like the original lens and possesses a 1¼-inch diameter barrel to allow the camera to be inserted into a standard telescope focuser, ensuring perfect alignment with the telescope optics.

With the introduction of the ToUcam, Philips eliminated a very useful feature for planetary imaging that had been included in the Philips Vetsa webcam software driver. Earlier versions of the camera driver, when set to 240 x 320 resolution, recorded a full resolution "partial scan" of the central ¼ of the image. This allowed one to record the full size disk of a planet at normal resolution while discarding a majority of the empty frame around the planet. This also allowed either higher frame rates, because the smaller image size enabled more frames per second to pass through the USB 1.1 connection bandwidth, or smaller resulting AVI file sizes. The "updated" ToUcam and SPC900NC driver shrinks the entire image to a lower resolution and thus lunar or planetary details are recorded on fewer pixels and are much smaller. The new driver reduces image resolution by reading the full 640 x 480 frame, then discarding both every other scan line and every other pixel in the scan lines retained. Because this records only ¼ of the original image data, upsampling the image to 640 x 480 will not restore image detail.

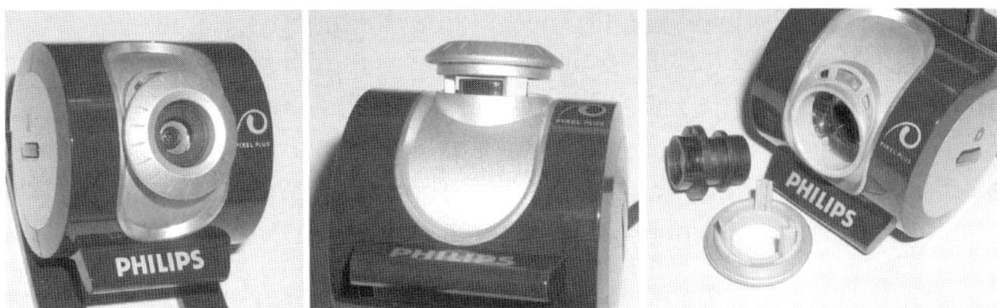

Fig. 1.13 *The Philips SPC900NC (left) camera has replaced the ToUcam. How to remove the apparently fixed lens bezel in order to install a telescope adapter is not obvious. A flat tool can be used to gently pry up on both sides of the lens bezel (center). The three hooks securing the bezel will pop loose allowing the bezel to be withdrawn. The lens can then be unscrewed (right) to reveal the threaded lens collar attached to the camera's front printed circuit board above the CCD sensor. Photos by Robert Reeves.*

The ToUcam 840 and SPC900NC receive all of their power through the USB cable that connects them to the computer. The ToUcam 840 uses 1.2 watts of power while the SPC900NC uses only 75 milliwatts (23 milliwatts in standby mode) while in operation. Both cameras have negligible impact on laptop battery life and both produce a high-quality 24-bit color image. The control software is capable of correcting the white balance in light conditions ranging from 2500K to 7500K.

The ToUcam 840 is capable of producing frame rates as high as 30 per second, but astronomical users are forewarned that because the camera is USB 1.1 any frame rate above five per second utilizes image compression to download data through the 12 Mb per second bandwidth available with USB 1.1. The newer SPC900NC is advertised as USB 2 compliant, but early examination of the camera's electronic components reveals little operational difference with the older USB 1.1 ToUcam. The camera is advertised as being capable of up to 90 frames per second, but consensus among imagers who have examined the electronic workings of the SPC900NC is that it is still basically USB 1.1 and thus applies some image compression to high frame rates.

Both cameras also have audio capture capability through a built-in digital microphone. The user may wish to use this feature to record verbal notes about the video being captured, but this is discouraged for two reasons. First, the audio signal uses part of the USB bandwidth and it is desirable to reserve as much of that as possible for the video signal (to prevent USB 1.1 cameras from invoking any form of image compression in order to deliver the desired frame rate through the available data pipeline). Second, some image processing programs have trouble with audio data attached to a video file. The popular RegiStax program is reported to encounter difficulty with audio files, although I didn't see evidence of this when I accidentally left the sound on during a number of lunar videos.

The faster frame rate available with the SPC900NC is attractive to planetary imagers who desire large amounts of video frames in a short period in order to limit image smearing from planetary rotation. However, the higher frame rate inherently limits the shutter speed to being equal to the inverse of the frame rate. In other words, it is not possible to expose each frame for a period longer than the individual frame scan and read time. For example, if the camera is taking 75 frames per second, the slowest available shutter speed will automatically be 1/75 of a second. This inherent shutter speed limitation may reduce the frame rate available when imaging dim high-magnification targets.

Of great interest to advanced webcam astrophotographers who "hack" the camera's electronics to apply long-exposure modifications is the fact that modifications previously applied to the ToUcam will also work with the SPC900NC. In fact, there is such a high degree of commonality between the old Vesta, the ToUcam, and the new SPC900NC that software drivers for the ToUcam will work with the older Vesta cameras, and installing SPC900NC firmware into a ToUcam will make it respond like an SPC900NC.

For the experienced webcam imager, the SPC900NC continues the Philips camera tradition of good performance and ease of use with astrophotography. Complaints about it center around the camera's obscenely bright LCD "camera on" status light. The earlier Philips cameras had a dim red LED that astronomers could live with. The SPC900NC possesses a white LED so bright the author has used it to illuminate nighttime close-up targets photographed by the camera. The bright LED can be quashed with the application of black paint or fingernail polish.

The default video output format for most webcams is AVI. Although this produces huge video files because each frame is saved as a full-sized image, astrophotographers desire this format because it preserves all the image detail, and the popular image-stacking and processing programs look for this format as the video source. When the SPC900NC is operated through its native Philips camera operating software, it produces compressed MPEG video instead of AVI. Fortunately for the astrophotographer, when the SPC900NC is operated through the K3CCD Tools program, it reverts to the astronomically desirable uncompressed AVI output.

1.10 An Astrophoto Philosophy

One of the questions often asked about amateur astrophotography is, "Why do it? Don't the professional observatories and NASA have facilities that can take much better images of the sky?" This may be true, but then again, one would never question why it is so fascinating to see the sky through an amateur telescope when professional astronomers can see it much better through their giant instruments. We look through telescopes for fun, we do it because we are curious about the universe around us, we do it because the objects we see in space can be beautiful, and we do it because of the challenge in finding faint astronomical objects in the dark night sky. Similarly, we enjoy astrophotography because a camera can extend the range of human vision, allowing us to see the normally unseen. It also allows us to share the universe with others. Many enjoy the technical challenge of adapting imaging equipment to a telescope and creating exciting views of celestial objects that rival the results achieved by professional astronomers and NASA probes. We take justifiable pride in displaying a beautiful picture of Saturn or a spacewalk-like view of the Moon and knowing "I did this. This is my picture."

If astrophotography is one of your interests, it is best to develop a philosophy that can guide you toward your goals in this pursuit. It should be your own personal justification for pursuing an activity not commonly practiced by the average person. Your philosophy should explain why you do astrophotography, how you do it, and how far you are willing to pursue it to achieve results you are satisfied with.

I do astrophotography to satisfy just one person—me! I want to record the beauty of the constellations, chart the river of stars that makes up the Milky Way, to see the unseen, and record these vistas to enjoy again later. If I am satisfied with what I do, there is a trickle-down effect. Others around me are exposed to new ideas and new worlds through my enthusiasm for this activity.

To succeed at astrophotography, your philosophy should dictate that you pursue projects that are attainable with your own equipment, whether it is a modest telescope or a top-of-the-line beauty. It should not preclude experimenting see what is possible with your given equipment. In fact, your philosophy should encourage pushing the limits of your equipment in quest of better and better results. But realistic goals must be maintained to avoid disappointment.

Overreaching is a sure way to become discouraged in a new endeavor. It is my hope that this book will help beginners gain the confidence and skills needed to achieve their full astrophotographic capability. But simply spending money on fancy equipment does not guarantee success. Skills and techniques need to be mastered by the individual. An analogy with a legendary race driver is appropriate: sometimes it is the driver's skill and

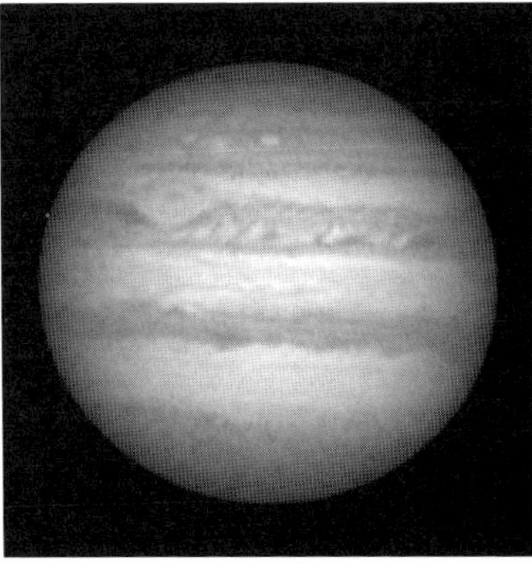

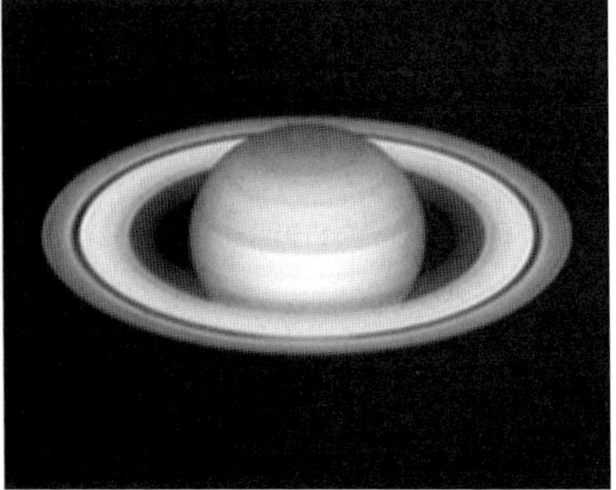

Fig. 1.14 *Astrophotography should never be considered a competition with other astronomers. The idea is to have fun and achieve the best image your equipment is capable of producing. By careful application of astrophoto techniques, Steve Foster captured these excellent images of Jupiter and Saturn using a ToUcam 740 and a 12.5-inch Newtonian telescope operating at f/25. Photos by Steve Foster and Frank Hinson.*

not just horsepower that wins the race. Webcams are a lot like that too.

A critical analysis of one's results should also be part of an astrophoto philosophy. If something doesn't work, the problem must be solved in order to achieve success on another try. Bad results can be very instructive if carefully analyzed. Was it an equipment problem, a procedural error, or attempting more than the equipment was capable of? Recognizing that one can make mistakes is important. Simply blaming the camera or telescope for poor results will perpetuate the problem if it was "operator error." Problems will occur, the learning curve can be steep at first, and frustrations will happen either through photographic problems or poor weather. Even the most experienced sky-shooters make their share of mistakes. My own would fill a separate book in their own right! But it is important to persevere.

Even the best astrophoto is useless if it is not seen. A good astrophoto philosophy dictates that good results be shared. Today, that means showing them to local astronomy club members, posting new images on personal web pages, and submitting the best ones to popular astronomical publications. Do not worry about having your images "ripped off." That is not likely to happen, because the only realistic venue for them is publication in the relatively small astronomy magazine and book market, the economics of which dictate little monetary reward.

A final note about personal philosophy is to recognize that astrophotography is a natural ambassador for astronomy and science in general. I have yet to see someone who wasn't moved by a beautiful portrait of the

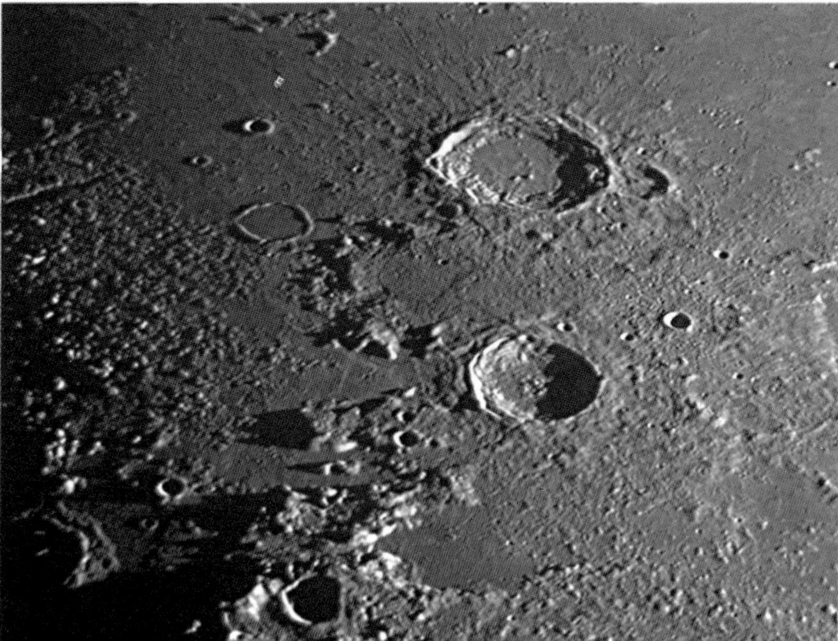

Fig. 1.15 *Lunar images such as this are easily achieved with webcam style cameras attached to modest amateur telescopes. The craters Aristoteles and Eudoxus were imaged by combining 600 separate video frames taken through a Celestron-8 at f/20 using an ATK-2HS camera. Photo by Robert Reeves.*

Orion Nebula or the Andromeda Galaxy. Our efforts should thus be geared toward capturing views of the universe and spreading the faith by sharing them with others.

Now that we see astrophotography is more than just a hobby, let's see what is involved in taking great images of the sky with the deceptively simple but amazingly capable webcam.

Chapter 2
Basics of Webcam Astrophotography

For the astronomer, the webcam has quickly evolved from being just an affordable computer peripheral to a viable, versatile alternative for producing images that are competitive with high-end CCD cameras. But not all webcams are created equal, especially when it comes to the kind of sensors used to make images. Most of the low-priced webcams use Complementary Metal Oxide Semiconductor, or CMOS, sensors that are less sensitive to dim astronomical targets than the CCD-equipped models. Also, newer USB 2.0 cameras capable of high-speed imaging, such as the Creative Live Ultra, have fully automatic exposure controls, rendering them incapable of the exposure extremes encountered in astrophotography. The astronomy webcam imager must sort out the myriad models available and determine which ones are truly useful to his or her personal imaging goals.

In addition to the webcam-style imaging devices available from major telescope suppliers and independent manufacturers like Atik, Orion, and SAC, there are dozens of basic commercial models available in different price ranges, from amazingly inexpensive to about $150 for the bare camera. Most produce adequate results on bright objects like the Moon, but a few rise above the rest: the Philips Vesta, ToUcam 840, and SPC900NC. The CCD-equipped Logitech QuickCam 4000 used to make this list, but recent manufacturing changes—notably a switch from a Sony CCD to a Sharp CMOS sensor—has significantly reduced the low-light sensitivity of the QuickCam. The only way to tell which sensor is installed in a particular QuickCam is to remove the lens and look at the sensor.

The older Vesta cameras have another issue that astro-imagers need to be aware of: they use two different versions of CCDs. All Philips webcams use either a Sony ICX098AK or ICX098BQ sensor. The "BQ," however, is newer and has better sensitivity and fewer hot pixels than the AK. Generally, the older Vesta cameras use the "AK" chip, but there is a production crossover point where either chip was installed in both the Vesta and ToUcam models (the newer ToUcam and SPC 900NC cameras use the

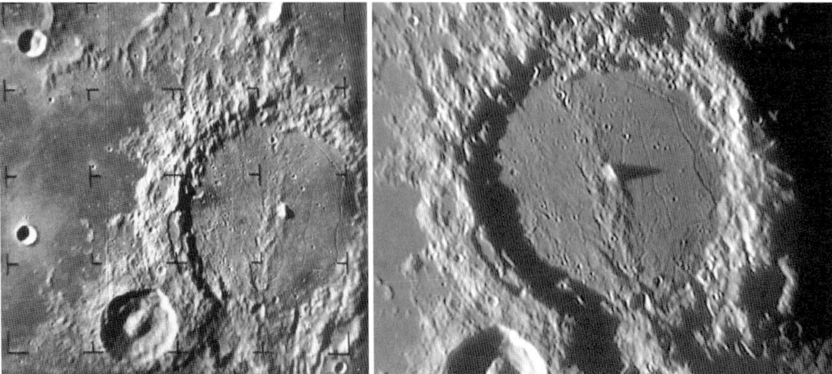

Fig. 2.1 *Lunar photos taken with webcam techniques rival images returned by NASA's early Moon probes. At left is an image of the crater Alphonsus taken by the Ranger 9 spacecraft in 1965. At right is the same crater taken with a Philips* ToUcam Pro II 840K *and a 12-inch reflecting telescope operating at f/29. Ranger 9 image courtesy of NASA, webcam image by Zac Pujic.*

"BQ" chip). An Internet site that shows you what to look for in determining which sensor is present is:

http://homepage.ntlworld.com/molyned/images_of_web-cams.htm.

Fortunately, the AK and BQ sensors have differing appearances. Using a magnifying lens, examine the edge of the light-sensitive portion of the CCD. If it is slightly brownish-purple, it is the older AK sensor. If the edge is bluish, it is a newer, more sensitive BQ CCD.

The sensitivity of an unmodified webcam is usually stated in units called lux. The lux is an internationally recognized standard for measuring the brightness of illumination and is considered to be part of the Metric measurement system. One lux is equal to one lumen per square meter. Rather than getting bogged down in explaining the definition or origin of these various scientific units, I think it more practical to present "real world" comparisons of the relative brightness of various lux levels. For instance:

- Sunlight on an average day varies from 32,000 to 100,000 lux.
- A TV studio is illuminated at about 1000 lux.
- A bright office has about 400 lux.
- Bright moonlight is about one lux.

The footcandle is an older measurement of illumination sometimes still used in the United States, primarily in building codes. For comparison, a footcandle is equal to 10.76 lux.

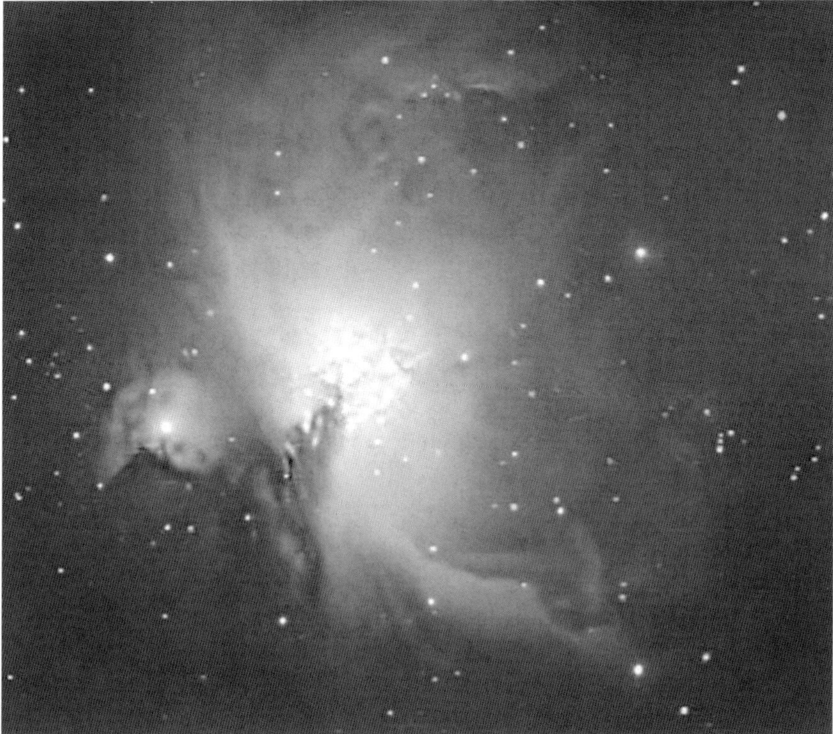

Fig. 2.2 *Many of the bright "showpiece" deep-sky objects can be imaged from urban settings using long exposure-capable webcam-style cameras and small backyard telescopes. A Meade Deep Sky Imager and Orion 80ED refractor were used to take this image of M42. Photo by Dave Street.*

2.1 Pixel Counts

This is a good place to clear up a confusing point about pixels. There are several different methods of measuring the number of pixels on a sensor, but for the astrophotographer, what really counts is the number of actual non-interpolated pixels in the raw image. I'll use Sony's ICX098BQ chip as an example. This chip, which is sensitive to light levels as low a one lux, has a total of 692 x 504 pixels and a raw output of 659 x 494 pixels. After internal camera processing, however, only 640 x 480 pixels are used to form the image. Why? The difference between the number of pixels on the sensor and the number of pixels in the final image stems from the fact that the top two, bottom eight, left two, and right 31 rows of pixels are covered with a black dye to create what is called "video signal shading" to give the camera's operating system a true black so it can take a dark-current reading needed to calculate what is completely black. Within this shaded frame is the sensor area actually exposed to light when an image is made.

2.2 Exposure

Webcams produce amazing astrophoto results, but the camera settings must be tweaked for proper exposure. When the camera is turned on for the first time, the control program will likely be set to "auto." This is a good setting for the kind of normal terrestrial imaging the camera was designed for, but it usually fails miserably when imaging an astronomical target because of the extremes in contrast. An image of a planet, for instance, will be surrounded by vast expanses of black space that will fool the auto-exposure controls into overexposing the planet. By clicking off the "auto" option and manually adjusting the shutter speed and image controls, the planet will become recognizable. The same is true for a lunar image along the shadow-filled terminator.

It is acceptable to boost brightness and gain settings in order to find and focus dim targets, but when adjusting the image-capture exposure for photography of solar system objects, leave the brightness setting in the middle of the selection range where it has no effect on the data being recorded. Altering the brightness range tends to clip image highlights and reduce the dynamic range of the image. Image-capture exposure must be set by adjusting exposure speed and gain settings, not the brightness. Also, remember that changing the frame rate will also shift the exposure settings, so recheck them after any frame rate changes.

To achieve the best results from a webcam, use the shortest possible shutter speed to freeze the effects of atmospheric blurring and preserve lunar and planetary detail. This means it is necessary to pass the maximum possible amount of light from the telescope to the camera. For this reason, if possible, do not use a star diagonal with a webcam. Any such additional component in the optical path will reduce the total light throughput the telescope and thus require a slower shutter speed to compensate.

For lunar and planetary imaging, webcams prove to be the system to beat. No other low-cost commercially available camera achieves the performance on bright solar system objects of a webcam coupled to a good telescope. And for deep-sky imaging, home-modified and commercially-modified astronomical webcams are amazing in their ability to scoop up starlight. For example, the Atik IIs that I use can image not one, but two stars within the center of M57, the Ring Nebula, with just a seven-second exposure through an 8-inch SCT.

Webcams are most sensitive in the red portion of the spectrum and less sensitive in the blue. This is good for astrophotography because the cameras' peak sensitivity is near the wavelength of the hydrogen-alpha line, one of the dominant emissions of nebulae. Thus, we must take care not to filter it out with infrared rejection filters. Fortunately, good brand-name infrared filters take this into account.

Any webcam can image the Moon and bright planets. But only modified webcams can image deep-sky objects. There are several approaches to deep-sky astrophotography with a modified webcam: snapshot, tracked-but-not-guided and guided.

- **Snapshot.** It is possible to take 30-second, 60-second, and even two-minute or longer exposures with modified webcams. The limiting factor is image noise. The cooler the temperature the less noise, so winter astrophotography has an advantage. Single shots are noisier than a stack of images, but snapshots can be effective for a supernovae survey, asteroid search, and variable star monitoring (with image calibration). A photographic Messier marathon is even possible at certain times of the year. Fast focal length Newtonians, short tube refractors, or an SCT with a focal reducer will work well for such activities. With 60-second exposures and focal lengths of *f*/5.6 or longer, city light pollution and moonlight are of little consequence, so snapshots can be done from urban backyards. Noise is also very controllable using dark-frame subtraction.
- **Tracked-but-not-guided.** If no means are available to guide a telescope during a long exposure but the mount tracks well and is polar aligned, the tracked-but-not-guided technique allows one to synthesize a long exposure by accumulating multiple short ones. This method is limited by the ability of the telescope mount to maintain good unguided tracking for the duration of the exposure—usually no more than 30 seconds. This technique, coupled with today's widely available image stacking software, has relieved imagers of the manual guiding burden faced by the previous generation of film photographers.

  Stars fainter than those visible on many star charts can be detected with this technique, but this is naturally constrained by how deep the accumulated exposure can reach. The dimmest object each individual exposure can record is limited by the faintest photon event recorded by the CCD photosite. A stack of 120 separate 30-second exposures may not display as deep an image as twelve 5-minute exposures because the faintest detail may not accumulate sufficient photons on a CCD photosite in 30 seconds to trigger a photon event.

  Satisfying images of star clusters and showpiece objects can be taken with the tracked-but-not-guided technique, but if the goal is faint deep-sky targets, then guided, multiple long exposures from a dark sky location will be needed.
- **Guided exposures.** These can encompass everything from single 30-second exposures with a telescope drive that is not accurate enough for the tracked-but-not-guided technique, to multiple exposures of several

minutes duration each, when using a low-noise camera under dark skies. Lengthier exposures will not only record dimmer stars but also begin to show faint detail in nebulae and galaxies.

Proper image exposure is a balance between both exposure time and the gain setting. Short exposure times can be offset by boosting the gain setting (similar to using a fast film with shorter exposures in a conventional camera). However, too much gain can introduce excess noise that will degrade the image (similar to the coarse grain of high-speed film). With solar system targets, you can monitor the exposure meter provided with K3CCD Tools and adjust the exposure to keep it around ⅔ to ¾ maximum. Do not adjust the "brightness" setting to alter the apparent exposure. For astronomical targets, it is best to leave the brightness setting at its default middle level, or an incorrect gain setting will be applied leading to poor exposure or the "onion ring" effect around a planetary limb. Exposures long enough to peak the exposure meter mean some pixels are saturated and detail will be lost.

"Integration time" is CCD-speak for exposure time, or the length of each separate exposure in a series. Depending upon the speed and frame rate selected, the actual shutter speeds exhibited by Philips webcams and the selected shutter speed may be two different parameters. In practice, any difference between the selected and the actual shutter speeds is not critical because variations in exposure that cause image degradation can be corrected on the spot. However, it is useful to have a working knowledge of how the Philips cameras actually respond to given settings.

Interestingly, the Philips Vesta, ToUcam, and newer SPC900NC cameras actually expose longer at slow shutter speeds than the setting parameter would indicate. This is helpful with imaging planets at high magnification since we need all the light-gathering power we can get. Matthias Meijer tested these three cameras at settings normally used by lunar and planetary photographers. His tests showed that shutter speed did not vary between 320 x 240, 353 x 288, and 640 x 480 resolution settings. The ToUcam, operating with Version 1.0 of the Philips ToUcam webcam driver, exhibited accurate shutter speeds only at 15 frames per second and 1⁄33- and 1⁄50-second shutter speeds. The older Vesta camera remained accurate through all but the slowest and fastest speeds. The actual shutter speeds, displayed for the 640 x 480 resolution settings, are shown in milliseconds in Table 2.1.

The Philips SPC900NC displays a different user interface that does not define the selected shutter speed in fractions of a second. Instead, a movable slider with eleven click-stop positions allows the setting of the

Table 2.1
Actual Vesta and ToUcam Shutter Speed at 640 x 480 Resolution

	Vesta 675			ToUcam Pro			
Speed	**5 fps**	**10fps**	**15fps**	**Speed**	**5 fps**	**10fps**	**15fps**
Auto	200ms	100ms	66ms	x	x	x	x
1/25	100ms	40ms	40ms	1/25	200ms	100ms	66ms
1/33	28ms	28ms	28ms	1/33	100ms	50ms	33ms
1/50	20ms	20ms	20ms	1/50	40ms	32ms	20ms
1/100	10ms	10ms	x	1/100	20ms	12ms	x
1/250	3.5ms	3.5ms	x	1/250	10ms	6ms	x
1/500	1ms	x	x	1/500	5ms	4ms	x

Table 2.2
Actual SPC900NC Shutter Speed at 640 x 480 Resolution

Position	5fps	10fps	15fps
1	200ms	100 ms	64ms
2	100ms	50ms	32ms
3	40ms	32ms	20ms
4	20ms	13ms	10ms
5	10ms	6ms	4.8ms
6	4.8ms	3ms	3.2ms
7	1.4ms	1.2ms	1.2ms
8	x	0.5ms	0.5ms
9	0.4ms	0.4ms	0.4ms
10	0.4ms	0.18ms	0.18ms
11	x	0.12ms	0.08ms

shutter speed at arbitrary positions between slow and fast. Table 2.2 shows the actual speeds at each of the eleven positions, again in milliseconds.

Similar shutter speed tests can be conducted with any camera by attaching a bar to a stepper motor or low-speed electric motor that rotates at a known speed. By calculating the sweep of the bar during the exposure, the true shutter speed can be deduced.

2.3 Gain

A webcam's gain setting is similar to changing the ISO speed setting on a conventional digital camera. Higher gain settings increase the amount of amplification applied to the voltage signal produced by electron events at each pixel's photosite. This amplification is applied before the analog-to-digital conversion of the signal. The number of electron events is propor-

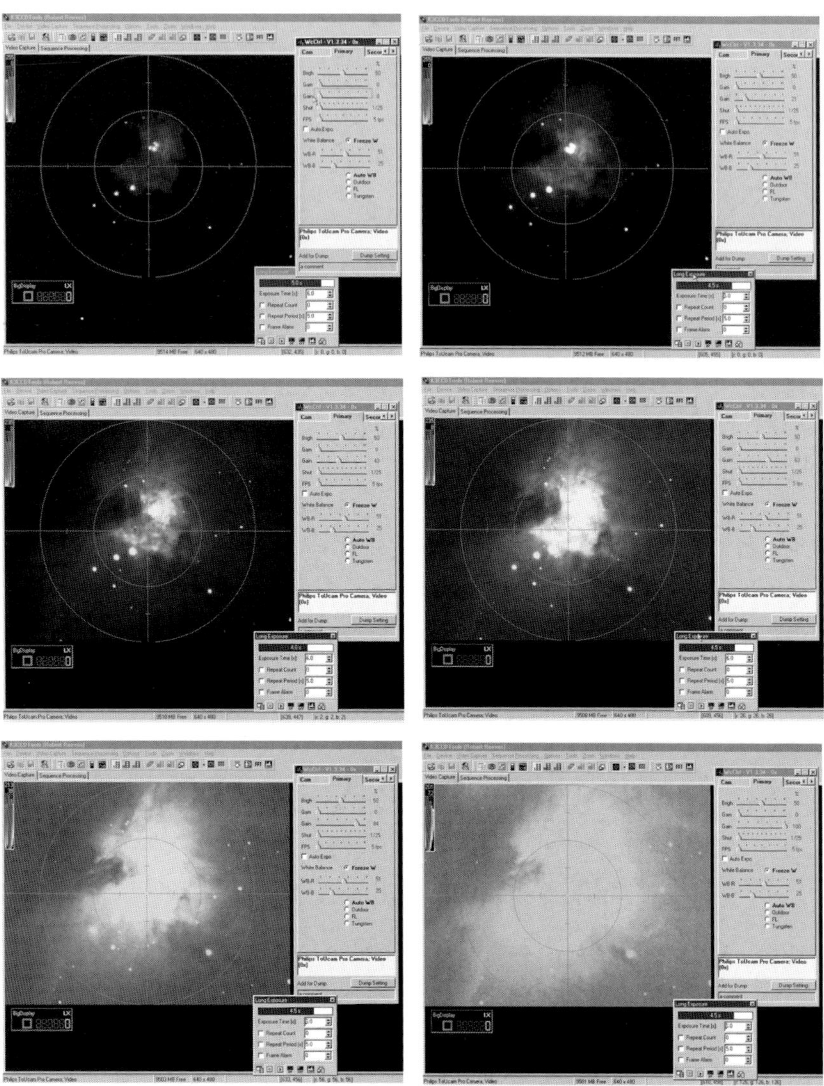

Fig. 2.3 *The effect of varying gain settings can be seen in this series of 6-second exposures of M42 through a Celestron-8 telescope using an ATK-2HS camera with a 0.3 focal reducer. Gain is increased from 0 to 20%, 40%, 60%, 80%, and 100%. At high gain settings, sensitivity is high, but noise overpowers the image. At 100% gain, M42 is almost unrecognizable. Photos by Robert Reeves.*

tional to the intensity of the incoming photon flux. If the incoming light level is low, a higher amount of amplification, or gain, is applied to increase sensitivity. Unlike conventional digital cameras that apply increased gain with electronics inside the camera, webcams use a software driver in the camera's host computer to control such functions. The apparent increase in sensitivity eventually reaches a point of diminishing returns where additional gain does not significantly improve the desired image but instead overwhelmingly amplifies the undesired noise always present in a digital sensor.

A high-gain setting and slow shutter speed will help find dim targets, such as Saturn magnified by a Barlow, or dim stars to be used for autoguiding. Elevated gain setting can also reveal small targets to aid in focusing, like Jupiter's moons, but at the same time planetary detail will be overly saturated and Jupiter will look like a bright featureless ball. Gain-magnified noise in the image appears coarse, similar to the way high-speed photographic film is grainier than slower films. The webcam mantra is thus "gain equals grain." Once the target is located and centered in the field of view, reduce the gain setting to 40 percent or less. Image noise will also be limited and the result will be a much cleaner image. It is best to increase the exposure time to compensate for lower gain settings.

There is not a linear mathematical relationship between the gain and shutter speed, so no science can be applied to balance gain and exposure. These factors vary from target to target and for different optical systems. The proper balance has to be found individually for each set of circumstances to produce what looks best. The tradeoff is that high gain and short exposures accentuate the grainy background while low gain and longer exposures produce "cleaner" images but will be sensitive to blurring caused by poor seeing. Fortunately, the user can partially remedy the "graininess" from high-gain settings by using software like RegiStax to average many frames together and smooth out noise.

2.4 Amp

While gain deals with the software processing of the video signal from the webcam, amp relates to the functions of the CCD chip itself. CCD sensors contain a vital component called an amplifier to turn the voltage signals from their photosites into something that the camera electronics can process. In long-exposure-modified cameras, this amplifier creates a small glow in the normally dark portions of the image because in normal operation the "amp" is always switched on. This glow adds noise and tends to spoil the image. A point to remember is that the relative brightness of amp

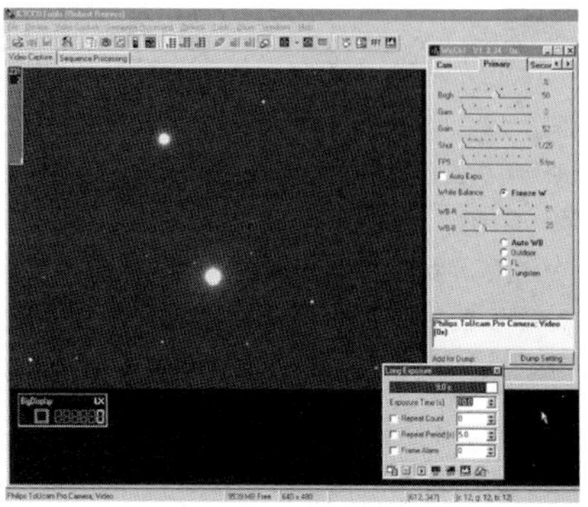

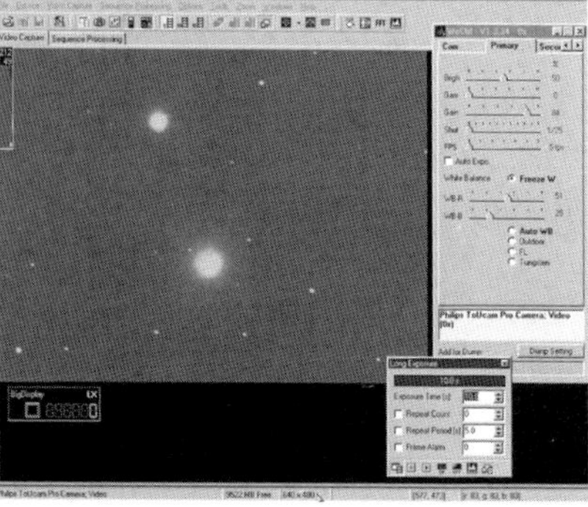

Fig. 2.4 *With webcam imaging, higher gain equals higher grain. These two 10-second exposures of a star field show the difference between 52% gain and 84% gain. Increasing the gain to higher levels creates noise in the image, giving it a grainy appearance. Photos by Robert Reeves.*

glow compared to the target is always going to be the same. High gain is always going to show more amp glow because the glow is amplified along with the rest of the image signal. Reducing exposure or gain will show less amp glow, but the brightness of the target will also be reduced. This places a limit on the exposure length of a webcam that lacks additional modifications.

The term "amp-off" refers to long-exposure webcams that have been further modified so the amplifier is turned off during the exposure and then switched back on briefly when the image is "read" from the CCD. This reduces, or eliminates, the intrusive amp glow. Although it is critical to the proper function of the CCD, the amplifier need not be on throughout an exposure.

2.5 Frame Rate

The goal in planetary imaging is to capture as many video frames as you can as quickly as possible. But this goal clashes with the rule of thumb stating that older USB 1.1 camera connections should use 5 frames per second when capturing astronomical videos. Anything faster will degrade the image quality of individual frames. While some users claim good results at frame rates as high as 15 per second, the fact is that the bandwidth of the USB 1.1 connection cannot carry the full amount of image data at higher frame rates. At high frame rates, some degree of image compression is used to pass a rapid sequence of images through the camera's 12 Mb/sec USB 1.1 connection. This will ultimately introduce artifacts and degrade the final image. Image stacking programs need the best input possible to create the final combined image. Newer USB 2.0 connections do not have this bottleneck, but it is a limitation we must accept with older USB 1.1 cameras. The number of frame drops (discussed in Section 2.6) increases dramatically at higher frames rates because the USB 1.1 data bandwidth cannot keep up with the increased data flow. The compression used with higher frame rates is "lossy," in that it discards as much image data as necessary to accomplish the image transfer. This is shown in Figure 2.5 where 10 frames-per-second produces a less detailed image than 5 frames-per-second. Since webcams are designed for "normal" pictorial photography of extended objects with relatively smooth tonal gradations, the loss of image information is rarely noticed on a "live" terrestrial image. Astronomical imaging, however, often involves fine detail in a relatively small area, such as the globe of a planet or very detailed high-contrast areas of the lunar surface.

A computer's basic processing speed, or the CPU's clock speed in megahertz, measures how fast it can process incoming data. Manufacturers list the minimum computer system requirements for operating a webcam. Generally, older Pentium II 300 MHz computers are adequate for the job with USB 1.1 cameras. Of course, the higher the amount of onboard RAM, the better it is for running all programs, including a webcam.

The initial aiming and focusing can be done at higher frame rates to avoid image flicker and jerkiness, because the eye will average out the detail in the rapid succession of frames. High frame rates at this point help with focusing when the atmosphere is turbulent as well as with acquiring small targets like planets. (Slow, flickering, frame rates make it hard to judge the point of best focus when operating the focuser.) When slewing the telescope to search for a planet that is just out of the field of view, high frames rates are necessary because the planet may skip unnoticed through the field if the frame rate is updating the image display too slowly. Once focus is achieved, the frame rate must be slowed down for actual imaging.

With planetary exposures limited to five minutes, the user is faced with the decision to capture either 1500 higher-quality 5 fps frames or 3000 lower quality 10 fps frames. The theory is that while the 10 fps frames may be of lower average quality, the greater number of frames increases the chance that atmospheric seeing will allow more of the individual frames to be of higher quality, than the fewer frames available at 5 fps. Which works best with your particular imaging system can only be determined with some experimentation. My preference, when using my own optical and computer system, is 5 fps.

The beauty of webcam imaging lies in the ability to cull the best-focused and steadiest images from a series and combine them to create a

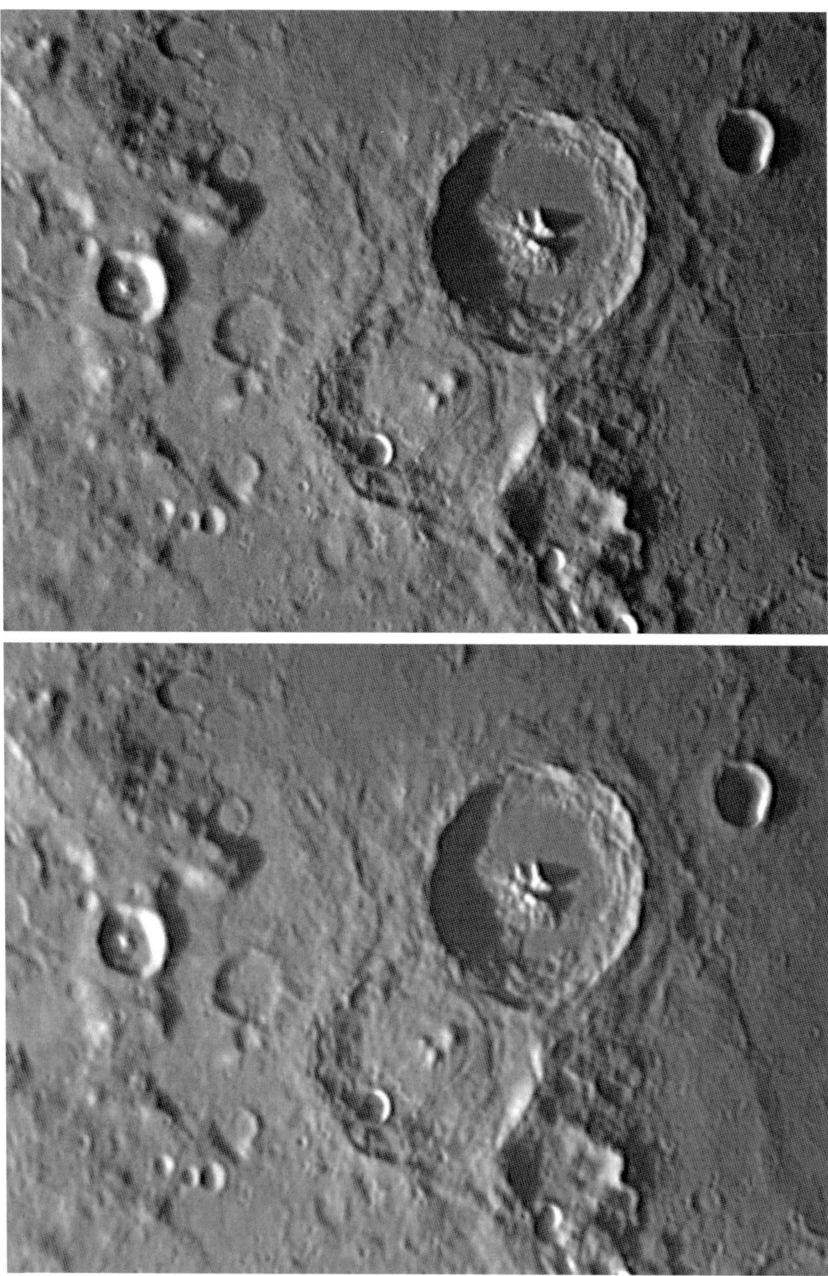

Fig. 2.5 *The top image was exposed with a ToUcam 840 through a Celestron-8 telescope at f/20 using a frame rate of 5 frames per second. The bottom image was taken immediately afterward under similar seeing conditions using 10 frames per second. The faster frame rate produced a noticeably degraded image because of the higher image compression required to pass more video data through the limited USB 1.1 bandwidth. Photos by Robert Reeves.*

detailed master image. Frames blurred by poor seeing can be filtered out by software and only the very best ones used to create the final image stack. The more frames gathered, the higher the number of good quality frames that will be available for stacking. Hence, the need to gather as many as practical within a given time becomes apparent.

A final note to remember about frame rates is that they are not independent of the exposure time. The Philips ToUcam driver software will automatically adjust the exposure or gain setting when the frame rate is changed. This is done to keep the exposure level constant. For instance, if the camera is operating at 5 frames per second at a 1⁄25-second shutter speed, we know from Table 2.1 that the camera is really operating at 1⁄5-second. Thus when the frame rate is changed to 10 per second, the camera has to increase the shutter speed to a minimum of 1⁄10-second. To maintain constant overall exposure, the gain setting must automatically increase. We can, of course, manually lower the gain if we choose, but the software will initially automatically change it when the frame rate is increased.

2.6 Dropped Frames

Frame dropping occurs when the camera produces more data than the USB connection can transfer, or the computer's hard drive can accept. Older USB 1.1 connections are inherently slower than the newer USB 2.0 standard. The length of the USB cable also affects its ability to carry high data rates. The three- to four-foot cables that come with a webcam may be inconveniently short at the telescope, but adding an extension cable may slow the data rate and increase the likelihood of frame dropping. At ten frames per second most USB 1.1 users experience frame drop. All the frames that are recorded are complete, but not every frame taken by the camera makes it into the recorded AVI file.

There are a number of things that can be done to speed up the data transfer and reduce or eliminate dropped frames. Use WDM (Windows Driver Mode) mode for video capture. This requires the use of K3CCD Tools but using this software for video capture is a good idea anyway. Using VFW (Video for Windows) mode at the highest quality compression will often lock up the computer when using K3CCD Tools. If the compression is lowered to 80 percent, dropped frames cease to be a problem, but compression artifacts are visible in the images.

A slow hard drive interface will impede the flow of data from the camera. If your hard drive supports it, make sure the Direct Memory Access (DMA) option for your hard drive is turned on. Sometimes it is off by default. Enabling DMA can dramatically reduce the rate of frame drop.

If the hard drive on your computer is an older, naturally slow unit, an external hard drive with a USB 2.0 or IEEE1394/Firewire interface may be an option to speed performance while imaging. A 7200 RPM IDE hard drive operating with a moderately powerful CPU (700 MHz or faster) will usually have no frame drop at 5 frames per second even though the USB 2.0 connection to the external hard drive defaults back to USB 1.1 speeds when older USB 1.1 cameras are in use.

A 7200-RPM hard drive will perform better for video recording than an older 5000-RPM model. The condition of the hard drive also plays a major role in how fast the incoming video stream can be recorded without having to reject video frames. If the drive needs defragging, the video stream will not be recorded in a continuous series of tracks on the drive. The drive will have to hop between free tracks on different areas of the drive, slowing its ability to keep up with the torrent of video data, and the computer will start to drop frames in an effort to keep up with as much of the video stream as it can. Creating a separate drive partition just for imaging can simplify maintenance of the hard drive.

In older cameras where the 320 x 240 resolution is a true full-resolution partial scan of the full frame, reducing the resolution from 640 x 480 to 320 x 240 will eliminate frame drop and allow higher frame rates. High-magnification planetary imaging does not use the outer parts of the full-resolution frame anyway, so those data aren't needed. Eliminating this portion of the frame reduces the frame file size by 75 percent without harming the resolution of the planetary image. Unfortunately, the newer driver supplied with the ToUcam and SPC900NC does not crop to the central part of the field of view and thus newer cameras cannot benefit from this option. The new camera driver reduces the resolution of the entire image when in 320 x 240 resolution. However, something to consider when choosing higher full-scan frame rates is that planetary images have large areas of black sky background while lunar images generally do not. The large black background in planetary images compresses more efficiently than do large areas of detailed lunar landscape and thus fewer compression artifacts will be seen in planetary images taken at a higher frame rate.

Still, some users argue that even if there is frame drop at higher frame-per-second rates, there are more frames for stacking than if the video were taken at lower rates. For example, a 60-second video taken at 5 frames per second and experiencing no frame drop will result in 300 frames available for stacking. If the frame rate is boosted to 10 frames per second for the same 60-second video length, but experiences frame drop one third of the time, the result is still 400 video frames, or 100 more than recorded at 5 frames per second with no frame drop. On the surface, this argument looks good, but in reality frame rates higher than five per second create so much data that in order to pass it through a USB 1.1 connection the camera is forced to compress it. This should be avoided if possible because compression can create image artifacts that will affect the resolution of the final stacked image.

Some things that may help reduce the rate of frame drop are:

- Reducing the frame rate from 10 to 5 per second.
- Do not use a USB extension cable between the camera's native USB cable and the computer unless it is a logistical necessity. USB connections up to 10 feet are supposed to function well, but in reality, any extension beyond the camera's own cable reduces the speed of video data transfer.
- Turn off any real-time monitoring in your anti-virus software. When imaging, also disable automatic virus updates that try to connect to the Internet via a wireless connection at preset intervals. If an antivirus program begins a scheduled scan while the webcam is operating it may turn on during a video capture sequence and tie up enough computer resources to create frame drops or even freeze the computer.
- Shut down all unnecessary programs running in the background such as email programs that gather messages at set intervals.
- Make sure the frame rate in both the native camera control software (such as VRecord with a ToUcam) and aftermarket camera control software (such as K3CCDTools) are both set to the same rate.
- Disconnect any other USB peripherals while the camera is in use.
- Do not move a USB mouse on an USB 1.1 system during a video capture. Since both camera and mouse share USB bandwidth, mouse movements may induce dropped frames.
- Change the video format to IYUV and make sure WDM capture is active instead of VFW.
- Defrag the hard drive and, if possible, create an "astro-imaging" partition dedicated to optimized recording of video data onto sequential hard drive data tracks.
- Disable laptop power-saving options to prevent the hard drive going into standby mode or the processor from shifting to lower speed if no keyboard inputs are detected after a certain period.
- On older laptops with more limited CPU processing power, see if there is a utility to keep the cooling fan on continuously. If the fan comes on during video capture, it may induce frame drop.

Some of the above mentioned items might not be inducing frame drop with the video system on your computer. If you have applied several of these suggestions and your frame drop problem disappears, reapply each item one at a time until you find which one is the actual culprit affecting your video recording.

2.7 Noise

One of the major advantages digital imaging has over film, especially in low light, is the efficiency of the CCD sensor in capturing faint images. But just as film astrophotography has its inherent limitations and problems with low-light levels, specifically reciprocity law failure, digital imaging has its own indigenous problems that we have to deal with in long-exposure low-light conditions. This situation manifests itself as electronic noise in the image, or the snowy, coarse appearance seen in webcam images when the gain setting is raised to high levels.

There are two components to every image taken with an electronic sensor: signal and noise. Signal represents the desired variations in pixel brightness due to the features in the target being imaged; the greater the variations in brightness the greater the variations in contrast. But there are also undesired variations in brightness caused by noise. A good image needs a high ratio of "signal-to-noise" so the target will be clear and distinct. If high amounts of noise are present, the image lacks contrast and may display speckles and artifacts.

Modern CCD sensors used in cooled astronomical cameras exhibit relatively little noise. Webcam sensors, however, are inexpensive devices manufactured to less stringent tolerances and can be considered inherently noisy compared to true astronomical CCDs. Noise is usually of little consequence in short lunar and planetary exposures because it is averaged out in the process of stacking many video images to create a single still image. Long-exposure capable cameras will display some degree of noise. Understanding the source of noise will help understand how to deal with it.

In addition to commonly discussed causes such as bias, thermal noise, and sensor inconsistencies like hot or cold pixels, less well-known phenomena including photon noise, detection noise, and readout noise contribute their share of degradation to the desired image signal.

Photon noise is actually independent of the image sensor itself. It is noise in the input light reaching the sensor and is the result of inherent statistical variations in the arrival rate of photons incident upon it. When the arrival rate of photons is low, such as in the low light levels common with deep-sky astrophotography, the variation in arriving photons is greater and causes more noise.

Detection noise is caused by the sensor's inability to record all incoming photons. Not all photons that strike a photosite on the sensor result in electron events that build a charge, or signal. Because the detective quantum efficiency (DQE) of a sensor is always less than 100 percent, there are fewer photoelectrons than incoming photons, meaning that some of the photons are lost, increasing the variation in the arrival of effective photons and increasing noise.

Readout noise is caused by the inability of the camera electronics to perfectly detect all the photoelectrons stored at each photosite, convert them into a voltage, and then convert the analog voltage into a digital signal. Each of these steps can introduce noise into the signal. Readout noise is important in webcam astrophotography because some camera models have more efficient electronics than others, thus reducing its effects. For instance, single-frame images of the Moon taken at the prime focus of a given telescope with an Atik IIhs are much "cleaner" and contrasty than those taken through a standard ToUcam, even though the latter has smaller pixels and should resolve finer detail. The difference is in the larger, more sensitive 7.4-micron pixel black-and-white CCD used in the Atik camera instead of the smaller 5.6-micron pixel color CCD used in the ToUcam.

Bias, sometimes called "offset," is basically a pixel's built-in electrical charge that is always present in the photosite even if there is no exposure at all. In advanced CCD imaging, the effect of bias is eliminated by taking what are called bias frames—a measurement of the pre-exposure electrical charge at each photosite. This measurement is then subtracted from each photosite's output after an actual exposure. In short-exposure webcam photography of solar system objects, we ignore bias because it is an inconsequential portion of the noise present. For the vast majority of long-exposure webcam astrophotography, such as that devoted to taking pictorial images, bias is again ignored unless the camera's output must be calibrated for scientific measurements. Bias only becomes an issue when precisely calibrated images are needed, but such work is beyond the scope of most webcam users.

In addition to bias, thermal noise or dark current will be present at each photosite even if no light falls on the sensor during the exposure. This additional electrical charge will artificially raise the pixel values by an amount proportional to the length of the exposure. Thus the longer the exposure, the more noise will affect the image. Thermal noise looks like "static" in a weak broadcast TV image (something the new generation of cable TV viewers may never have seen!). Thermal noise increases with temperature, approximately doubling for every 7°C increase in sensor temperature. There are a number of strategies to reduce noise caused by the variations in pixel electrical charge. These include cooling the sensor, stacking multiple images, and filtering the final image with additional software. The image calibration process removes the detrimental effects of pixel brightening due to this additional electrical charge.

The inconsistent sensitivity of one photosite compared to another on a sensor, or external factors such as light leaks in the optical path to the sensor, vignetting, dust on the sensor, or internal reflections lead to what are called flat-field inhomogeneities. Basically, this means some portions of the sensor are more or less sensitive to incoming photons than others. Hot or stuck pixels (always on maximum brightness) or cold or dead pixels (always dark) are the most noticeable forms of inhomogeneities. These defects will cause undesired bright or dark areas in the image. These are controlled through the image calibration process by taking flat frames that map out the defects affecting sensitivity.

In daily terrestrial use, when compared with typical low-cost webcams, the ToUcam stands out as very sensitive and noise-free. However, when used for low-light astrophotography, there will still be a lot of noise. Because of increased image compression at higher frame rates through the ToUcam's USB 1.1 connection, faster frame rates will display more noise in their individual frames. This presents the solar system astrophotographer with a choice: use a low frame rate and stack fewer but cleaner individual frames; or use a higher frame rate and stack more but noisier individual frames, while allowing the stacking process to smooth out the noise. At first, the idea of using faster frame rates with higher inherent noise may seen counterproductive since it also uses greater amounts of hard drive space to store the video. But the higher frame rate also uses a higher shutter speed and thus freezes the effects of poor seeing that may plague slower shutter speeds.

Noise is rarely a serious problem with unmodified webcams because their inherent short exposure times limit their use to bright objects. Unless the gain settings are significantly boosted to image dim targets such as Jupiter and Saturn at long focal lengths with their associated high *f*/ratios, the fraction of a second exposure times used with lunar and planetary imaging produce little noise.

A standard webcam like the Philips ToUcam will get noticeably warm during extended operation. Modified long-exposure-capable webcams are susceptible to thermal noise and users have to employ various cooling methods to limit thermal noise in the imaging sensor. Some people resort to the simplest and most basic way to cool a webcam for long exposure imaging—put it in a freezer before use. This technique, however, has several flaws, some of them detrimental to the camera. First, the camera has limited "cool time" and quickly warms back up under ambient conditions. Second, there is a high risk that condensation will not only coat the

Table 2.3
Video Frame Stacking Noise Reduction Factor

Number of Video Frames	Noise Reduction Factor	Number of Video Frames	Noise Reduction Factor
4	2	81	9
9	3	100	10
16	4	200	14
25	5	300	17
36	6	400	20
49	7	600	24.5
64	8	1000	31.6

cover over the CCD, thus blurring the image, but may also short out the electronics on the camera's circuit board.

The most common cooling method for modified webcams is a small fan, such as a low-amperage, ball bearing-mounted computer CPU cooling fan that can be powered via the camera's USB connection. Cooling the webcam's sensor will allow higher gain settings, and thus more sensitivity without additional noise. A more advanced and effective method is to employ a Peltier cooler to reduce the sensor's temperature. Such cooling can provide a significant reduction in long-exposure thermal noise. Reducing the temperature 7°C will halve the noise, reducing it 14°C with halve it again to one-quarter of its original level, while reducing it by 21°C will reduce noise to one-eighth its uncooled level. Achieving a 20°C reduction with off-the-shelf Peltier coolers adapted to small webcam imaging circuits is not an unreasonable goal.

Stacking, or averaging, multiple images of the same target can also reduce noise. Noise is reduced by a factor equal to the square root of the number of images being stacked together. For instance, if four images are stacked, the noise in the combined image is reduced by half. If 100 are stacked, the noise in the combined image is reduced to 1⁄10 that of a single one. If 1000 images are stacked, noise is reduced to approximately 1⁄32 that of a single image. Since high-gain short- or long-exposure video frames are somewhat noisy to begin with, webcam images realize great benefit from stacking hundreds, or even 1000 separate frames.

Filtering the stacked image with proper software can further reduce noise. Some software packages such as RegiStax perform low pass, or Fast Fourier Transform (FFT), filtration that reduces the fine detail part (high frequency) of the image. This type of filtering smoothes the image, making noise less visible, but at the expense of image detail and sharpness. This type of filtering is good for normally diffuse objects like nebulae and galaxies. However, as applied in RegiStax on lunar and planetary images, the FFT filter still performs an amazing job of improving the image.

A median filter smoothes noise by replacing each pixel with density values equal to the median value of adjacent pixels. This type of filter can maintain good sharpness while removing noise from the image. Various astronomical image-processing programs such as Images Plus, AIP4Win v2.0, and Astroart as well as Photoshop allow the application and even the custom creation of various types of filters.

Grain reduction programs such as SGBNR from Pleiades Software and Neat Image also do a credible job of reducing the effects of image noise even though these programs were actually designed to reduce the coarse appearance of grain in photographic film.

As a general rule, with short-exposure webcams it is good practice to use the lowest possible gain setting and to capture at least two minutes of video at 5 frames per second: that is, 600 frames. From among these frames, enough good images will be available for the auto selection routines in RegiStax or Astrostack to choose the best frames for stacking and noise reduction.

For the very best results using videos that are exceptionally steady, manually select the best frames for stacking. The autoselect function in some programs measures image contrast to determine the best frames and thus the process can select frames that have atmospheric distortion in them. Distorted frames may have adequate contrast to pass the selection criteria, but detail in such frames will not align with other frames and will detune the sharpness of the completed stack.

Once image stacking is complete, it is important not to over-apply wavelet and other sharpening filters or these may accentuate the noise that we have worked hard to eliminate.

2.8 White Balance

All light sources have an inherent colorcast associated with them. The human eye has an amazing capacity to accommodate wide variations in lighting color. If it didn't, the world would seem very bluish at noontime and very reddish at sunset. In film photography we are used to considering the "color temperature" of the light source when choosing which film to use. We have the choice of "daylight" or "tungsten" films that are balanced respectively, for bluish sources or reddish artificial light sources. If the color of the light source does not match the sensitivity of the film, we use color-correcting filters to prevent significant color shifts in the resulting images. This color shift is most prominent for white-colored objects. For instance, to render colors correctly on daylight film while shooting under incandescent lighting, we use a strong blue filter to compensate for the yellowish tint of the light.

Digital cameras, including color webcams, have what is called the "white balance," a parameter that gives them a large advantage over film cameras. Since digital cameras cannot change sensors to accommodate a different color of illumination, we have to modify the sensor's perception of the off-color light in order to remove any colorcasts present. The white balance is a digital camera's electronic compensation for varying colorcast in lighting. Compensation is achieved by changing the gain for the red, green, and blue channels so that white objects are truly white and there is no colorcast in the image.

For astronomical photography, we get the best results when using films balanced for normal daylight illumination. The same is true with digital cameras. For the most part, we can leave the white balance setting on "auto" and let the webcam perform the white balance selection automatically. If desired, the camera's software also allows the user to override the control and set the white balance manually. Cameras do a reasonably good job of automatically setting the white balance, but in some cases they can be fooled, resulting in off-color images. This is especially true with astronomical subjects surrounded by large areas of black sky. For most users, the true white balance is a judgment call.

Remember, however, that it is always better to perform as much correction as possible at the image source rather than later using an image-processing program where data are lost during each step of the correction process. For that reason if the auto white balance does not produce good planetary colors use the following procedure: In daylight, point the telescope at a white card and allow the camera's auto-white balance to set itself, and then unclick it in the camera controls. This procedure should be done in daylight because illuminating the card with a flashlight or other artificial light will create a colorcast. The camera's operating system should remember the white balance settings during the later nighttime astrophotography session.

2.9 Video Time Length for Lunar and Planetary Imaging

Another consideration when trying to achieve the very best planetary resolution, is how long should the AVI capture time length be before the finest details, as defined by the Dawes Limit, become smeared by a planet's rotation? Peter Campbell offers the following equation to answer that question:

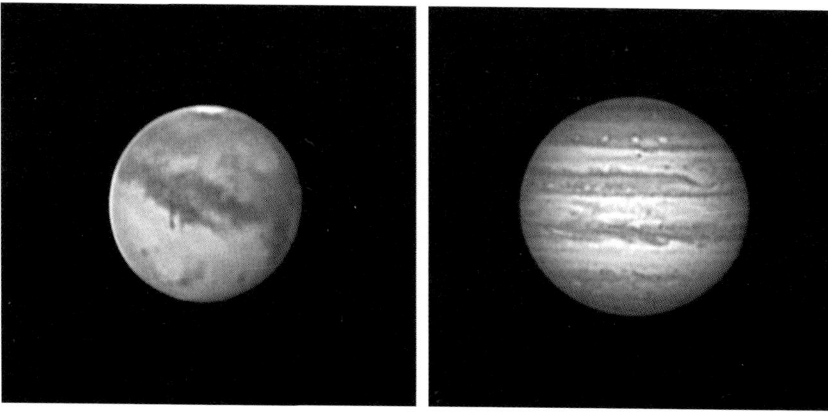

Fig. 2.6 *The rotation of Mars and Jupiter puts limitations on how long a video of the planet can be before details from the first and last frames no longer register together. A practical video time limit for planets with most astrophoto combinations is about five minutes. Photos by Steve Foster and Frank Hinson.*

$$T = \frac{S \times R \times 3600}{D} / 3.14$$

where

T = Total capture time in seconds
S = How much smearing, in arc seconds, you can accept
R = Planet's rotation period, in hours.
D = Apparent diameter of the planet in arcseconds.

For the sake of argument, Peter assumes the maximum amount of detail smearing allowed will be half the Dawes Limit for any particular telescope. Thus when Mars is imaged through a 6-inch aperture telescope, the maximum smearing allowed will be 0.38 arcseconds. Since Mars at favorable opposition is about 25 arcseconds in diameter and rotates in approximately 24.6 hours, the formula calculates a maximum exposure of 428 seconds. This is long enough to capture over 2000 frames at 5 frames per second.

For a faster-rotating planet like Jupiter imaged through a 12-inch telescope, the resulting exposure limit will be shorter. The planet rotates in about 9.9 hours and at opposition reaches an apparent diameter of 45 arcseconds. The formula thus dictates a maximum exposure of 48 seconds, limiting the user to a little under 250 frames at 5 frames per second. For most imaging situations where seeing is less than optimal and the planet is not at opposition, these rules are a little more lax. In practice, Jupiter can usually be imaged for up to five minutes with little noticeable image degradation from rotation. Saturn rotates almost as fast as Jupiter but the planet's surface is blander and lacks the minute surface detail that both Mars and Jupiter display. It is possible to exposure video sequences of Saturn for about 10 minutes without losing significant detail. Peter Campbell's equation does, however, offer a guideline for achieving optimal results when both seeing and telescope performance are at their best. But nothing in astrophotography is concrete, and as we saw in the Master Webcam Imagers section of this book, stacked and image-processed webcam images routinely achieve greater detail than the Dawes Limit suggests is possible.

The Moon is a relatively static target in that it always presents the same general face toward us. Surface features remain stationary although there is a limited amount of tilting and shifting of the lunar globe from day to day due to the effects of libration. We are thus not significantly limited in our video capture time when imaging the Moon. From hour to hour it is possible to see changes in shadow lengths along the terminator. The higher elevation crater and mountain features along the dark edge of the terminator will slowly appear at "sunrise" or disappear at "sunset." But over all, there is no practical astronomical limitation to the length of a video capture for lunar features. Even along the terminator, a ten-minute video at 5 frames per second is no problem as long as the image is properly tracked, there is sufficient hard drive space to record it, and the video file size does not exceed limitations imposed by the computer operating system. This however represents a "longest case" scenario for lunar imaging. In practice, a 600-frame video is more than adequate for a majority of lunar imaging.

2.10 AVI File Size Limit

The above discussion helps us determine how long, time-wise, an AVI imaging sequence can be, but there is another limitation to the length of an AVI video sequence: the file size as stored on the computer hard drive. Depending on which version of the Windows operating system you use and which type of file management system you have, the maximum size of an AVI file can be as "small" as one gigabyte or a maximum of four gigabytes. In other words, not all AVIs are created equal. One gigabyte of data may seem huge, but consider that a two-minute webcam video sequence taken at 10 frames per second will create a file this size.

Most computers using Windows 95, 98, 98SE, and ME no longer use the older FAT16 (File Allocation Table) file management system, but old laptops used as telescope computers may employ it. The FAT16 system cannot store more than two gigabytes in a single file. Windows NT 4.0 utilizing FAT16 can store four-gigabyte files, but few computers used by astrophotographers utilize the NT operating system. Windows 98, 98SE, ME, 2K, and XP utilizing the newer FAT32 system can store four-gigabyte files. Technically, the maximum file size is actually four gigabytes, minus one byte, or (2^{32} – 1 bytes). Today, Windows 2000 and XP utilize the New Technology File System. NTFS can store files up to 64 terabytes in size.

One might ask, if computers are now capable of storing individual files larger than four gigabytes, why are AVIs taken by webcams still limited to four gigabytes? Indeed, on some systems, the AVI file size "barrier" seems to be only one or two gigabytes. There are a number of reasons for this. Basically, the multimedia system built into Windows 95 is incapable of handling files over one gigabyte. Fortunately, Windows 95 is essentially obsolete now. Also, the initial AVI file size limit was only one gigabyte when using Media Control Interface (MCI) based software. This was increased to two gigabytes when using Video for Windows. Today, most AVI parsers still use something called signed arithmetic when creating the video file. This forces a file storage limit of two gigabytes when using Windows 98. Even if the AVI creation program allows larger file sizes, the Windows 2000 and XP operating systems are still limited to four gigabytes because AVI files use 32 bit pointers. Thirty-two bit pointers cannot address more than four gigabytes.

Users who want to capture as many video frames as possible while using computers or software packages that invoke an AVI file size limit run the risk of exceeding that limit. This results in the dreaded "failed-to-decompress-.AVI" message when the video is opened for processing. Fortunately, most video-capture software packages allow specifying the maximum capture file size. This option automatically stops the video capture before the appropriate one-, two-, or four-gigabyte file size is exceeded.

In practical terms, the limitations of AVI file sizes will be of little consequence to the average webcam astrophotographer. As we saw earlier in this chapter, there is a point of diminishing returns past which acquiring more than a certain number of video frames does little to improve the final image. AVIs with 1000 video frames are within the capability of any Windows operating system and are, on average, very adequate for imaging the Moon and planets. Increasing the AVI size to 2000, or even 4000, will statistically improve the final result, but is often not worth the extra time and effort needed to acquire and process the larger video file.

2.11 Dust on the CCD

"Dust donuts" are artifacts that will eventually appear on images from any astronomical digital imaging device that features a removable lens. Basically, the donuts are shadows of almost invisible dust specks on the imaging sensor. It is virtually impossible to keep dust out of the camera after repeated handling with the lens removed. Some of this dust inevitably lands on the cover over the imaging sensor. Although they are nicknamed donuts, these shadows have no holes in unobstructed optical systems. True dust donuts complete with holes are present only in Schmidt-Cassegrain, Cassegrain, and Newtonian systems in which the secondary mirror appears in the center of the dust speck's shadow. Removing dust from a camera sensor will become a normal part of basic set-up just like tweaking the collimation of a reflecting telescope before a high-resolution imaging session.

It is surprising that something as tiny as a speck of dust could produce such a large spot on an astronomical image, but the physics of light works to accentuate such particles if they are present on the imaging sensor or a filter near the sensor. For instance, in a typical *f*/10 optical system, a mote only 100 microns in diameter—slightly wider than a human hair—lying on a filter one-half inch above the imaging sensor will cast a shadow 1200 microns in diameter. The reason the shadow is so large is because the long focal length brings light rays into a nearly parallel state. Thus, the tiny dust spot is no longer hidden by diffused light coming from many directions. CCD expert Richard Berry described this as an "anti-pinhole" effect because the physics are similar to a pinhole camera, except that the dust speck is an obstruction instead of an orifice. The light loss inside the shadow of the speck is only 0.6 percent compared to the unshadowed portion of the image and is essentially undetectable. The problem arises when image processing of astronomical targets increases the contrast and sharpness of details, accentuating dust donuts until they are visible on the lighter portions of the image.

2.12 Cleaning the CCD Sensor

Dust on the main telescope mirror or lens is usually not a problem. A surprisingly large amount of dust can accumulate on the instrument's primary optics with little effect on the image other than a slight reduction in contrast. But dust and other foreign objects on the eyepiece, Barlow lens, camera filter, or on the imaging sensor itself can be a problem. Any dust or dirt particles that accumulate on the sensor will be more noticeable on lunar images than on planetary or deep-sky images in which much of the background is black. Since astrophotography is usually done outdoors where there is often a breeze, dust will eventually migrate onto the sensor.

Some tips for combating dust on a camera's imaging sensor:

- Minimize the time the camera is open with no lens or body cap.

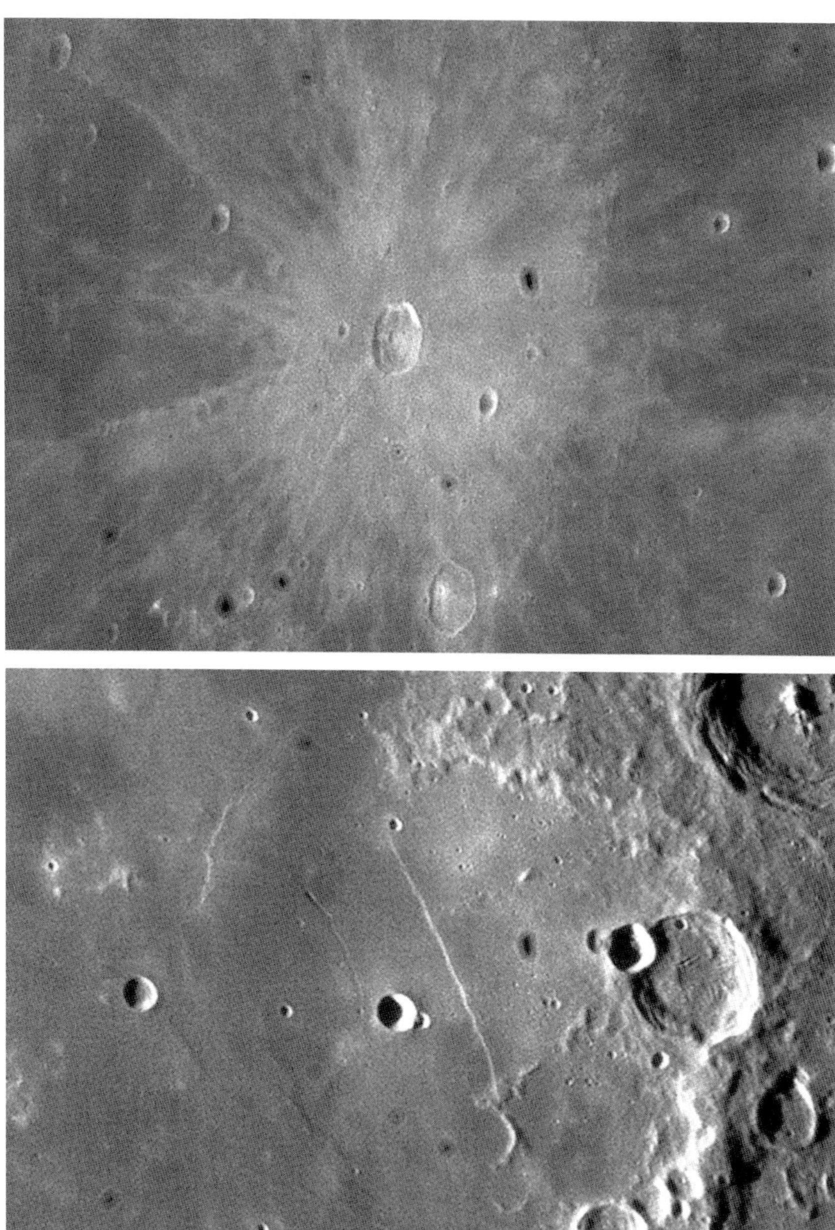

Fig. 2.7 *Dust particles on the glass cover of a CCD will cast a shadow on the sensor, creating black blobs visible on lighter image backgrounds. Dust shadows are invisible against the black background of deep sky-objects, but readily show up on lunar images that cover the entire frame. Photos by Robert Reeves.*

- Since electrical charges may attract dust, some people power down their cameras when they have to be removed from an optical component that has been protecting the sensor.
- If the "normal" lens has been removed from the webcam, always leave the eyepiece adapter installed and place a cap over the open end.
- If a cap is not available, keep an IR filter installed on the eyepiece adapter.
- Blow out the inside of the dust cap with compressed air before installing it on the camera.
- When a lens or accessory must be changed on the camera, hold the camera aimed downward to minimize particles falling into its opening.
- If possible, when imaging without any filter (such as when autoguiding) with an open optical system like a Newtonian reflector, rotate the telescope tube so the camera is aimed downward while it is in use.

Technically, dust particles do not actually lie on the sensor itself, but on the glass cover or infrared filter that lies just in front of the imaging sensor. Dust on the sensor does not create any physical damage, like scratching a negative with film images, but it does call for spot removal from each image with an image-processing program. Eventually this will become a drudgery and more direct action will be needed to eliminate the offending dust blob.

Do not remove the lens or cover and peer inside the camera to look for dust. This act is guaranteed to simply put more dust inside it. The best policy for cleaning the sensor on a webcam is to keep it clean to begin with. The fewer times it needs to be cleaned, the better. But in time, dust will eventually end up on the imaging sensor no matter how careful the user is. Most dust is too small to be seen by the unaided eye and the best way to find it is to record it on an image taken with the camera.

"Charting" the location of any dust or small grit on the CCD is simple. Run the webcam with only an eyepiece adapter and no IR filter using your usual camera control program and set it to about 1⁄500-second exposure and a low gain setting. Aim an ordinary television remote control into the camera from about a foot away while viewing the camera output on screen. Vary the distance between the remote and camera until the camera sensor is about half saturated with the IR signal from the remote control. The small point-like IR source will cause any dust spots to show as shadows.

Eastman Kodak, one of the world's largest manufacturers of electronic imaging sensors, reminds us that the glass cover over an imaging sensor can never be perfectly clean, but its contamination can be lowered to acceptable levels. Kodak further stresses that cleaning operations should be done in an electrostatic discharge (ESD) protected workstation under a clean hood

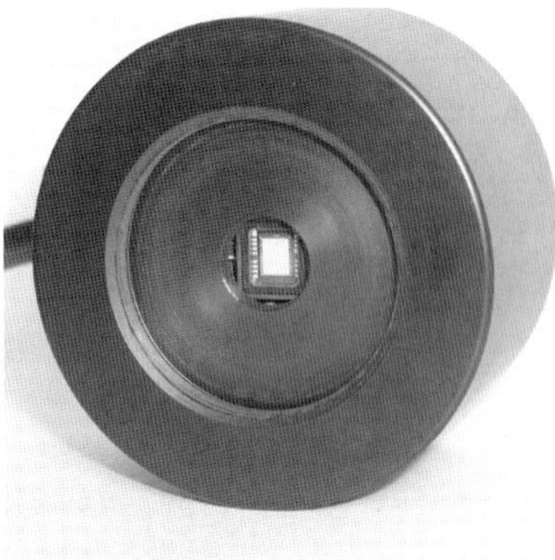

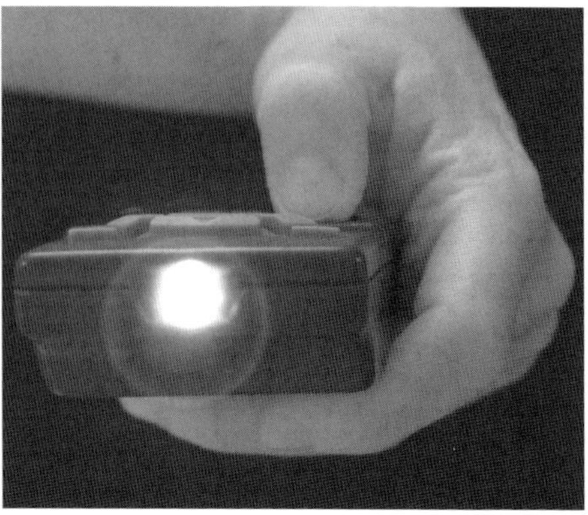

Fig. 2.8 *Eventually dust and small debris will fall on the face of a webcam CCD sensor. Care must be exercised when cleaning the glass cover of a CCD sensor to prevent scratching or smearing. A webcam CCD is very sensitive to infrared radiation and thus a TV remote control is a good "point source" of infrared light to illuminate a CCD to check for dust on the sensor. Emission from the remote control is invisible to eye, but is quite bright in the infrared. Photos by Robert Reeves.*

with a nearby air ionizer. A grounded anti-static wrist strap should also be used. Kodak recommends that nothing other than lens cleaning paper and solvent touch the sensor cover. Finger grease can etch the glass and mechanical contact can scratch it. If gloves are used, they should be powder-free and antistatic. Most latex gloves have powder to aid in putting them on and most gloves create rather than dissipate static electricity.

Usually, dust is very mobile and if it can be easily moved it can be easily removed. It is rare that a particle of debris sticks tightly to the cover of an imaging sensor. Examine the sensor with a 6x or10x magnifier under good light. If an objectionable amount of dust appears on it, there are a number of simple ways to clean it. First, attempt to blow the dust off with a hand squeeze bulb blower without allowing the blower tip to touch the image sensor. If this is not sufficient enough to dislodge the dust, more powerful foot-operated blow nozzle devices are available that allow better manual control of a more forceful stream of air. Foot-operated pumps are available for inflating air mattresses. Try to blow as parallel to the sensor surface as possible to avoid gas pressure driving the particle harder onto the glass. If the sensor is mounted in a recess, care must be taken to avoid pushing contaminants onto it from other places within the camera.

Kodak recommends that the gas used to clean a sensor not have an electrical charge and therefore does not recommend the use of the popular "canned air" blowers. The coldness of the gas and possible fluid discharge can cause electrostatic events that can be detrimental to the imaging sensor. With the official Kodak warning noted, a majority of webcam astrophotographers have used these devices with the proper precautions. These blowers contain pressurized volatile liquid that evaporates to produce the gas blown out of the can's nozzle. It is not unusual for them to spray out some liquid if they are shaken or tipped too far from an upright position. Always test spray on a household mirror before spraying a canned air blower onto an important optical device. If it should spray some liquid onto the optical surface, it may cause more harm than good. Gas escapes at high velocity from the plastic extension tube on a canned air blower; so be careful not to place the nozzle too close to delicate camera parts.

If simply blowing on the sensor does not remove the dust, it is time to actually wipe off the sensor. Kodak recommends that only lens cleaning paper be used to contact the glass cover. Lens paper that is not already folded and mounted on a stick or swab can be clamped in a hemostat to avoid contamination from fingers. Fold the paper so it is the same width as the sensor cover. The cleaning operation should be done in one swipe that covers the entire sensor.

Isopropal alcohol or 200 proof ethyl alcohol is recommended as a sensor cleaning solvent. Methanol is discouraged because not only is it toxic, it does not clean as well. Acetone should never be used because it can dissolve resins that attach the glass cover to the imaging sensor. Before using alcohol on the sensor, be sure to first disconnect the camera from the computer to prevent the possibility of fire. Pure alcohol for cleaning imaging sensors is available from outlets that serve digital photographers' needs.

Use enough solvent on the lens cleaning paper to insure it is wet, but don't soak it so much that it drips. Shake the paper to insure there is not excess solvent on it, but do not allow it to dry too much. There must be enough solvent to enable the paper to slide across the sensor. Use each piece of cleaning paper for one swipe, and then discard it. Always swipe the sensor in the same direction. There will be a short trail of evaporating solvent behind the paper as it crosses the sensor. If it is still not clean, repeat the process. However, if the sensor is not clean after three swipes, there is the possibility that its cover has been damaged.

Debris large enough to be seen with the naked eye can be picked off the sensor with a SpeckGRABER, a device borrowed from the world of digital SLR photography. Marketed by Kinetronics, a SpeckGRABER is a thin pencil-like device with a soft washable pad on the tip that provides a high-adhesion surface that sticks to contaminating particles more strongly than the particles stick to the surface of the imaging sensor. The offending particle can be lifted off and the SpeckGRABER washed with soap and water for later reuse.

Chapter 3
Astrophotography Techniques and Optical Limitations

3.1 Introduction

In this chapter we will look at the various types of astrophotography that can be accomplished with webcams and examine the factors that govern the resolution achieved with these devices. The different techniques can be sorted into several categories, each with its strengths and weaknesses. The categories are further divided by whether a standard webcam or a long-exposure modified camera is required. The standard short-exposure webcam is, of course, limited to bright objects, but long exposure-modified cameras are capable of doing all types of astrophotography.

The different categories are:

1. Fixed-tripod photography with a camera mounted directly on a tripod for stability.
2. Piggyback photography with the camera mounted atop a guided telescope to allow longer exposures without star trails.
3. Prime-focus photography with a camera body attached to the focuser of a telescope (i.e., no eyepiece used).
4. Barlow projection photograph with a camera body receiving a highly magnified image projected from a Barlow lens in a telescope.
5. Eyepiece projection photography with a camera body receiving a highly magnified image projected from the eyepiece of a telescope.

Skyshooters familiar with traditional digital camera imaging methods will notice the absence of the afocal imaging technique in which the camera uses its own infinity-focused lens to view the target through an infinity-focused telescope. This system of photography is ignored with webcams because it is cheaper, easier, and more convenient to mount a lightweight webcam at prime focus using an adapter rather than attaching it to a bracket that supports it above the eyepiece.

With the night sky filled with everything in the universe, it may seem odd to ask what is there to photograph? Other than the Moon and planets, matching a specific celestial target to your camera may be a bit of a mystery to the novice skyshooter. But the upside is that with digital photography, there is no cost for experiments that do not work and experience is the best teacher. Therefore it is important to keep a log of the objects you shoot, the camera settings, and the optical configuration used. Keeping such records may seem like drudgery at the time, but take it from me that it is worth the effort. I wish I had been more diligent in my record keeping because such information is vital to the learning process and later astrophoto success.

3.2 Fixed-Tripod Photography Introduces the Camera to the Sky

As the name implies, this activity involves simply placing the camera on a tripod to hold it steady during an exposure. This technique can be used for short exposures with an unmodified camera to show events like sunset, moonrise, or conjunctions of bright planets. The crescent Moon and Venus in twilight is always a pretty picture, especially if you place some objects such as recognizable buildings or trees in the foreground. This renders an ordinary image into a nocturnal landscape.

Similarly, the bright constellations can be recorded with a long-exposure modified camera using exposures of up to 30 seconds. If it is possible to still use the camera's "normal" lens, wide views of constellations can be achieved. The stars will remain pinpoints because the exposure is brief enough that the apparent motion of the night sky will not cause the stars to streak. Again, placing familiar landscape or architectural objects in the foreground will add interest to an otherwise starkly technical image. Try to photograph bright constellations such as Orion near the celestial equator, or the asterism of the Big Dipper in the northern sky.

If the camera has been long-exposure modified and remounted into a custom case, the use of wide-angle camera lenses is an option. Be aware that even a 20mm camera lens will act like a telephoto because of a webcam's small imaging area, so the field of view will be limited to asterisms or portions of a constellation.

Photography of Earth satellites is also an interesting offshoot of fixed-tripod astrophotography with long-exposure-capable cameras. A photographic record of your observation of the passage of the International Space Station, the Hubble Space Telescope, or perhaps the bright flare of an Iridium communications satellite will make an interesting keepsake. The technique for photographing various bright satellites is as simple photographing constellations with a fixed camera. The trick is predicting when a certain satellite will appear and what part of the sky will it pass through so that the camera can be aimed at that area to await its passage. Once the satellite is visible, or the Iridium flare begins to happen, the exposure is started and runs for the duration of the event.

Fortunately, the Heavens-Above web site (www.heavens-above.com) simplifies the satellite pass predictions. Once you log onto the web site, you can register your exact location on the Earth. This places a harmless cookie in your web browser so that on later visits to the site, it will generate satellite pass predictions for your location without having to reenter your map coordinates.

3.3 Piggyback Adds Tracking Capability

Piggyback photography with a conventional camera lens turns a modified webcam into a powerful astronomical instrument because it can now record objects in beautiful astronomical scenes that are invisible to the unaided eye. Telephoto lenses allow us to get close-ups of star clusters and nebulae that approximate the same field of view as seen through a low-power telescope eyepiece.

Modified webcams allow unlimited exposures, although electronic noise often limits their exposure times to the 30- to 60-second range. But this is still long enough to record objects that are too dim for the eye to see. However, these longer exposures present the problem of having the camera track the moving stars in order to prevent star trailing. Because a webcam's small imaging sensor restricts the field of view, it is popular to employ a conventional camera telephoto lens to achieve the same field of view we employ to capture through a telescope using a full-frame 35mm camera. For instance, placing a Philips ToUcam on a 200mm telephoto lens approximates the field of view seen with a 35mm camera on an 8-inch *f*/10 SCT, although with a very reduced image scale. Tracking the moving stars using such magnification is most easily accomplished by the piggyback method of astrophotography. This involves attaching a camera and lens to an equatorially mounted, polar-aligned telescope with a bracket that either bolts to the tube or is secured with clamps around the tube. The camera uses its own lens to image the sky and uses the telescope and its clock drive solely as a tracking platform to follow the stars as Earth rotates.

Guiding is not critical because modified webcams operate in the 30- to 60-second exposure range. If properly polar-aligned, most telescope drives are accurate enough to allow unguided, trail-free exposures in that short time. Since a webcam's output is a digital image, longer equivalent

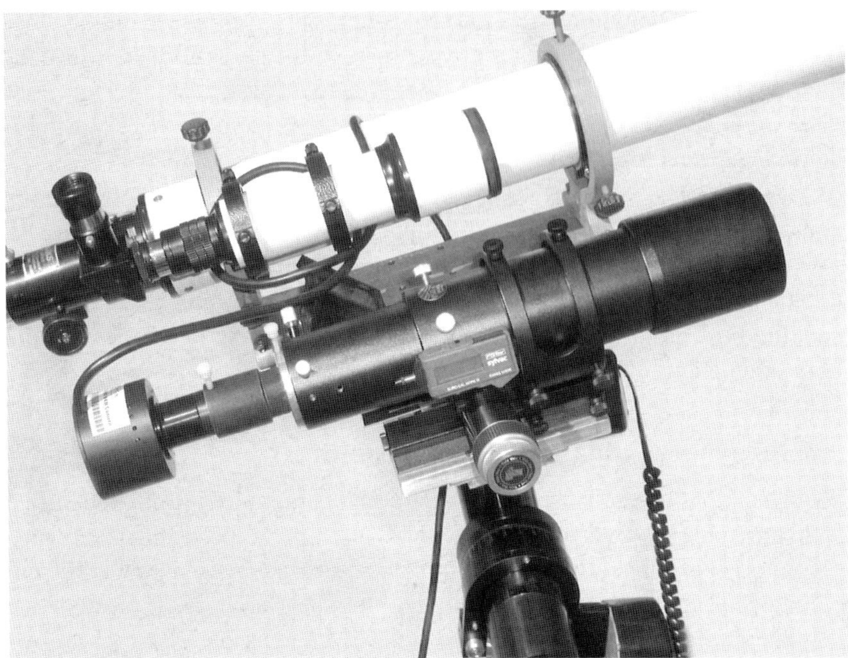

Fig. 3.1 *Attaching a webcam-style camera either piggyback on an equatorially-mounted telescope, or parallel with the guidescope as shown here allows it to track moving celestial targets. Photo by Robert Reeves.*

exposures can be achieved by digitally combining repeated short exposures with an image-processing program (this technique will be explored in Appendix B).

Since the response of a webcam's digital sensor is linear, it records at the same rate throughout the exposure and multiple images can be combined to create the effect of a much longer exposure. Thus, instead of taking a single half-hour exposure like a film astrophotographer would, the digital astrophotographer would take 30 separate one-minute exposures of the same object. Breaking the exposure into segments has two advantages: First any single exposure with a serious guiding problem can be eliminated from the final image stack. Only the best-guided images can be used to create the final image. Second, many of today's good electronically-controlled telescope mounts are capable of accurately tracking an object for a minute without the need for constant guiding corrections if the mount is well aligned to the celestial pole.

The technique of stacking multiple unguided 30-second exposures to emulate longer exposures has given rise to a new term in astrophotography…tracked, but not guided. This technique is useful with telephoto lens piggyback webcam photography on a polar-aligned telescope because over the course of a series of automated 30-second exposures, a telescope mount will typically track well for a few minutes, then encounter a brief time when the drive displays periodic error. The result is several images are well-tracked, then one is slightly trailed when periodic error is encountered, then several more are well-tracked before the periodic error repeats in a fixed cycle. With enough automated exposures in hand, the percentage of trailed images can be deleted from those stacked to create a final unguided but tracked image.

A word of warning: piggyback photography requires an accurately polar-aligned equatorial mounting which pivots the entire telescope tube assembly around an imaginary line that is parallel to the celestial pole. Not all telescopes that track the stars across the sky use this method. Some recent models of computerized telescopes operate on an alt-azimuth mounting that uses electronics and control motors to automatically follow the motion of the stars across the sky. These mounts have a small computer built into them to continuously calculate the location of the stars relative to the telescope's location on Earth and the local time. The computer then commands motors on the telescope's altitude and azimuth axes to slew the instrument in order to follow the sky. This will accurately track a celestial object and keep it in the field of an eyepiece, but in reality, the telescope is simply following the object through a series of small up or down and side-to-side zigzags. The telescope may tip up and down while tracking but its orientation remains constant relative to the Earth. The top of the tube will always remain "up." This will cause the field of view through the eyepiece to slowly rotate as the sky seemingly pivots around the celestial pole while the telescope remains level relative to Earth. Hence, the image in a camera attached to such a telescope will also exhibit field rotation. The resulting image would resemble a traditional polar star-trail image taken with a tripod-mounted camera, except the star pattern would not match those around the true celestial pole.

3.4 Prime Focus Telescopic Imaging

The simplest method for achieving high magnification with a webcam is to mount it at the prime focus of a telescope. Here, the camera lens is removed and the camera body is attached to the telescope in place of an eyepiece and the telescope's optics focus the image directly on the camera's focal plane. Essentially, the telescope becomes a giant telephoto lens. While it is technically true that the term "prime focus" only refers to the focal plane of a primary optic, such as the mirror of a Newtonian reflector

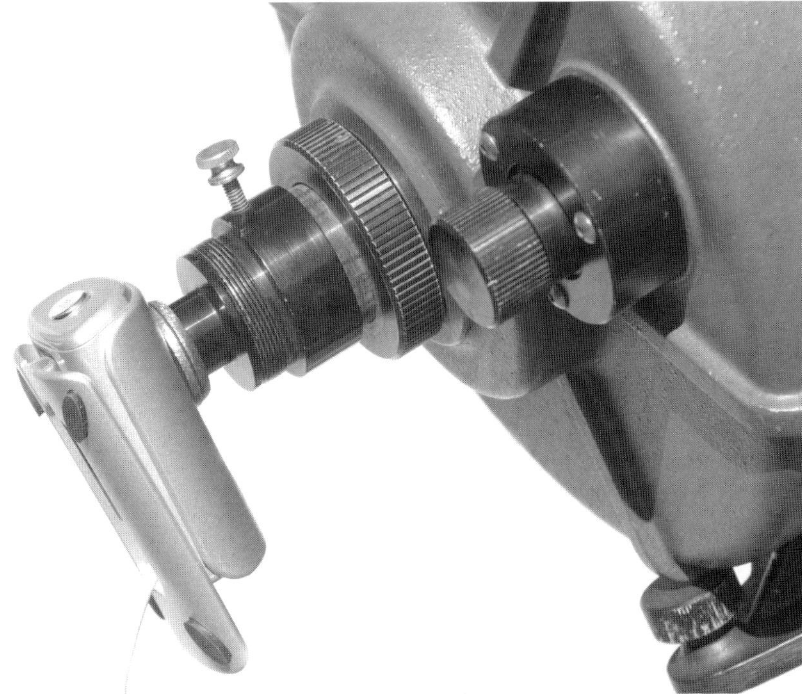

Fig. 3.2 *A webcam attached at the prime focus of a telescope becomes a powerful photographic instrument. A ToUcam at the focus of a Celestron-8 telescope produces a view similar to that through a 400-power eyepiece. Photo by Robert Reeves.*

or the objective lens of a refractor, it is loosely applied to imaging at the focus of any optical configuration not using an eyepiece. (For the sake of scientific clarity, a similar imaging technique with one of the popular SCT-style telescopes should be called "Cassegrain focus"). To do prime-focus imaging, a camera is mounted on an eyepiece adapter and inserted into the visual back on the rear cell of a Schmidt-Cassegrain telescope or into the focuser of a Newtonian or refractor. With this method the focal length and *f*/ratio of the telescope is the focal length of the photographic system, so a standard 8-inch *f*/10 Schmidt-Cassegrain telescope becomes a 2000mm focal length *f*/10 telephoto lens.

The approximate image size achieved at prime focus as compared to a webcam's "normal" view is easy to calculate. Simply divide the telescope's focal length in millimeters by 5, as the small sensor size of a webcam provides a field of view approximating that seen with a 5mm eyepiece. Thus, a telescope with a 2000mm focal length used with a webcam yields a field of view that is about the same size as that seen with a 5mm ocular on a 8-inch SCT (400 power), allowing amazing close-ups of

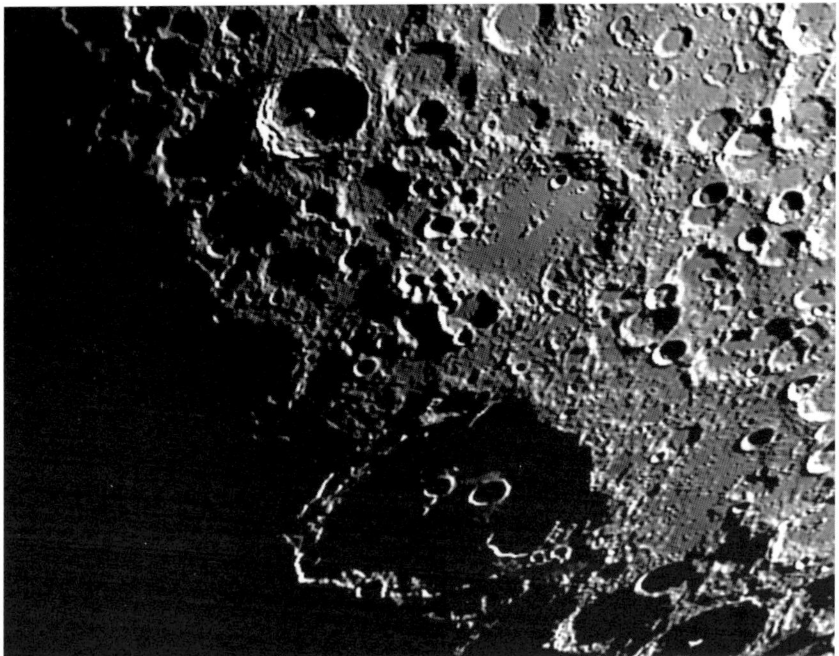

Fig. 3.3 *Because of its physically small dimensions a webcam CCD sensor sees a small field of view when imaging at the telescope's prime focus compared to a conventional camera. Photo by Robert Reeves.*

the Moon and planets.

The field of view at prime focus can be calculated using the formula:

$$X = \frac{D}{F} \times 57.3$$

where:

X = Field size, in degrees,
D = Image diameter at focal plane, and
F = Focal length of telescope.

For example, using a standard webcam imaging sensor with 5.6-micron pixels yields an imaging surface of 3.6 x 2.7 millimeters. Using the 3.6 millimeter width of the imaging sensor with an 8-inch *f*/10 SCT yields the following results:

$$X = \frac{3.5}{2000} \times 57.3 = 0.103 .$$

Thus, the 3.6-millimeter diameter webcam sensor at prime focus of a

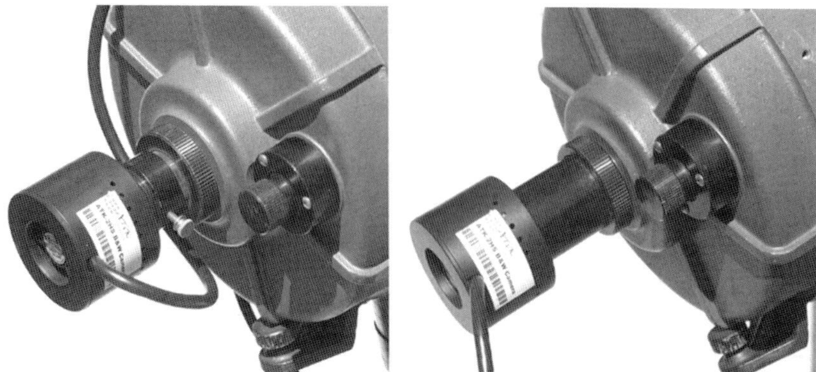

Fig. 3.4 *All webcam-style cameras attach to a telescope drawtube with a 1¼-inch adapter. Adapters secured with a setscrew (left) should be checked often because in cold weather, metal shrinks and setscrews can accidentally release the camera. Cameras that accept T-adapter accessories (right) can be "hard coupled" to Schmidt-Cassegrain telescopes using threaded T-adapter photo accessories. Photos by Robert Reeves.*

2000-millimeter focal-length telescope will have a field of view slightly over one-tenth of a degree.

A webcam's inherently small field of view is often too small at prime focus to image deep-sky objects. To attain a larger field of view, place a focal reducer between the telescope and camera. The larger field size is calculated by multiplying the telescope focal length by the reducer's reduction factor. For example, if a 0.6 focal reducer is used, an 8-inch *f*/10 telescope becomes a 1200-millimeter-focal-length *f*/6 instrument and field size is determined using these figures with the above formula. The imaging procedure is the same as that used with prime focus.

Lunar close-ups and planetary imaging are good prime focus targets for unmodified webcams, while deep-sky objects like galaxies, star clusters and nebulae are the principal targets of prime-focus work using long exposure-modified cameras. However, the high magnification attendant with this technique dictates that care be given to the accurate guiding and tracking of the telescope mount. Any imperfection in the tracking accuracy will trail star images into egg-shaped blobs. Fortunately, the same factors that help the piggyback technique—inherent short exposures and today's accurate telescope mounts—allow the webcam imager to achieve good success with modest equipment.

If you haven't a lot of polar alignment or guiding experience under your belt, I suggest you practice before attempting prime focus work. While small errors are nearly inconsequential in piggyback work, every deviation of the star image looms large in a prime focus shot. Because the

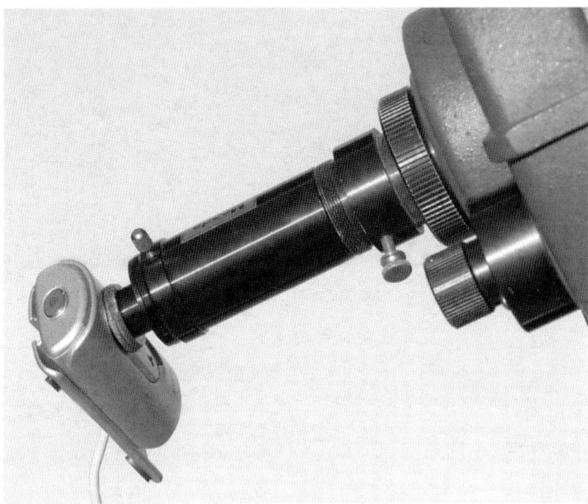

Fig. 3.5 *The easiest way to gain focal length for solar system photography is to use a Barlow lens. The Barlow inserts into the telescope focuser and the camera in turn inserts into the Barlow. Photo by Robert Reeves.*

camera is attached where the eyepiece normally resides, we cannot directly view through the telescope in order to guide it during the exposure. There are two ways to get around this limitation; a separate guidescope attached piggyback to the main telescope, and a device called an off-axis guider (OAG). These are discussed in Chapter 9.

3.5 Barlow Projection Telescopic Imaging

The use of a Barlow lens or Tele Vue Powermate is a common way to increase the focal length of a given telescope to achieve higher webcam imaging magnifications than possible with prime focus alone. Barlows are available in various powers that will multiply a telescope's focal length by 2x, 2.5x, 3x, or 4x and are used by inserting the device into a telescope's eyepiece holder in place of an eyepiece, then inserting the camera into the Barlow lens barrel.

Barlows are of little use in deep-sky imaging because they increase the magnification of a telescope too much for large dim targets (except, perhaps, small planetary nebulae). With increased focal length comes increased focal ratio, and thus the photographic speed of the telescope is slowed by the magnification factor. For example, an *f*/10 telescope with a 2x Barlow becomes an *f*/20 system and the use of a 3x Barlow results in an *f*/30 system. However, *f*/20 to *f*/30 works well with bright lunar and planetary objects under good seeing conditions and Barlows achieve their

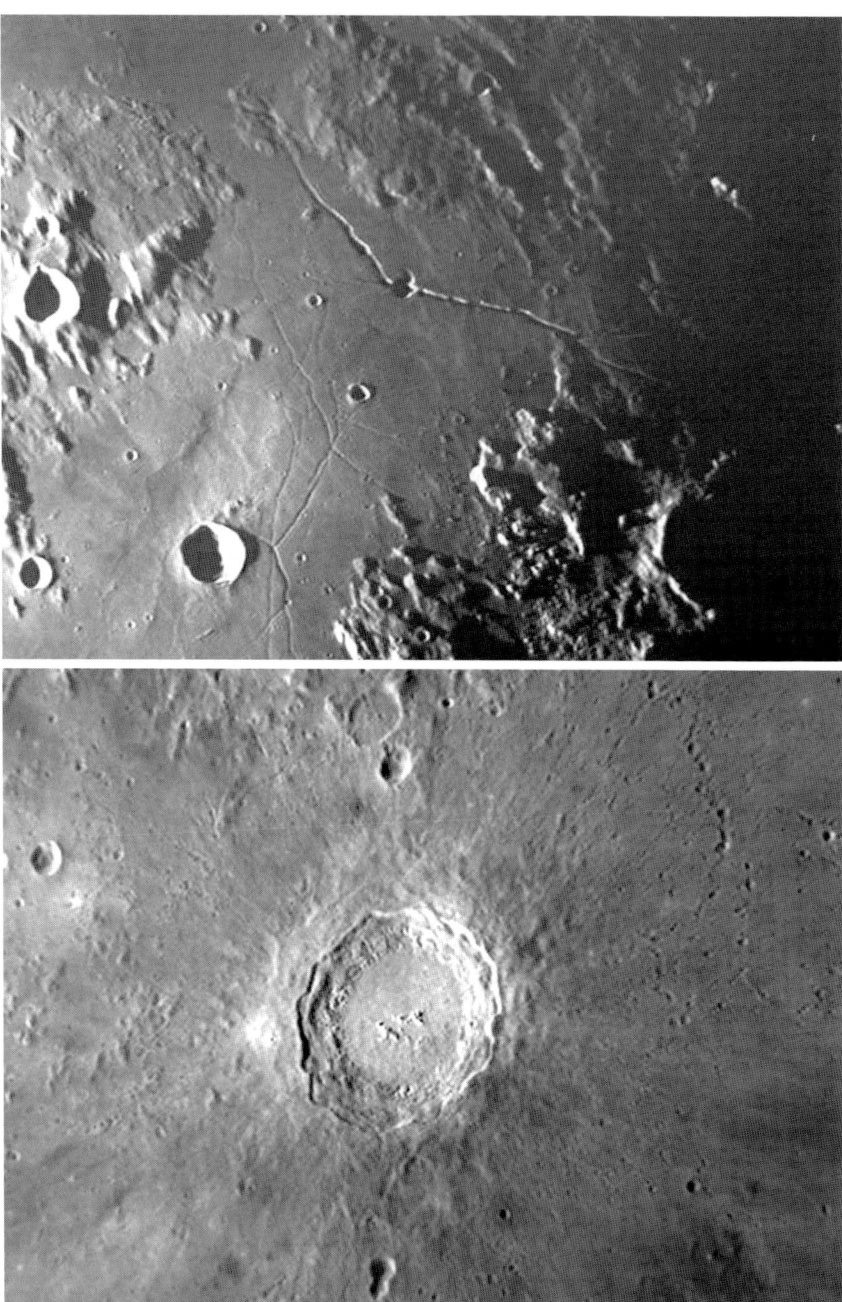

Fig. 3.6 *A 2x Barlow used with an ATK-2HS camera with 7.4-micron pixels on an 8-inch f/10 Schmidt-Cassegrain telescope will yield a field of view similar to that seen through a 600-power eyepiece. At the top are the Hyginus and Triesnecker Rille systems on the Moon. Below, detail in the terraced walls of the crater Copernicus is visible. Photos by Robert Reeves.*

optimum performance with such targets.

The Powermate is similar to a Barlow in that it increases the focal length of an optical system. Powermates are available in up to 5x versions and as explained in Section 4.8, they have inherent advantages for photography over a Barlow.

3.6 Eyepiece Projection Telescopic Imaging

Eyepiece projection is an alternative technique for achieving higher magnifications than possible with prime focus. With this method, the camera is mounted to the telescope with an adapter that contains an eyepiece at one end and a tele-extender on the other. (In this case, the term "tele-extender" should not be confused with the similarly-named Barlow-like device used to multiply the focal length of SLR and DSLR camera lenses.) Traditionally a tele-extender positions the camera at a certain distance from the eyepiece and provides T-adapter threads on which to mount the camera. The most common tele-extender device is a six-inch long projection tube used with Schmidt-Cassegrain telescopes. The principle is that the image projected from the eyepiece is enlarged on the focal plane in much the same manner as a photographic slide is enlarged when projected onto a screen. The further away the camera is from the eyepiece, the bigger the projected image is on the focal plane. With increased projection distance, however, comes reduced photographic speed as the image dims with greater enlargement. The effective focal length of eyepiece projection systems with average amateur telescopes can approach 15,000 to 20,000mm, depending upon the eyepiece used. However, photographic *f*/ratios increase to between *f*/50 and *f*/100, resulting in exposures lasting up to several seconds on the Moon and bright planets.

The following three formulae describe the basic photographic parameters when using the eyepiece projection technique:

$$M = \frac{D - F_2}{F_2}$$

$$T = F_1 \times M$$

$$F = \frac{T}{A}$$

where

A = telescope aperture,
D = distance of projection from eyepiece to focal plane,

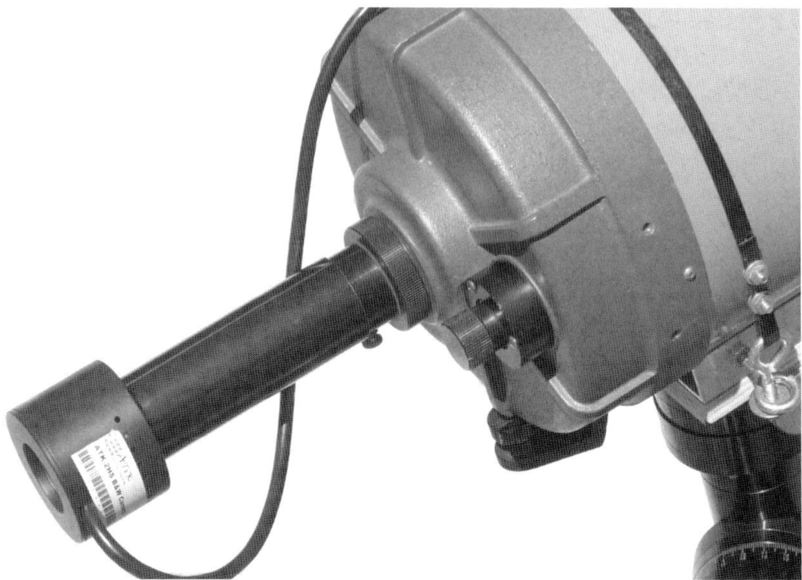

Fig. 3.7 *Any webcam-style camera that accepts T-thread accessories can be used with eyepiece projection photography systems designed for conventional film and digital cameras. Extremely high focal ratios can be achieved by using short focal length eyepieces in the projection system. Photo by Robert Reeves.*

F = focal ratio of projection system,
F_1= focal length of the telescope,
F_2= focal length of the eyepiece,
M= projection magnification factor, and
T = total projection system focal length.

If we plug in the numbers for a common 8-inch *f*/10 Schmidt-Cassegrain telescope using a 6-inch projection tube and a 25mm eyepiece, we come up with the following (converting all dimensions to millimeters):

$$M \text{ (projection magnification factor)} = \frac{150 - 25}{25} = 5$$

$$T \text{ (total projection system focal length)} = (2000 \times 5) = 10000\text{mm}$$

$$F = \frac{10000}{200} = 50\,.$$

We find that our 8-inch Schmidt-Cassegrain telescope using a 25mm eyepiece with 6-inches of projection distance now operates as a 10,000mm focal-length *f*/50 photographic system.

The enormous magnification and small field of view with eyepiece

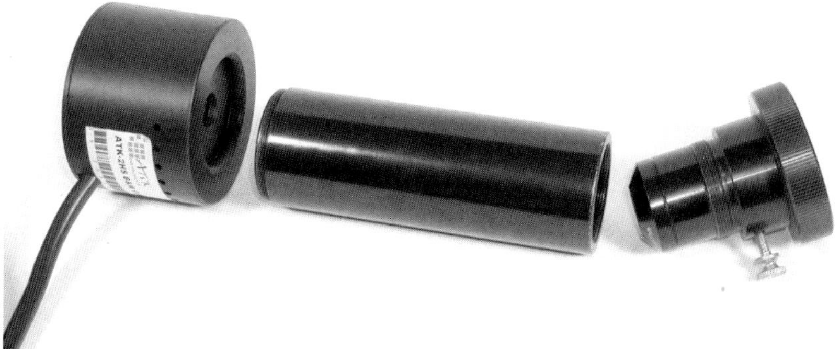

Fig. 3.8 *The principle of eyepiece projection astrophotography is similar to a 35mm slide projector; the greater the distance from the eyepiece to the camera focal plane, the larger the image scale on the focal plane. The 6-inch projection tube shown here threads into the visual back of an SCT telescope and also has T-threads to couple directly to an ATK-2HS camera. Photo by Robert Reeves.*

projection and resulting high *f*/ratios require accurate telescope tracking even though the exposures are far shorter than those used on deep-sky objects. The object of such photography is to record minute lunar and planetary detail. Even the slightest tracking drift can soften the focus of very high magnification views of these subjects. There is no practical way to guide a telescope during eyepiece projection photography, so accurate polar alignment and a good tracking rate are a must.

It should be noted that the afocal projection technique, where a camera's infinity-focused normal lens is used to image the view seen through a telescope eyepiece, is possible with a webcam. However, this style of photography is more complicated than the simpler method of using prime focus or a Barlow at prime focus. The wide availability of inexpensive eyepiece adapters to couple a camera to a telescope has made webcam afocal photography simply not worth the effort.

3.7 Resolution

Now that we understand the different imaging methods available to the webcam astrophotographer, lets look at some of the parameters that will allow us either to record all the detail available or conspire to limit the amount of detail the camera can record. The term "resolution" has various meanings in astronomy, computers, and photography. In this book, we will concentrate on resolution as it applies to digital astrophotography through a telescope. Here, resolution refers to the number of arcseconds per pixel resolved by a particular optical system. Obviously, the greater number of pixels covered per arcsecond of image field, the greater the image detail that can be resolved. But as we shall see, this applies only up to a certain point.

Other definitions of resolution are also used in celestial imaging. With computers, resolution typically refers to how many pixels the display monitor uses to show the image on the screen, such as 1024 x 768 or 1600 x 1200. In photography it usually refers to how many pixels are used to show a particular field of view. This can be expressed either by the total number of pixels on the imaging sensor, such as 5-, 6-, or 8-megapixels, or it can refer to the number of pixels used to create a print of a picture taken by the camera, such as 3072 x 2048. In visual astronomy, high resolution is historically referred to as the ability to perceive fine detail in an eyepiece, as when the components of a 0.75-arcsecond double star are split.

Some webcams that have 640 x 480 pixel sensors advertise that they can take 1280 x 960 pixel images. These cameras have an option that allows taking snapshot still images, and the 1280 x 960 resolution is actually an interpolated resolution scaled up from the camera's native 640 x 480 video resolution.

To get the most performance from a webcam and your telescope, the proper focal length must be used in order to record all the available image data. It is important to understand that individual pixels merely show variations of brightness, therefore in order to resolve any detail there must be at least 2 pixels of differing brightness. A single pixel cannot show any detail, it will only show a particular brightness density. This concept remains true with star pictures as well as with lunar and planetary images. A dim star can be imaged by a single pixel, but without the surrounding pixels showing the darker sky background, the star will not be recognized. The same is true for lunar and planetary detail. There must be at least 2 pixels, a bright one and a dark one, for any detail to be recognized.

In digital astrophotography, the term "sampling" refers to how many pixels are used to display individual details, such as a single star. Individual pixels on an imaging sensor are usually square and thus appear in the image as little squares with varying tonal density. If too few pixels are used to produce an image of a celestial object, stars can take on a square appearance, and lunar details can have a jagged, stair-step appearance. Such images suffer from under-sampling.

The number of pixels needed to adequately show image detail is related to the focal length of the telescope and the size of the pixels on the imaging sensor. If the focal length is too short, some detail will be lost because it is under-sampled and not spread over enough pixels to be recorded. Conversely, if the focal length is too long for a given aperture,

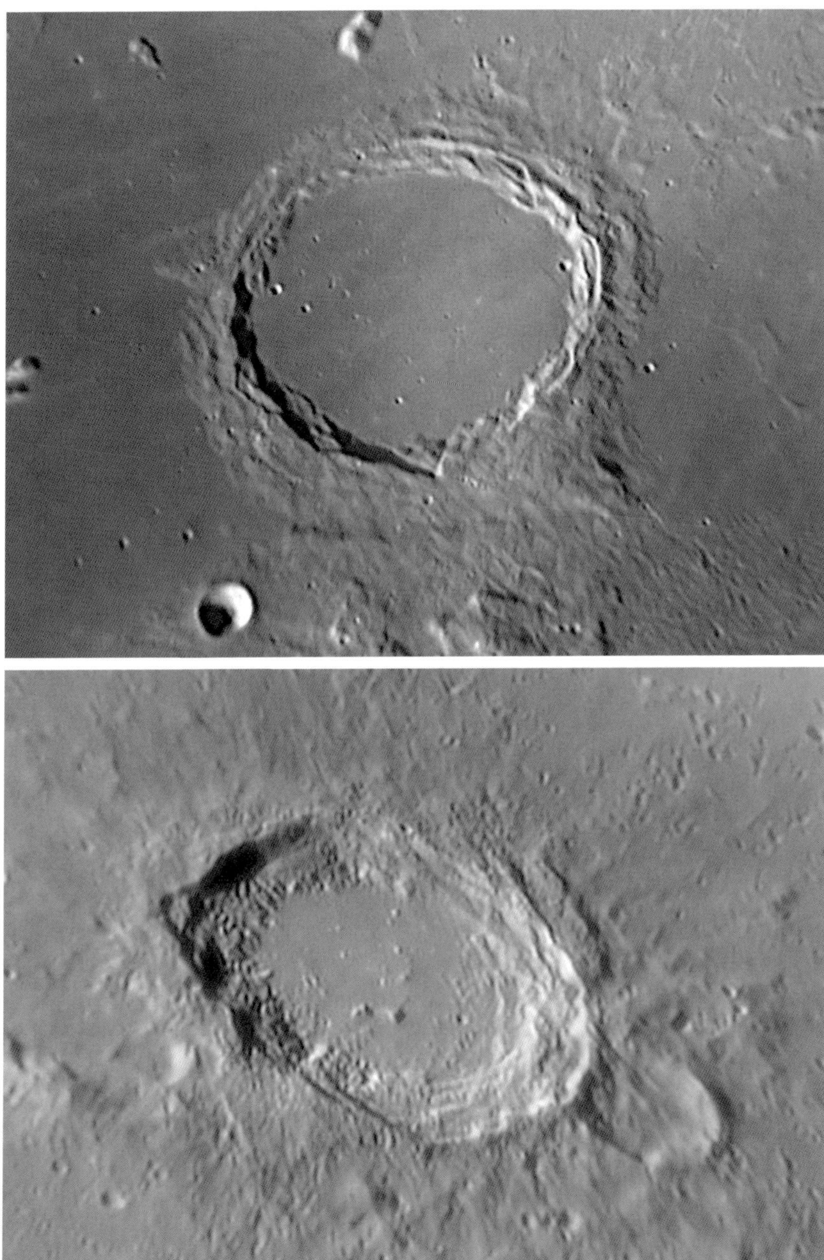

Fig. 3.9 *The above images of the craters Archimedes (top) and Aristoteles (bottom) were taken with a 12-inch reflector operating at f/28. The theoretical Dawes Limit resolution of a 12-inch instrument is 0.38 arcseconds while the f/28 effective focal length of 336 inches produces an image scale of about 0.13 arcseconds per pixel with a ToUcam 840. Oversampling the image by a factor of three allowed the ToUcam to record all the available image detail. Photos by Zac Pujic.*

the limited image detail is spread over too wide an area on the imaging sensor resulting in "empty magnification." In the latter case, the exposure also must be longer to compensate for the reduced image brightness created by the larger scale image, and guiding must be more precise because of the higher inherent magnification. Longer exposure also allows the detrimental effects of poor seeing to blur the image.

For lunar and planetary imaging it is best to achieve "full sampling," that is, to choose a magnification that places the smallest resolvable image detail on 2 pixels at the camera's focal plane. When imaging the stars, this is related to the telescope's Dawes Limit. Traditionally, we refer to the Dawes Limit to define the smallest detail, in arc seconds, that can be resolved by that telescope (see Table 3.1), then strive to achieve a focal length that projects that detail onto two pixels at the camera's focal plane.

It is, of course, possible to "over-enlarge" an image with too much focal length and in doing so gain no additional detail because either local seeing conditions blur the image too much or the optical system is not capable of resolving additional detail. For this reason, we strive in telescopic digital astrophotography to match the resolution capability of the telescope with the imaging capabilities of the camera.

There are a number of factors that limit a webcam's resolution: focus, atmospheric seeing, the quality of the telescope optics, the Dawes Limit for that aperture of telescope, and its focal length versus the spacing of the pixels on the imaging sensor. We can control some of these factors, but not all of them. There is nothing we can do when atmospheric seeing is poor except wait for a better night. If poor optics limits resolution, we can try again with a better telescope. If telescope focus is not adequate, we can refocus. But the Dawes Limit and the pixel spacing on the imaging sensor represent fixed points that we have to work with in trying to achieve the highest resolution possible with a given telescope aperture and webcam.

Another limitation in resolving fine detail is collimation. If the optics are not aligned well, an instrument cannot properly focus the target even if the seeing is perfect. The optics must also be thermally stable, that is, there must be adequate telescope cool-down after dark. Following the star testing guidelines in Harold Suiter's book, *Star Testing Astronomical Telescopes* or using the freeware program "Aberrator," written by Cor Berrevoets, can check a telescope's optical quality. With Schmidt-Cassegrain instruments, two common problems are optical roughness and spherical aberration, both of which create problems when imaging the Moon and planets. When testing a telescope, use an artificial star if the seeing never gets good enough to reveal the diffraction pattern on images of real stars.

Table 3.1
Dawes Limit vs. Focal Length

Aperture		Resolution
(Inches)	(Millimeters)	(Arcseconds)
1	25	4.6
2	50	2.3
3	75	1.5
4	100	1.2
5	125	0.9
6	150	0.8
8	200	0.6
10	250	0.5
12	300	0.4
20	500	0.2

3.8 More on Dawes Limit

The Dawes Limit for telescope resolution was formulated over a century ago by British astronomer William Rutter Dawes (1799–1868). Although it is traditionally used to define the visual resolving power of an instrument, this resolution parameter is actually valid only for pinpoint objects such as stars. It does not apply to high-contrast wide-area targets like the Moon or planets. The effects of contrast in lunar and planetary images often allow the detection of details up to ten times smaller than indicated by the Dawes Limit. For instance, the Encke Division in Saturn's A ring is only 0.05 arcseconds wide, yet we can often image this feature with 10-inch backyard telescopes under good seeing conditions. In the case of linear features, the detail-limiting factor is diffraction. If the feature being resolved is narrower than the diffraction limit of the telescope, it is impossible to resolve. But an understanding of the Dawes Limit will give us a reference point to help us recognize both the resolution potential, and limitations, of our telescope.

To understand how we can work with the Dawes Limit and pixel spacing, we need to understand what it is and how it interacts with the pixel spacing on the camera sensor. The Dawes Limit is a mathematical expression derived empirically for the smallest possible image detail visible in a given size telescope. Basically, because of the wave nature of light, rays tend to spread out after passing through an aperture and this phenomenon, called diffraction, blurs fine details. A telescope theoretically cannot resolve two stars of equal brightness whose angular separation is less than the figure stated by the Dawes Limit. The limit is calculated with the formula

$$R = \frac{115.8}{D}$$

where R is the separation in arcseconds and D is the diameter of the telescope objective in millimeters. As we can see, the resolving power of a telescope is dependent upon aperture, the larger the aperture, the higher the potential resolution.

The Dawes Limit is not an absolute figure. Poor optics will not live up to the calculated resolution. On the other hand with good optics, proper magnification of the image at the camera focal plane, and modern image-processing and stacking techniques, it is possible to resolve detail somewhat smaller than that predicted by this benchmark. See Table 3.1 for representative examples of the Dawes Limit resolution for various apertures.

3.9 Determining Image Sampling Needs

Remembering that it takes two pixels to resolve any individual detail, a rule of thumb to prevent under-sampling is to divide the local seeing in half and match that number to the pixel size of the camera to be used. For instance, if seeing is poor, (limiting resolution to, say, four arcseconds), choose a focal length that provides two arcseconds-per-pixel resolution at the camera focal plane. If seeing is exceptional, (allowing resolution of about one arcsecond), choose a focal length that provides one-half arcsecond per pixel resolution at the camera focal plane. Two formulae that can help calculate the desired sampling in arcseconds are:

$$\text{Sampling in arcsec (Metric)} = \frac{206.265}{(\text{focal length in mm}) \times (\text{pixel size in microns})}$$

and,

$$\text{Sampling in arcsec (English)} = \frac{8.12}{(\text{focal length in inches}) \times (\text{pixel size in inches})}.$$

To get maximum lunar and planetary resolution under conditions of excellent seeing, it is best to match the pixel size of the camera with a focal length that produces between 0.5 and 0.25 arcseconds per pixel. This usually calls for more focal length than provided by common portable amateur telescopes. Using a webcam with a typical 8-inch SCT calls for a focal length in the 120- to 200-inch range (3048 to 5080mm). The additional focal length is gained through the use of a high-quality Barlow or Power-

Table 3.2
Resolution per Arcsecond vs. Focal Length

Focal Length	Arcsec per Pixel	
Inches	5.6 micron	7.4 micron
20	2.27	3.00
40	1.13	1.50
60	0.75	1.00
80	0.56	0.75
120	0.39	0.50
180	0.25	0.33
240	0.19	0.25
320	0.14	0.19
400	0.11	0.15

mate, as described in Section 4.8.

Although this 0.5- to 0.25-arcsecond resolution target exceeds the theoretical Dawes Limit for an 8-inch telescope, users will find that under excellent seeing conditions with a well-collimated and focused telescope, images processed in programs such as RegiStax will display planetary details that exceed this visual resolution "barrier." This is because the contrast features in planetary images allow fine details to be seen. Indeed, under excellent seeing conditions, an 8-inch telescope using a webcam with a pixel size of 5.6 microns such as that used in the Philips ToUcam 840 or SPC900NC is capable of producing stunning images that rival lunar close-ups achieved by NASA's Moon probes in the 1960's. Such high resolution is an attainable goal in webcam imaging. But in most cases we will find that seeing is going to limit telescope resolution to about one arcsecond.

Table 3.2 shows the pixel resolution in arcseconds for both 5.6- and 7.4-micron pixel sensors used in a variety of standard webcams and commercially-modified long-exposure webcams at a variety of focal lengths typically used with popular 8-inch SCTs coupled to various focal reducers or Barlows. Table 3.3 shows the best application for a particular arcsecond resolution per pixel.

One would think that if the pixel spacing were 5.6 microns, then enlarging the image so that the telescope's Dawes Limit exceeds this figure at the focal plane would automatically result in the best resolution possible. However, a number of factors come into play that can degrade focus. The first is the fact that as focal length increases so does the depth of focus, or the zone in which the image seems to be in focus. This expanded zone

Table 3.3
Pixel Resolution vs. Seeing Conditions

Pixel Resolution in arcseconds	Recommended Application
Greater than 2.0	Undersamples image in most cases. OK for deep-sky with very poor seeing
1.4 to 2.0	OK for poor seeing or where sensitivity outweighs resolution
0.65 to 1.4	OK for all uses in typical seeing
0.5 to 0.65	OK for solar system anytime and deep-sky under good seeing
0.25 to 0.5	OK for solar system anytime and deep-sky under best seeing
Less than 0.25	High resolution solar system where sensitivity not needed

of focus can make it harder to determine the point of absolute best focus. The quality of the optics used with the webcam also plays a critical part in the resolution achieved. Any optical flaws will only degrade the image, not improve it, and users of refractors may also have to contend with chromatic aberration.

Perhaps most important in the chase to match the focal length with pixel size is the fact that enlarging the image will also dim it. Using a 2x Barlow will dim the image by a factor of four. This will call for either longer exposures or higher gain settings on the camera, both of which will degrade the image through increased atmospheric turbulence or increased image noise. Increased magnification and exposure time also call for a steadier telescope that accurately tracks the target, lest the image smear on the focal plane during the exposure.

A final point should be mentioned about the use of increased focal length to properly match the Dawes Limit with the pixel spacing on the image sensor. The need for sensor cleanliness increases with longer focal length because the light cone exiting the optics and converging on the focal plane is narrower and creates sharper shadows on the sensor if any dirt or dust is present.

3.10 Effect of Seeing on Aperture

The above discussion of a telescope's resolution capability, as dictated by the Dawes Limit, does not take into account a critical factor: seeing. It is rare that the atmosphere is steady enough to allow a telescope to actually perform at the Dawes Limit. Atmospheric seeing varies enormously due to location, weather, and time of day. One of the prime observing locations

Table 3.4
Effect of Seeing on 8-inch Aperture Telescope

Actual Seeing (arcseconds)	Equivalent Aperture Governing Resolution Under Perfect Seeing, in inches
4	1.14
3	1.52
2	2.28
1	4.55
0.5	9.12

on the planet is the summit of Mauna Kea in Hawaii. Here, at an altitude of nearly 14,000 feet, the seeing averages about half an arcsecond. The legendary good seeing on Mount Wilson in California averages about one arcsecond. However, the typical seeing at an average urban location can degrade to between one and four arcseconds. If the atmosphere is so turbulent that it distorts the incoming wavefront to the extent that objects cannot be imaged with better than 4-arcsecond resolution, much of the effectiveness of a telescope's aperture is lost. In effect, an 8-inch instrument still gathers light with its full aperture, but poor seeing degrades its resolving power to that of a smaller telescope.

To illustrate the effect of poor seeing on the resolving power of a telescope, the Dawes Limit formula on page 76 can be rearranged to solve for the aperture D using a given resolution in arcseconds:

$$D = \frac{115.8}{R}.$$

Table 3.4 approximates the effects on a given telescope of various degrees of poor seeing. Here we see that an 8-inch telescope operating under 4-arcsecond seeing is performing no better than an instrument of just over one inch aperture would under perfect conditions. Remember, however, that the 8-inch telescope is still gathering light with its full aperture and will detect objects that are dimmer than a smaller scope can detect, regardless of turbulence. Seeing is constantly variable, and even if it averages between two and four arcseconds on a given night, there are often brief periods of calm conditions that allow a telescope to deliver its best resolution.

3.11 Increasing Air Mass at Lower Altitudes

Even if seeing conditions at the zenith are excellent, as we get closer to the horizon, we must look through increasingly greater amounts of air mass

Table 3.5
Air Mass as a Function of Altitude above the Horizon

Altitude	Air mass	Altitude	Air mass
90	1.00	45	1.41
85	1.00	40	1.56
80	1.02	35	1.74
75	1.04	30	2.00
70	1.06	25	2.37
65	1.10	20	2.92
60	1.16	15	3.86
55	1.22	10	5.76
50	1.31	5	11.42

and potentially more turbulence. Stellar magnitude extinction also becomes a factor at lower altitudes. Extinction remains less than one magnitude at an altitude of 20 degrees, but climbs rapidly to nearly 10 magnitudes at the true horizon. The calculation of the equivalent air mass at any given altitude above the horizon is performed with the formula

$$\text{Air mass} = \frac{1}{\cos z}$$

where $\cos z$ is the cosine of the altitude above the horizon, expressed in degrees.

Table 3.5 shows the air mass factor for various astronomically useful altitudes above the horizon.

At the zenith, we are looking through the thinnest portion of the atmosphere. Here, a column of air, if it were of uniform density, would be about five miles high. This is considered to be one air mass. At about 60 degrees altitude, our line of sight takes us through the equivalent of about 1.2 air masses. At 45 degrees, the equivalent air mass increases to 1.4. At a 30-degree altitude, the equivalent air mass increases to 2. Below 30 degrees, the equivalent air mass quickly increases until at the horizon we view a star through 38 air masses. Because of the rapid progression of increasing air mass with lower altitude, most high-resolution imaging is done above 30 degrees altitude to minimize the deleterious effects of turbulence.

3.12 Atmospheric Dispersion

Dispersion is the refraction, or bending, of light as a function of wavelength. This effect will be present even under perfect seeing conditions. This phenomenon creates problems for the astrophotographer when shoot-

Table 3.6
Atmospheric Dispersion at Various Altitudes

Altitude (in degrees)	Arcsec of dispersion
90	0.00
60	0.35
45	0.60
30	1.04
15	2.24

ing at low altitudes because atmospheric dispersion refracts blue wavelengths to a greater degree than red wavelengths. The result is the familiar "rainbow" effect we see with bright planets, where there is a blue fringe on one side of the planet and a red fringe on the other. Much of the time we dismiss this effect as the result of poor optics, but in fact, even the best telescopes suffer from atmospheric dispersion when viewing at low elevations because it is a natural physical phenomenon.

Atmospheric dispersion was not much of a problem with planetary or high-resolution astrophotography in the past because other factors such as inherently slow shutter speeds and coarse resolution masked the phenomenon. Today, webcam planetary photography is routinely achieving high enough resolution that dispersion is now a factor that must be controlled or image quality will noticeably suffer when imaging at altitudes lower than 40 degrees. Table 3.6 shows that atmospheric dispersion can be a highly limiting factor when imaging the inner planets at low altitude, and small stars will be noticeably elongated at altitudes below 30 degrees.

Fortunately, planetary imagers can control the effects of dispersion to a great extent with the use of an Adirondack Planetary Atmospheric Dispersion Corrector (PADC). This device is discussed in Section 4.11.

Chapter 4
Accessories for Webcam Astrophotography

There is more to astrophotography than just a camera and telescope. Numerous accessories are needed not only for comfort and convenience while imaging, but some of them are absolute necessities without which this activity would be impossible. Webcam imaging is no different. Here we will discuss some of the accessories that are important for success in this endeavor.

4.1 Webcam Adapters

The single most useful accessory for using a webcam on a telescope or other optical device is an adapter to mount the camera directly into the telescope eyepiece holder or onto any of several popular brands of 35mm camera lenses. Today, high quality adapters are available from numerous commercial sources.

Until several years ago, the accepted method by which to mount a webcam to a telescope was to cut the bottom off a plastic 35mm film canister and glue the canister to the webcam body. This stopgap measure worked well enough to prove that webcams were powerful celestial imaging devices, but in practice a flimsy film canister is not rigid enough for accurate imaging.

Early in the rise of webcam astrophotography Steven Mogg of Australia recognized the need for well-machined adapters to mount various brands of webcams to both telescopes and camera lenses. Mogg's adapters have become the standard design adopted by manufacturers for attaching these cameras to a wide variety of optical instruments. Sylvain Weiller on the QCUIAG email group provided the original idea for the now-universal webcam adapter. After much discussion between QCUIAG members, the dimensions and thread size for the adapter were worked out. Steven Mogg had both a lathe and time on his hands and decided to start producing the adapter. The result was today's standard webcam adapter featuring a 1¼-

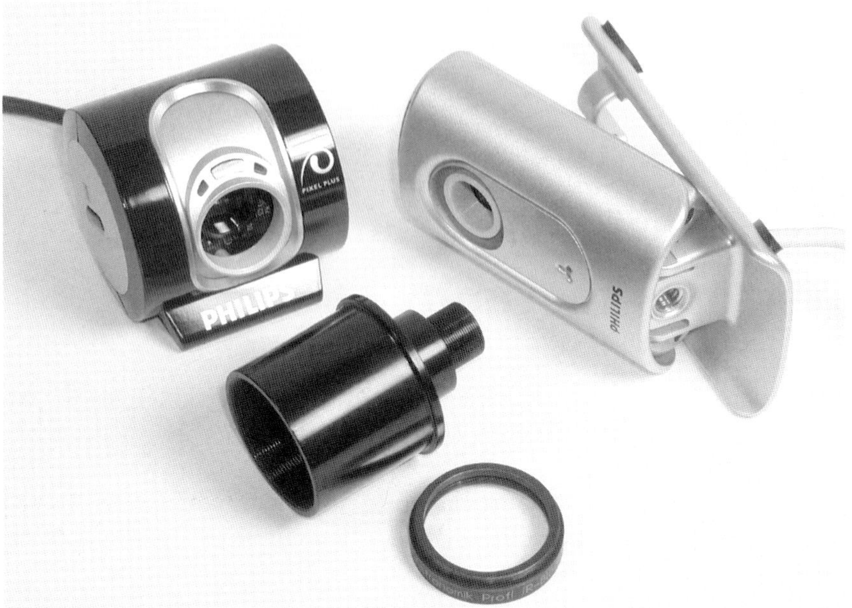

Fig. 4.1 *An adapter with a 1¼-inch outside diameter is required to mount a webcam in a telescope focuser. The adapter screws into the webcam's lens mounting thread. An infrared filter screws into the end of the adapter and should be considered a universal accessory for webcam imaging. Photo by Robert Reeves.*

inch barrel at one end which fits the telescope eyepiece focuser and M12 x 0.5 threads on the other end. The metric designation denotes threads with a 12-millimeter diameter and a pitch of 0.5 threads per millimeter for screwing the adapter into a webcam.

Mogg's adapters allowed the telescopic webcam revolution to accelerate by providing an easy means of adapting cameras to various optical devices without the need for a lot of tinkering. Today, only ATM purists who desire to make all of their own components need bother with making their own webcam adapters. If an astronomy retailer does not have the proper adapter for your webcam, visit Steven Mogg's website for a vast assortment of webcam mounting applications for both telescope eyepiece holders and camera lenses.

4.2 Filters for Webcam Astrophotography

In large cities, the background glow of the night sky can be 50 or more times brighter than in a rural setting. In urban areas, sky glow caused by city lights and light reflected back to the ground by dust particles, water vapor, smog, or industrial soot and smoke will cause the sky to be too bright for exposures of more than 30 to 60 seconds' duration. This sky glow will fog the image to the point where the stars and nebulae are no longer visible. In areas away from the city, where there are minimal lights, exposure times can increase and may be limited only by the capability of your camera. As we shall see, there are ways to suppress urban sky glow by using filters to block out portions of the spectrum where street-lights are dominant, and image processing steps that can be done with digital images to reduce the effects of sky glow.

4.2.1 Filters and the Wave Nature of Light

To understand how filters work with astrophotography, we have to understand the wave nature of light. Each color of the visible spectrum has its own unique wavelength expressed in units called Angstroms, one Angstrom equaling 0.0000000001 meters, or a ten-billionth of a meter. In more practical terms, however, wavelengths can also be expressed in nanometers, one nanometer being equivalent to one billionth of a meter.

The visible spectrum ranges from red light, which has a wavelength of 700 nanometers, through orange, yellow, green, blue, and indigo to violet, which has a wavelength of 400 nanometers (Figure 4.3). Wavelengths just beyond the longest and shortest ends of the visible spectrum are called, respectively, infrared and ultraviolet. Although the eye cannot perceive these wavelengths, webcams are sensitive to them and we must use special filters to prevent them from striking the sensor. Ultraviolet and infrared wavelengths will not harm the sensor, but these unseen wavelengths will alter image coloration, and since they do not focus at the same point as visible light they will create bloated halos around stars.

Table 4.1 summarizes the characteristics of a variety of astronomically useful Kodak Wratten filters we can use to block unwanted wavelengths or accentuate desired wavelengths. The wavelength is considered to be "cut off" if transmission at that wavelength is 50 percent or less. The last column refers to the percent of hydrogen-alpha radiation transmitted by the filter.

There are several different types of filters. A cutting filter will not pass a wavelength shorter than the band it was designed to transmit. An example is a red hydrogen-alpha filter designed to block everything except the crimson light of hydrogen-alpha emission. A band-pass filter transmits a limited band of wavelengths. Lumicon's Deep Sky and the Hutech IDAS Light Pollution Suppression filter are examples of this type. Monochromatic, or interference, filters transmit a single wavelength of light. An example is the hydrogen-alpha solar filter used to view solar prominences.

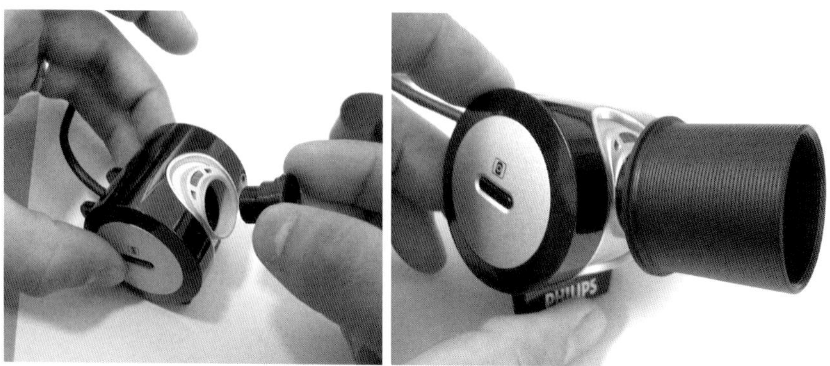

Fig. 4.2 *The Philips SPC900NC camera has a recessed lens mount, so the telescope adapter designed for the ToUcam will not fit unless it is remachined to the 900NC lens bore. BrightStar, the same Portuguese company that manufactures components for the Atik series of webcam-style cameras, produces a special insert that allows adapters from other cameras to be used with the SPC900NC. Photos by Pedro Pereira.*

Table 4.1
Filter Spectral Characteristics

Filter	Color	Lower Cut-off	Upper Cut-off	Passes % of H-alpha
Wratten #1A	UV	380 nm		91.0
Lumicon M-V	UV	420 nm		96.0
Wratten #2B	Yellow	390 nm		91.5
Wratten #2A	Yellow	405 nm		91.0
Wratten #2E	Yellow	415 nm		91.0
Wratten #8	Yellow	495 nm		91.0
Wratten #12	Yellow	520 nm		91.0
Wratten #15	Yellow	530 nm		91.0
Wratten #21E	Orange	505 nm		90.3
Wratten #47A	Blue	415 nm	475 nm	0.0
Wratten #47B	Blue	430 nm	430 nm	0.0
Wratten #47	Blue	440 nm	440 nm	0.0
Wratten #80A	Blue	495 nm		18.0
Wratten #44	Blue	476 nm	510 nm	0.0
Wratten #25	Red	600 nm		87.9
Wratten #29	Red	620 nm		88.5
Wratten #92	Red	640 nm		76.4
Lumicon H-alpha	Red	640 nm		90.0
Schott RG 645	Red	650 nm		60.0

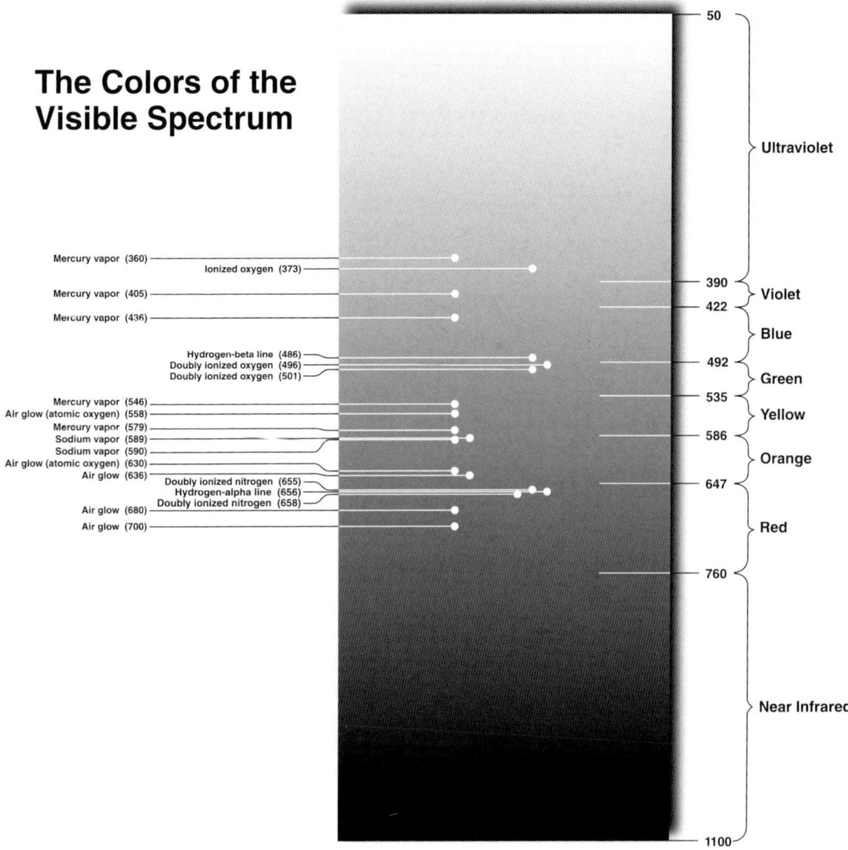

Fig. 4.3 *At left, the important light pollution emission lines and astronomical emission lines are displayed. At right, the wavelengths of colors in the visible spectrum are shown.*

This ultra-narrow band hydrogen-alpha solar filter should not be confused with the wider bandpass nighttime hydrogen-alpha filter.

Filters affect the focus of the optical system they are used on by an amount dependent on the refractive properties of the filter itself and the wavelength of light passed by it. If a filter is placed in front of a lens, the glass of which it is composed has no effect on the lens's focus. The color of the filter, however, may shift the focus. If the filter is positioned inside the focused beam of the lens, the amount of focus shift caused by the filter glass itself is equal to the thickness of the filter multiplied by one and divided by the refractive index of the filter glass; that is, shift = filter thickness x (1/refractive index).

Telephoto camera lenses are often used with modified webcams for deep-sky imaging. Using a deep red or hydrogen-alpha filter with a camera

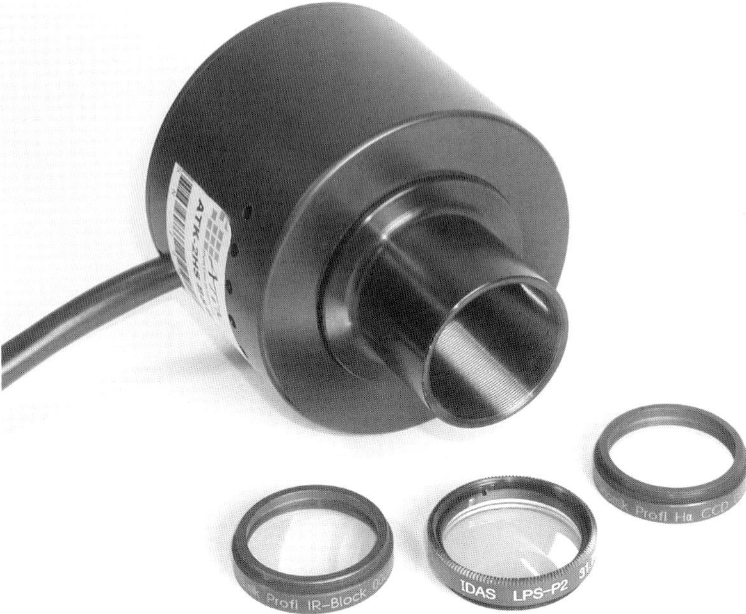

Fig. 4.4 *Just as conventional photography can benefit from the use of filters, webcam imaging can also be enhanced with the use of the proper filter. Webcam filters thread into the barrel of the telescope adapter. Some popular ones are: an infrared blocking filter, an IDAS light pollution control filter, and an H-alpha filter. Photo by Robert Reeves.*

lens will change its infinity focus setting to a point just short of its normal "white light" infinity setting and, hence, degrade star images if the focus shift is not compensated for. A narrow-wavelength hydrogen-alpha filter requires very precise focus. This is well worth the effort because most camera lenses are twice as sharp in red light as they are when admitting the entire visible bandwidth.

4.2.1.1 The Importance of Infrared Cutoff Filters in Webcam Astrophotography

An infrared filter is perhaps one of the most important types you can use with an astronomical webcam. In fact, it is virtually mandatory to use one with a refracting telescope. The need for such a filter comes from the fact that the imaging sensor in a webcam is sensitive to infrared wavelengths as well as visible light. Modern refractors benefiting from computer-aided optical design and exotic glasses to achieve near apochromatic performance present highly color-corrected, aberration-free visible light images. But many modern refractors are not corrected for these infrared wavelengths, which will contribute to star bloat because the infrared wave-

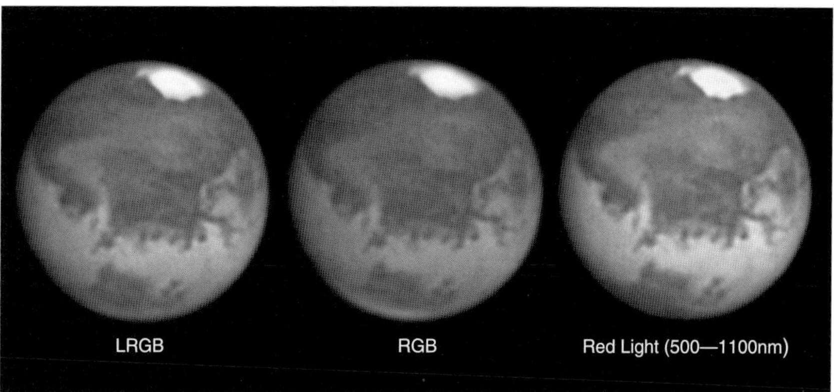

Fig. 4.5 *Using a high-detail infrared image of Mars as the luminance channel in an LRGB color composite improves the tonal density of the image while preserving surface detail. Photos by Damian Peach.*

lengths will lay an out of focus image over the focused visible light image. The resulting combined image will display soft lunar and planetary detail and halos around stars. In many cases, the infrared filter reduces the sensitivity of the camera, but it is essential for recording fine detail nonetheless.

Most webcams have a built-in infrared filter for "normal" terrestrial use. This filter is usually part of the lens, such as the rear glass in the 6mm *f*/2 lens that comes with the Philips ToUcam 840. The presence of this filter is evident by the pinkish reflective sheen of the glass. Other cameras may have the filter placed in front of the CCD sensor. The built-in filters used in current webcams do a good job of blocking infrared, but they are also aggressive in the deep red portion of the spectrum and block some of the important hydrogen-alpha wavelength. The older black-and-white Quickcam filter is absorbing glass and also blocks the far-red portion of the visible spectrum. If images without a filter are desired, do not mistake the window over the CCD imaging chip for a filter. This window is part of the chip itself and is there to protect the sensor.

A webcam's normal lens will be removed for most astronomical imaging applications. This will, in most cases, also remove the factory-installed infrared filter. To prevent the detrimental effects of infrared radiation on the image, another infrared filter will have to be reinstalled somewhere in the optical train. The most common method is to thread one into the end of the webcam eyepiece adapter.

As a general rule, Newtonian or Cassegrain telescopes do not need an infrared filter to achieve good webcam focus because their reflective optics, unlike even most apochromatic refractors, focus both visible and

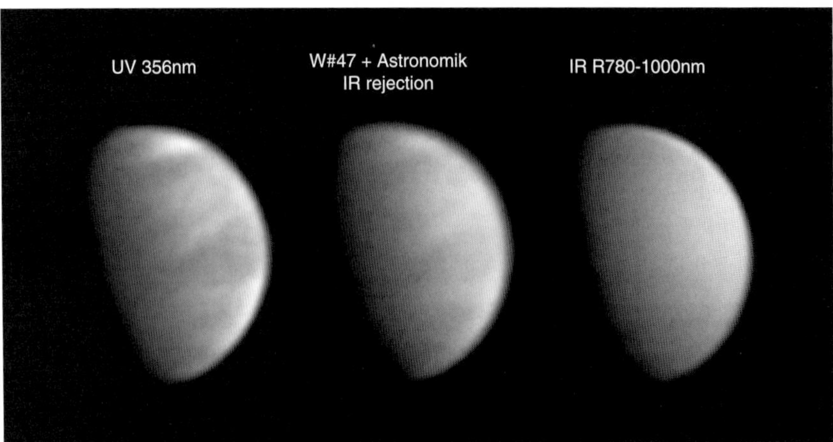

Fig. 4.6 *The use of a short-wavelength pass filter that blocks visible light while passing the ultraviolet allows a webcam to image the cloud features of Venus (left). The same view through an infrared pass filter (right) no longer reveals the cloud features. Photos by Damian Peach.*

infrared wavelengths at the same point. However, the vast majority of mirror-based telescopes are compound optical systems that use both mirrors and lens corrector plates. Examples are the popular Schmidt-Cassegrain design Meade or Celestron SCTs, or the Maksutov design like the Meade ETX. Therefore, as a general rule, Schmidt-Cassegrain and Maksutov telescopes do benefit from an infrared filter to some degree because the corrector plate acts like a lens.

The effect of infrared focus shift is not as pronounced with secondary optics, such as a focal reducer or Barlow lens, as it is with a refractor objective lens. However, it is still a good idea to use an infrared filter with these devices for maximum lunar and planetary detail.

If focus is not a problem because Newtonian optics are being used, planetary color imaging will still require the use of an infrared filter because of a phenomenon known as "IR bleed." This effect shifts the colorcast of a planet because the individual red, green, and blue Bayer mosaic pattern filters over each of the CCD sensor photosites, or pixels, are transparent to infrared. The excess infrared combines with the visible light colors and creates a color shift. Color correction is easier to do at the source of an image, rather than later during image processing, to eliminate the infrared.

Using an infrared filter also makes it easier to correct the effects of atmospheric dispersion. This phenomenon is easily seen with the naked eye when imaging Venus near the horizon. In this case, the atmosphere refracts differing wavelengths in unequal amounts, resulting in several overlapping multicolored planetary images. This affect also occurs to a lesser extent at higher elevations, and affects all astrophotos to some degree. Generally, any high-resolution imaging done at altitudes lower than 50 degrees should be done with an infrared blocking filter. The smaller the target, the greater the chromatic aberration, as tiny details will be smeared by atmospheric refraction making focusing more difficult. These effects can be corrected to a large extent with the use of the Adirondak Planetary Atmospheric Dispersion Corrector (PADC), as will be discussed in Section 4.11. But even with the PADC, truly sharp visible-light planetary detail will be impossible to attain without an infrared filter.

If the effects of chromatic aberration are significant, each color can be separated and realigned in an image-processing program such as RegiStax. An infrared filter will keep the individual color components of the image a purer color, making it easier to separate the blue from the red image. Without the filter, the blue component will also contain part of the infrared portion of the spectrum and will be irreparably smeared.

Even if your telescope design does focus both visible and infrared wavelengths at the same point, and a target such as the Moon is being imaged in black-and-white, it is a good idea to have a filter installed on the webcam simply as a dust cover. Infrared filters for webcams are designed to thread directly into the camera's eyepiece holder adapter and will prevent dirt and specks of dust from contaminating the sensor.

4.2.2 Popular Infrared Filters for Webcams

There are two types of infrared filters—absorbing glass and dichroic cutoff filters. Both work for visual use and webcam imaging. Absorbing glass filters have a slightly greenish tint and proportionately cut off more infrared at longer wavelengths. The more expensive coated glass dichroic filters have a pinkish reflecting surface and display a sharper cutoff at specific wavelengths. Photographic infrared filters are optically flat while others, such as those designed for heat rejection in a slide projector, are not flat. A photographic system must use an optically flat filter or image quality will suffer.

The Baader UV/IR Cut filter, manufactured in Germany, has a good reputation with webcam users. Baader claims the filter passes 98 percent of visible wavelengths, an important point when imaging dim or highly magnified objects. It is effective against wavelengths shorter than 400 nanometers, passes virtually all visible wavelengths, and blocks all wavelengths from 690 to 1200 nanometers. Beyond 1200 nanometers, the filter again begins to pass significant amounts of infrared, but lunar and planetary spectra are not strong in these regions and the infrared sensitivity of

CCDs diminishes significantly at 1200 nanometers. The Baader filter also has a unique knurled "crown" machined into its metal holder ring to facilitate easier handling and installation.

Astronomik, another German company, markets an infrared filter with transmission characteristics nearly identical to the Baader unit, but Astronomik makes no claims that their filter will block any UV wavelengths. I use this brand filter and have had good results with lunar and planetary targets. It passes all of the 656-nanometer hydrogen-alpha spectral line, then has a sharp cutoff for wavelengths between 675 and 1200 nanometers.

The popular glass filters discussed above are all available in the standard 32.8-millimeter diameter thread size used with the universal 1¼-inch webcam eyepiece adapter. The glass filters are not affected by humidity and are resistant to scratching.

4.2.3 Exceptions For Infrared Filter Use

For every rule, there are exceptions and this also applies to a webcam infrared filter. There are times when it is not desirable to have a filter present, like when imaging only in black-and-white through an all-mirror Newtonian telescope. In this case, the added sensitivity from utilizing a broader range of the spectrum allows shorter shutter speeds when imaging solar system objects in black-and-white. Also, when imaging deep-sky objects with a long-exposure modified camera, the added sensitivity gained by utilizing the infrared portion of the spectrum may allow imaging fainter objects than possible with the filter.

Most imagers who use a compound telescope for the latter activity regard the infrared filter as optional. The prevailing thinking is that although the lack of an infrared filter may affect image sharpness, the effects of atmospheric blurring over long exposures is greater and will naturally degrade the image anyway. In this case, the short exposures that are possible because more wavelengths are available without the filter are more beneficial to the image than removing the infrared wavelengths.

If a webcam is being used as an autoguider, critical star focus is not an issue while sensitivity to faint stars is. It is always advantageous to autoguide without an infrared filter, no matter what type of optical configuration is used, in order to detect fainter stars, especially if an unmodified camera is used. An unmodified ToUcam or SPC900NC working at the Cassegrain focus of an 8-inch *f*/10 SCT has no trouble imaging a 6th-magnitude star at the slowest shutter speed and maximum gain settings. The fact that a webcam "sees" by utilizing a significant portion of the infrared spectrum was demonstrated with my own ToUcam 840. While autoguiding on a fairly bright star, in this case about 3rd magnitude, the camera continued to image the star even when a layer of cirrus clouds blocked it from view through the telescope's finder. In this case, the IR wavelengths were penetrating the clouds and still registering on the infrared-sensitive CCD while the clouds blocked the shorter visible wavelengths. Although the visible exposure is interrupted, it can be quickly resumed without searching for a lost guide star when the clouds pass.

4.2.4 Infrared Pass Filters for Planetary Imaging

Do not confuse an infrared blocking filter with an infrared pass filter. A blocking filter passes visible light and blocks infrared, while a pass filter blocks visible light and passes the infrared. An infrared pass filter has its uses, such as imaging certain surface details on Mars or Jupiter, but the visible light view through an infrared pass filter is very dim because it is blocking the visible wavelengths. This may seemingly make it difficult to focus using such a filter, but remember that the webcam will "see" the infrared wavelengths the eye cannot, and thus with appropriate camera exposure settings the image on the computer monitor will still be bright.

Infrared pass filters are also useful for a new planetary imaging technique in which luminance data is gathered using through an IR pass filter, then combined with color data gathered through RGB filters. The Baader Infrared Pass Filter blocks all wavelengths shorter than 670 nanometers. At the long wavelengths visible through the filter, planetary details are less susceptible to atmospheric distortion and when the infrared luminance data are combined with other visible color data, the resulting planetary images are sharper.

For those who like to experiment, an inexpensive infrared pass filter can be made from Kodacolor 100 film exposed to fluorescent light for five seconds, then developed normally in C-41 chemistry (see Table 4.2). Any minilab can process this film. Photographic film is fragile, easily scratched, and does not respond well to high humidity, so it's best to sandwich it between glass or insert it into the tube of a Barlow lens to protect it.

4.3 Useful Filters For Imaging

One of the attractions of astrophotography with modern digital imaging is that the results are in color—but color imaging can also create image-processing challenges. Light pollution will not affect the colorcast of short exposures of the Moon and planets, but it can cause color shift problems when imaging deep-sky objects with long exposure-modified cameras. We therefore want to take steps to control light pollution and still maintain a

Table 4.2
Infrared Transmission of Fully Exposed Kodacolor 100 Film

Wavelength (nanometers)	Approx. Transmission %
Visible light <700	Below 3 percent
700 nm	3
750 nm	10
800 nm	70
850 nm	90
900 nm	88
950 nm	87
1000 nm	83
1050 nm	80
1100 nm	75

good color balance in the final image. Some filters that reject light pollution wavelengths may also create color tints that render the sky or a celestial object an unnatural color. This is not a problem with black-and-white imaging, but it limits our selection of filters that will work with color images.

One filter that can help digital imaging is an ultraviolet filter. Most stars emit significant amounts of UV radiation, and because most camera and telescope lenses are unable to focus wavelengths in this region of the spectrum at the same point as visible light, the results are fuzzy halos around otherwise well-focused stars. Lumicon markets their well-known Minus Violet filter that cuts all wavelengths shorter than 420 nanometers. Users may find the Baader UV-IR cut filter more useful because it solves both the UV and infrared "problems" with webcam imaging.

Another filter that is very useful for both planetary and deep-sky imaging through a refractor is the Baader Contrast Booster. This filter combines the beneficial characteristics of a UV cut filter with a high-contrast filter. Color aberration is reduced in achromatic refractors and contrast is increased with the elimination of the defocused blue haze created by short blue and UV wavelengths. The elimination of blue wavelengths lends a slight yellow tinge to the resulting image, but the increase in contrast more than offsets this effect. This filter has been found to be very effective when imaging Mars. The combination of Contrast Booster and the UV-IR cut filter together forms a powerful combination that eliminates unwanted wavelengths that degrade both planetary images and deep-sky work done under less than ideal sky conditions.

Other filters useful in webcam planetary photography are yellow, orange, and violet filters used to visually enhance planetary detail. A #21 orange filter will work well with Mars.

Venus is a brilliant but visually bland planet that shows almost no detail. The near-total cloud cover obstructs surface detail and the clouds themselves are devoid of visible light detail. Venus does display some cloud detail in the ultraviolet wavelengths that are beyond the range of human vision, and a webcam has some sensitivity in this portion of the spectrum. To image this cloud detail, a UV pass filter, such as the Baader-U Venus filter is required. This unusual filter passes 80 percent of the wavelengths between 300 and 400 nanometers, but virtually no visible light. To the eye, Venus is invisible through the filter but since the filter passes wavelengths that a webcam can detect, the result is a good image on the computer monitor. If a UV pass filter is not available, stacking a regular infrared filter with a Wratten #47 may reveal Venus cloud detail.

4.4 Broadband Light Pollution Rejection Filters

A number of broadband light pollution rejection filters is now available that reduce the detrimental effects of light pollution on color astrophotography. Light pollution filters are more complex and costly than single color filters because they must pass certain wavelengths across the spectrum while selectively blocking others. Because these filters must pass many important wavelengths, they are prone to also passing some of the light pollution wavelengths we are trying to avoid. The desired wavelengths needed for full-color photography, and the detrimental wavelengths of light pollution, are too close to each other to effectively filter out all the "bad" wavelengths while passing only the "good" astronomical ones. This limits the effectiveness of these broadband light pollution filters, but the results achieved with them are superior to those achieved without.

The most widely used broadband filters for telephoto lenses coupled to a webcam are Lumicon's Deep Sky filter and Hutech's IDAS Light Pollution Suppression (LPS) filter. Lumicon offers their filter in thread sizes up to 72mm while Hutech offers an 82mm size for large telephoto lenses. The larger Lumicon and Hutech filters are expensive, but much of that high price is due to manufacturing techniques mandated in the mid-1990's by Environmental Protection Agency policies. Both Hutech and Lumicon also market smaller versions of their Deep Sky and LPS filters that thread into 1¼-inch eyepieces and thus can be directly installed to the webcam's eyepiece adapter.

The German filter manufacturers Astronomik and Baader also produce lines of popular light pollution control filters that thread directly into an eyepiece and thus can be adapted to a webcam.

The Lumicon Deep Sky filter produces a slightly reddish sky background, but does allow longer exposures in slightly light-polluted areas. The filter renders nebulae in truer colors because it passes both red and blue wavelengths that are prominent in deep-sky objects while blocking yellow and green where most man-made light pollution exists. Since the Deep Sky filter is selective in the wavelengths it does pass, the nebulae look normal, but star colors may be changed. The Deep Sky filter does pass infrared radiation in the 850- to 900-nanometer range, but stacking an infrared-blocking filter with it will eliminate this. This filter typically allows exposures about three times longer than possible without filtration. Because they are not monochromatic, broadband filters have no effect on chromatic aberration and will not help in sharpening star images. Typically they pass about 90 percent of the light between 400 and 480 nanometers.

The IDAS filter is most effective against light pollution produced by mercury vapor lighting, the prevalent outdoor lighting used in Japan where the filter was developed. However, in the U.S. the switch to sodium vapor outdoor lighting in many urban areas renders the IDAS filter less effective. Still, some of the wavelengths associated with the orange sodium vapor do fall within the cutoff ability of the IDAS filter, so some benefit is nonetheless achieved. Since the mix of mercury/sodium vapor lighting in any urban area is unpredictable, it is not possible to state with conviction that the filter will achieve a certain level of performance in your area. However, a typical image can be exposed for twice as long when using the IDAS filter.

Light pollution reduction filters are most effective with long, multiminute exposures such as are common with film astrophotography. The effectiveness of the IDAS filter is therefore not as dramatic with digital imaging as it is with film because each exposure is inherently shorter due to electronic noise limitations. The elimination of light pollution wavelengths by the filter, however, does create a better signal-to-noise ratio, allowing images to be stacked achieving a deeper exposure without becoming overwhelmed by the sky background. The general consensus among users is that the IDAS LPS filter renders a more natural-looking sky.

Astronomik, Baader, and Lumicon also market narrow passband light-pollution filters that do a good job of increasing image contrast, but do so at the expense of eliminating some wavelengths entirely, and thus these filters may alter the color appearance of some celestial objects. The Astronomik CLS filter and the Baader UHC-S nebula filters both have similar transmission characteristics. Each passes wavelengths between 450 and 530 nanometers as well as more than 90 percent of the hydrogen-alpha wavelength. The Lumicon UHC is even more restrictive, passing only the 486-nanometer hydrogen-beta, and the 496- and 501-nanometer O-III wavelengths. All three filters are available in sizes that thread into eyepieces and thus can be adapted to webcams.

4.5 Hydrogen-Alpha Filters Block Light Pollution

Most deep-sky objects emit light in the red hydrogen-alpha wavelength, making this one of the most important wavelengths in astrophotography. Stars, galaxies and emission nebulae all shine brightly in this spectral region. Reflection and planetary nebulae radiate most of their energy in shorter wavelengths, but still emit enough hydrogen-alpha to be recorded at this wavelength. Therefore, by using a filter tailored to pass only light in the hydrogen-alpha region of the spectrum, we can eliminate most of the light-pollution sources in the blue, green and yellow wavelengths. This gives us a powerful means of blocking light pollution and sky glow, allowing exposures up to 10 times longer than without the hydrogen-alpha filter. Using deep red filters with a color-capable camera is a double-edged sword. They give us a good tool to combat light pollution, but unfortunately it turns a color camera into a monochrome camera and the resulting image must be converted to grayscale and viewed in black-and-white.

Astronomik and Lumicon make hydrogen-alpha filters designed to thread directly into the eyepiece holder adapter used on most webcams. The Astronomik H-alpha CCD filter has a narrow passband centered on this 656 nanometer wavelength and is opaque to longer infrared wavelengths between 675 to 1100 nanometers. Beyond that wavelength, the filter resumes infrared transmission. The Lumicon filter on the other hand begins transmission near 640 nanometers and remains transparent to all infrared beyond the 656 nanometer hydrogen-alpha wavelength. As of summer 2005, Baader and Hutech do not offer H-alpha filters in this size.

Enterprising people who modify their own webcams for long-exposure work often install the sensor and electrical circuits in a project box with custom modifications allowing the use of other filters designed for "conventional" astrophotography. In these cases, Kodak Wratten gel filters, available in 3-inch or 4-inch squares, can be used for celestial imaging. However, there is a reduced availability of these filters as some are out of production plus the number of camera stores that carry them is shrinking. The #25, #29, and #92 are all useful for red-light imaging as described above. The #25 is usually available as a threaded glass filter from most manufacturers. The deeper red #29 is available as the B + H 091 glass filter. The Lumicon and Kenko (marketed by Hutech) glass hydrogen-alpha filters are a close match for the #92 and statistically outperform it for celes-

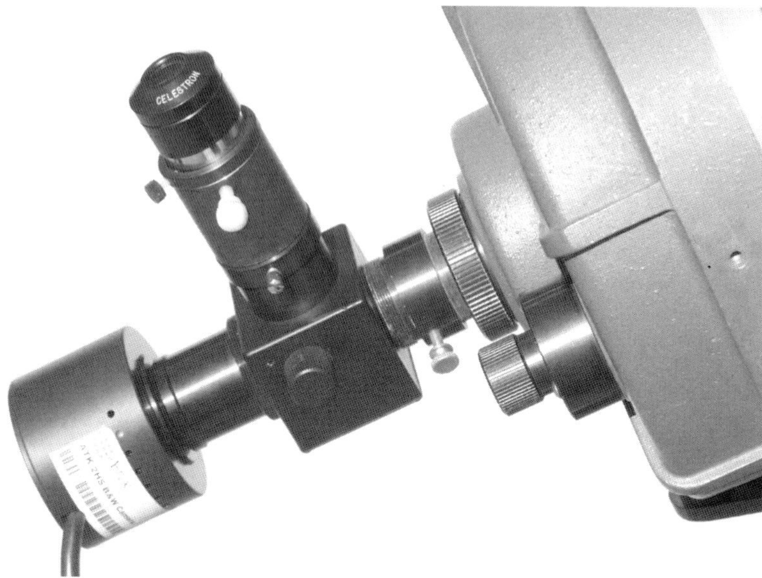

Fig. 4.7 *A large object like the Moon can be acquired with a webcam by using the telescope finder, but deep-sky objects, and especially high-magnification views of planets, require the use of a flip-mirror finder. The flip-mirror assembly installs between the telescope and camera. With the mirror in its down position, it temporarily diverts the view to a low-power eyepiece in order to center the target. Once centered, the mirror is flipped up and the camera can image the target. Photo by Robert Reeves.*

tial imaging. Glass filters cost more, but they are more durable than the soft, easily scratched, Wratten gel filters and astrophotographers will find them a better choice for imaging.

4.6 Flip Mirrors

While it is inherently simple to install a webcam in place of an eyepiece at the focus of a telescope, it must be realized that its field of view is very small compared to a 25mm or 40mm eyepiece. The field of view with the 4.5mm diagonal measure CCD sensor in a standard webcam is approximately equal to viewing through a 5mm eyepiece. That is 8 times narrower than the field of the 40mm eyepiece. Using a Barlow increases the difficulty in aiming the camera by further reducing the field of view. One quickly realizes that the main difficulty with planetary imaging is simply aiming the telescope and camera at the tiny target. An accurate finder scope is necessary and will be adequate by itself for lunar imaging. Some searching is usually necessary to center a planet in the webcam at prime focus while using only a finder scope, and if a Barlow is used, the photographer will have to go on a hunt to find the planet.

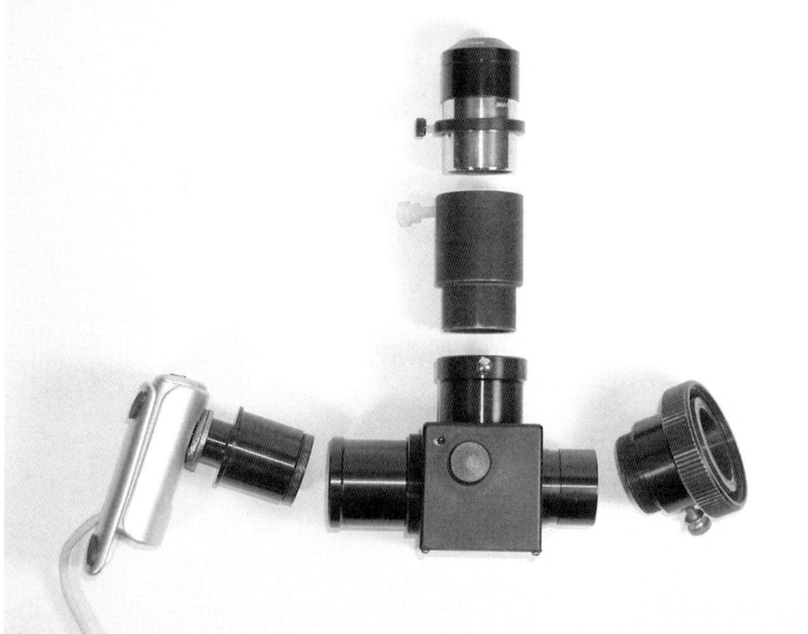

Fig. 4.8 *A flip-mirror assembly inserts into the telescope focuser or SCT visual back. An eyepiece extension tube is usually required to allow the finder eyepiece to be parfocal with the camera focal plane. Photo by Robert Reeves.*

An accessory that greatly simplifies aiming the telescope with a narrow-field webcam is a flip mirror. This device is often called a flip-mirror finder (FMF) because it is used to locate a small target. It attaches between the telescope and camera and uses a movable mirror to temporarily divert the light path to a wide-field eyepiece to allow centering the object in the field of view. Once the target is centered, the mirror is flipped out of the optical train and the light passes to the camera. A number of vendors offer a flip mirror. The Meade Model 644 is a high-quality model that works with 1¼-inch accessories and has a street price of about $150. True-Technology markets higher-priced FMF's that are equipped with a filter drawer. The added versatility of the filter drawer allows this unit to act as both a finder and filter holder for the camera.

It is possible to visually zero in on a tiny target by progressively inserting higher power eyepieces, then finally substituting the webcam. However, a steady telescope mount is needed to keep the target centered as each eyepiece and the camera are swapped out. This method is very time-consuming and one evening of trying to chase down Saturn while using a Barlow will quickly show the advantages of using a flip-mirror system.

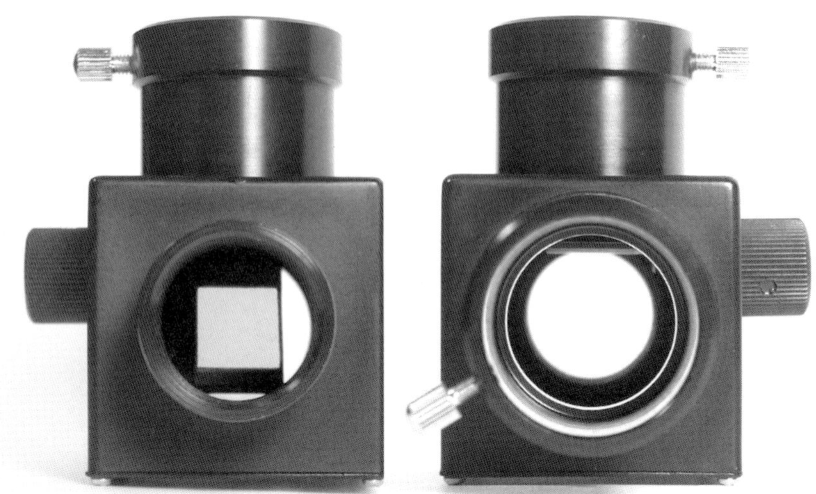

Fig. 4.9 *A flip-mirror assembly is shown with the mirror lowered (telescope-side view on left) to deflect the telescope view to the eyepiece holder on top. The mirror is flipped up (camera side view on the right) to allow the telescopic image to reach the camera. Note that inexpensive, but functionally adequate, flip-mirror assemblies may not completely block the light path when the mirror is lowered and thus can not serve as a dark shutter for taking dark frames. Photo by Robert Reeves.*

An economy option is to purchase a flip mirror manufactured by Astrovisioneer in China and marketed by the Telescope Warehouse in Las Vegas, Nevada. This unit inserts into any 1¼-inch eyepiece holder and accepts any 1¼-inch accessory. It works surprisingly well considering its cost is only ⅓ that of the Meade unit. I use one of these units and can attest that this flip mirror is well made and the finder eyepiece and camera fields of view are accurately aligned. The important thing is that the unit shows the same field of view in an eyepiece with the mirror down as the camera sees with the mirror up. By centering a planet in a dual crosshair 12.5mm guiding eyepiece, then lowering the mirror, I find a planetary target is within ⅛th of the diameter of the camera's field of view from the center—good enough for me! The barrel of the 1¼-inch outlet on the Astrovisioneer unit does not unthread to expose T-threads for attaching T-threaded options, but considering its price, this FMF does a good job as a webcam finder.

I found that to focus any of my eyepieces with the Astrovisioneer, the eyepiece needs to be slid out of its adapter tube so far that it exits the flip-mirror assembly before reaching focus. A similar problem is encountered with many other flip-mirror units, and when using an off-axis guider. Lumicon and Scopetronix market various lengths of eyepiece barrel and focusing-tube extensions for use with their off-axis guiders. Users will find that the appropriate length extension tube will fix the FMF eyepiece focus problem. I found a homemade 1¼-inch extension tube in my "astro junk box" that was actually too long. Rather than cut the tube down, I found that removing the barrel skirt from a swap meet "no-name" 26mm ocular allowed the upper part of the eyepiece containing the optics to slide far enough inside the adapter tube to reach focus. The field of view of the 26mm eyepiece is about six times that of the webcam, but it is also large enough that targets are easily visible within its field after using the finderscope. Once the target is in the 26mm eyepiece, a jog of the telescope controls places it in the center of the eyepiece field. Raising the flip mirror then reveals that the target is within the field of the webcam.

Deep-sky imagers who use long-exposure modified webcams will find that flip mirrors that totally block light from reaching the camera will be useful for making dark frame calibration exposures. Some inexpensive flip mirrors, such as the Astrovisioneer unit discussed above, use a mirror that is smaller than the optical path from the telescope to the camera. While this causes no real problem when visually locating targets with the mirror down, a small percentage of light is getting past the mirror and may reach the camera through reflections inside the flip-mirror box. Such light leakage makes the Astrovisioneer incapable of dark frame acquisition. For this activity, look for a model that either totally blocks light from passing while the mirror is down, or has a built-in dark-frame shutter that can be manually slid into place to shield the camera.

4.7 Parfocal Rings

Often when a webcam is installed on a "cold" telescope, that is, one that is not already accurately focused for the webcam's view, the target star or planet is so far out of focus that it is unrecognizable even if its image falls in the camera's field of view. It is possible for a webcam to view entirely inside the "donut hole" of the out-of-focus star. In this case, the camera can be aimed directly at the target star, but the stellar image is spread out so far that it cannot be detected.

A device useful for achieving a rapid, though coarse, webcam focus before installing the camera on a telescope is a parfocal eyepiece. A parfocal eyepiece will achieve focus at the same telescope focus setting as the camera. Any eyepiece can be "parfocalized" by using a small ring that slips over its barrel and clamps into place with a setscrew. The parfocal

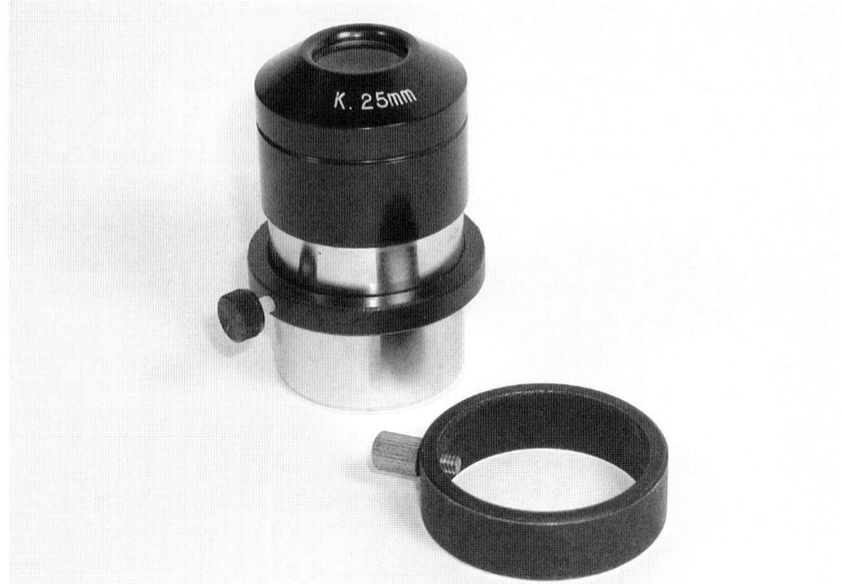

Fig. 4.10 *Parfocal rings, such as the Mogg ring on the eyepiece, or a homemade version, provide a positive stop to position a prefocused eyepiece at the focal point to match the camera focus. When the eyepiece is in focus, the camera is focused well enough to acquire the target. Photo by Robert Reeves.*

ring will allow the eyepiece to be inserted into the telescope in the same repeatable position once the webcam's focussed position has been determined. Rings are available from Mogg and any of the larger distributors of webcam adapters.

In practice, eyepieces are parfocalized with the camera by first accurately focusing the camera while its adapter is bottomed out in the barrel of the eyepiece focuser. Without disturbing the telescope's focus setting, remove the camera and focus the eyepiece by sliding it in and out of the eyepiece holder. When it is in good focus, lock the clamp on the parfocal ring while it is shouldered against the eyepiece focuser. The next time the telescope is focused using the parfocal eyepiece, the camera will also be in approximate focus when it is substituted for the eyepiece as long as both devices are bottomed out against the eyepiece focuser. The camera will initially be in good enough focus to allow finding the target, but a final critical focus will still have to be performed.

Similarly, if an FMF is used, the camera is first focused, and then the mirror is lowered to divert the view to the eyepiece. The eyepiece is slid in or out of the flip-mirror assembly until it is also in good focus. The parfocal ring is then shouldered against the eyepiece holder and tightened on the eyepiece. The parfocal eyepiece will now allow future coarse focus through the flip-mirror assembly. Additional tweaking of the camera's focus will be necessary, but the target object will be sharp enough to be easily acquired.

Another way to achieve coarse focus with telescopes that use a rack-and-pinion focuser is to make a series of template spacers the same length as the needed drawtube travel required to focus the camera. The spacer should measure the required distance between the camera body and the fixed, non-movable top of the focuser. Spacers of various lengths can be used to define the focuser travel needed for prime focus, or use of a Barlow, or focal reducer.

4.8 Barlows and Powermates

Barlows and Tele Vue Powermates are traditionally known as accessory optics that are inserted between the telescope focuser and an eyepiece or imaging device in order to increase the telescope's focal length and thereby its magnification. Use of these devices is the easiest way to gain additional focal length to match the telescope's expected resolution in arcseconds with the camera's pixel size. While it is possible to use a webcam with an eyepiece projection system like that used with a conventional camera, the resulting extreme focal lengths and high *f*-ratios are often impractical for the inherently small imaging area of a webcam. A much better and simpler way to boost magnification over that achieved at prime focus is to use a Barlow or Powermate. A webcam that adapts to an eyepiece focuser will also fit directly into either of these accessories.

A Barlow increases apparent focal length by using a negative lens to diverge the light rays as they exit the Barlow's top glass surface. Because of this, they do not work well with some optical configurations, as this will displace the exit pupil of an eyepiece farther out. This may increase viewing comfort with short focal-length eyepieces, but visual observers will find that the increased eye relief with a long-focus eyepiece may create vignetting with eyepieces that don't normally exhibit this problem. Vignetting happens when the eye relief is extended beyond the point intended by the optical designer, and the eyepiece lenses are too small to allow all the light rays through. The additional magnification may also increase focus degradation at the edge of the field, especially in the case of "Shorty" Barlows, the stronger negative element of which diverges light rays at a steeper angle.

Webcam imaging uses only the central portion of the field and thus the above-mentioned Barlow limitations—vignetting and soft focus at the

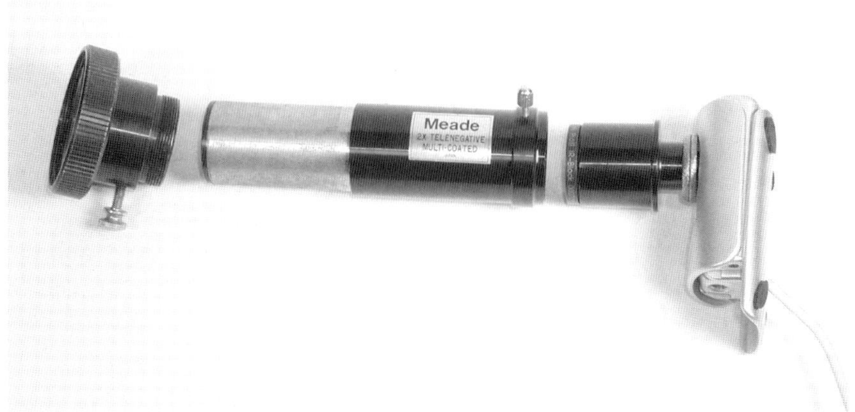

Fig. 4.11 *Barlow projection photography inserts a Barlow lens between the telescope eyepiece holder and the 1 ¼-inch webcam adapter to increase the telescope's effective focal length. Photo by Robert Reeves.*

edge of the field—do not come into play. They are outside the webcam's narrow field of view.

Barlows come in a variety of types featuring 2x, 2.5x, 3x, 4x and even 5x magnifications. A 2x Barlow will make the target twice as big at the focal plane. This is what we desire, but the use of a Barlow for increased magnification comes with a price. Simple math shows us that an object that formerly covered a single pixel on the camera's sensor now covers two pixels in each dimension. Now that the image is spread out it is also dimmer because of the inverse square law, and hence requires longer exposure time. Increased focal length and exposure time also allow any atmospheric turbulence or telescope vibration to degrade the image. The enlarged image results in a lower signal-to-noise ratio, but this can be partly compensated by taking more video frames for stacking and averaging to create the final image. Also, the resulting smaller field of view makes it harder to locate and center planetary targets. For these reasons it is advisable to use a Barlow or Powermate only under good seeing conditions.

The best combination of accessory optics to increase the focal length of a given telescope for lunar and planetary imaging depends upon the performance of the telescope itself and the pixel size of the camera used. The local atmospheric seeing is also a limiting factor in how much additional magnification can successfully employed. Generally, users of the popular *f*/10 Schmidt-Cassegrain telescopes, under good seeing conditions, achieve excellent lunar and planetary results between *f*/20 and *f*/40. Focal lengths as high as *f*/50 can be employed for objects with high surface

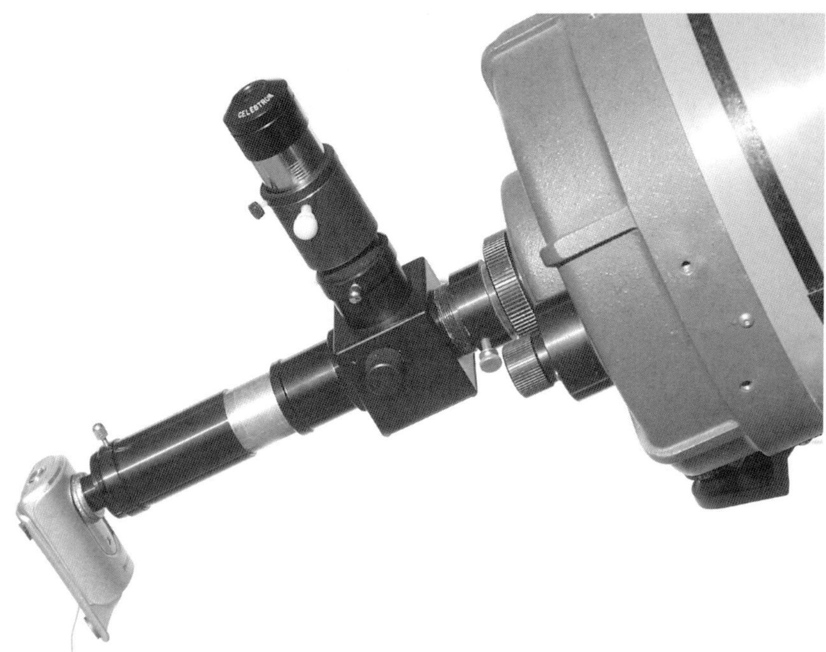

Fig. 4.12 *Imaging through a Barlow lens is the easiest way to achieve the focal length "sweet spot," usually between f/20 and f/30, required to obtain the best lunar and planetary performance from a telescope. Photo by Robert Reeves.*

brightness, such as Mars at opposition under steady seeing. Jupiter and Saturn have a much lower surface brightness and the "sweet spot" for imaging the giant planets is between *f*/25 and *f*/35. These focal lengths are not a hard rule and people who bend it during times of extraordinary seeing often take outstanding images.

Some people working under exceptionally steady skies have success stacking several Barlows for even greater magnification of the Moon, an object bright enough that exposure times remain reasonably short even with such equipment. A point to remember when stacking Barlows is that each one increases the apparent focal length by its stated magnification before the light path enters the next Barlow. Therefore two 2x Barlows increase the focal length four times, a 2x and a 3x Barlow increase focal length six times. While it is not advisable to do so because of the flexure of such a large optical stack, three 2x Barlows would increase focal length by eight times.

Another consideration when stacking Barlows is that introducing more glass elements into the optical train reduces the light throughput as each element scatters and reflects a certain portion of the light reaching it.

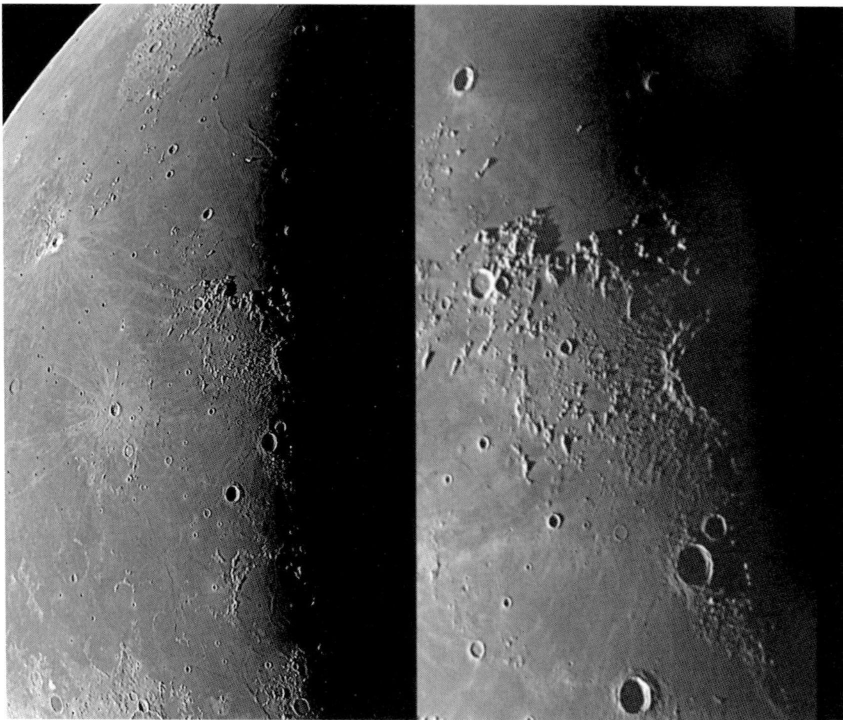

Fig. 4.13 *The improvement in detail derived from using a Barlow for solar system imaging is evident when comparing these views taken on the same night of sunset at the lunar crater Copernicus. Both were taken with an ATK-2HS camera on a Celestron-8 telescope and show the same pixel-for-pixel image scale at the camera focal plane. The left one was cropped from a mosaic taken at f/10, while the right image is a three-frame mosaic of the Copernicus area taken at f/20 using a 2X Barlow. Significantly finer detail is visible in the latter. Photos by Robert Reeves.*

One factor working in favor of the user is that modern optical coatings do a wonderful job of reducing reflections from each glass element, reducing light loss from about 5 percent to as low as 0.5 percent per glass-to-air surface. But it should be remembered that optical coatings are just that; coatings. They are not polished optical surfaces and are capable of introducing aberrations in the own right. When stacking Barlows, use high quality components for best results. After all, the system is only as good as its weakest link. Using cheap optics can result in mediocre images.

The magnification factor available from a particular Barlow design is not necessarily fixed. The advertised 2x, 2.5x, etc., magnification is achieved only when the focus of the eyepiece or imaging instrument is at a certain distance from the top of the Barlow. Increasing that distance increases the magnification factor, sometimes by a significant amount, and this can be used to benefit the webcam imager. The magnification of a 2x

Barlow increases to nearly 3.5x if the focus is 100 millimeters from the top of the Barlow. For a 3x unit the magnification increases to 5x at the same distance.

Because of the higher Barlow power at increased projection distance, some increased magnification can be realized by sliding the barrel of the webcam adapter out of the Barlow as far as it will go while still being firmly attached by the lock screw. There is not enough travel to achieve the magnification gain demonstrated at the 100-millimeter projection distance, but some additional Barlow power will be obtained.

Resourceful imagers can take advantage of a Barlow's varying magnification power by making custom-length extension tubes to position their camera focal plane at specific distances from the Barlow. The magnification of a Barlow is dependent on the distance F from the negative lens (remember that F is negative by definition), the longer the distance, the higher the magnification.

To calculate the increased magnification with distance, Nils Carlin provides the following formula:

$$M' = M - \frac{B}{F}$$

where

F = Barlow focal length,
M = Magnification, or "Barlow" factor,
B = Additional extension distance, and
M' = Larger magnification gained with additional extension.

This formula assumes one knows the focal length of the negative lens. Typically, however, the focal length of the negative lens is not known, but the Barlow's magnification factor is usually marked on its barrel. In this case, the focal length of the negative lens can be calculated with the following formula (again noting that F is negative):

$$F = \frac{C}{2 - M - \frac{1}{M}}$$

where

C= the height of the Barlow barrel protruding above the focuser.

You should be aware that it may be necessary to move the focuser inward a certain distance when using an additional drawtube for increased magnification. Schmidt-Cassegrain and Maksutov telescopes usually have a generous amount of mirror movement to accommodate focus so this is not an issue. However, the focuser travel on refractors, Newtonians, and true Cassegrains is more limited and may not permit focus to be reached. To find the required change in focuser position (P), use the following formula:

$$P = F\left(\frac{1}{M} - \frac{1}{M'}\right).$$

By increasing the magnification of a Barlow with increased projection distance, some additional spherical aberration is introduced, but since a webcam is imaging only the central portion of the field, this should not affect the image.

Because the projection distance behind a Barlow will affect its true magnification, be careful when placing a Barlow in an optical train that includes a star diagonal. A Barlow that normally doubles magnification when used alone may increase magnification by a factor of three or more if placed in front of a star diagonal because the diagonal increases the projection distance.

The Tele Vue Powermate is a derivative of the Barlow concept that eliminates the degrading effects the negative lens has on the performance of many long focal-length eyepieces. To do this, the Powermate uses a negative doublet to diverge the parallel light rays from the telescope optics in order to gain magnification, and then uses a positive doublet to restore their parallelism. For the visual observer, this has the advantages of providing true increased magnification without altered eye relief and associated vignetting or focus degradation.

Like traditional Barlows, Powermates also vary magnification as a function of distance from the top surface of the lens (i.e., the eye lens) to the focal plane but to a lesser degree. The magnification change with distance for 2x and 2.5x Powermates is negligible, with the 2.5x Powermate actually losing 0.25x over a 100 millimeter projection distance. The 4x Powermate gains 0.5x over this projection distance while the 5x unit yields an actual magnification factor of 7.7x at 100 millimeters (i.e., 2.7 additional). The latter is thus an excellent substitute for eyepiece projection photography and simplifies the attachment of any camera with the optional Tele Vue T-ring adapter which screws directly to the eyepiece, and in turn interfaces with a webcam T-adapter.

For the webcam imager, Tele Vue's Powermate offers several advantages over a standard Barlow:

- Telescentric operation, that is, the light rays are parallel when they exit the optics instead of diverging as they do with a Barlow. This is ideal for

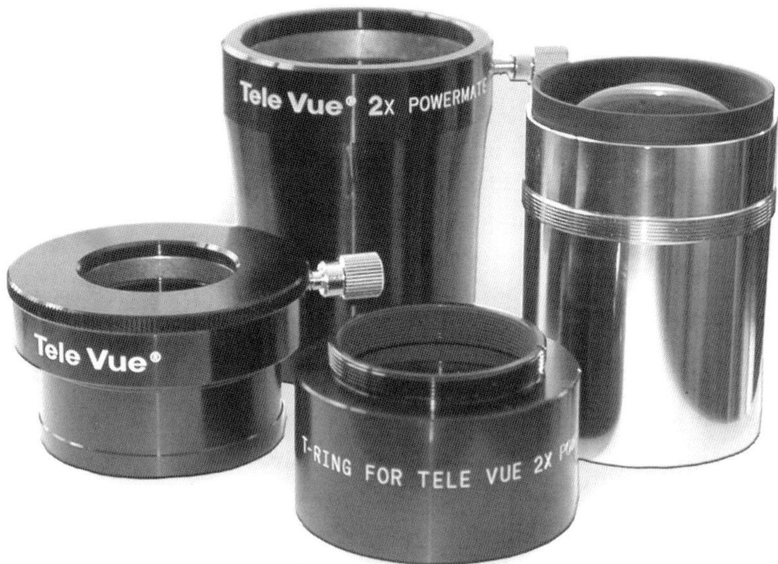

Fig. 4.14 *The Tele Vue Powermate is a versatile accessory for increasing telescope focal length. The 2-inch 2X Powermate shown here can accept 2-inch eyepieces, a 1¼-inch eyepiece adapter, or an adapter with T-threads to couple to a photographic device. Photo Courtesy of Tele Vue.*

use with solar H-alpha filters like the Daystar.

- The 2x, 2.5x, and 4x Powermates are nearly parfocal and provide relatively constant magnification regardless of the distance to the imaging focal plane. (An exception is the 5x Powermate which increases its magnification by 1x for every additional 35 millimeters to the imaging surface.)
- The 5x Powermate provides high image magnification without Barlow-stacking or eyepiece projection.
- Powermates do not invert the image.
- Powermates extend less than three inches from the focuser and accept all 1¼-inch eyepiece filters.
- Imaging with a Powermate is like imaging at prime focus with a refractor or Newtonian, or at Cassegrain focus, but with greater focal length.

When imaging with optics such as a Barlow, Powermate or other accessory optics like the Tele Vue Paracorr or some forms of focal reducer, check with the manufacturer to find the optimum back-focus distance needed to achieve best possible focus. For instance, the Powermate and Paracorr are designed to achieve best performance with a 56-millimeter back focus to accommodate the needs of popular 35mm and digital SLR cameras. Webcams do not need this much back focus and extension adapters may be needed to place the focal plane at the optimum 56-millimeter distance.

For those who enjoy tinkering with optical components, 2x "tele-extenders" for 35mm camera lenses can provide the optical elements to make a wide-aperture Barlow that can attach directly to the visual back of a Schmidt-Cassegrain telescope. Good units from brand names like Nikon and Pentax can often be found in the used equipment display at major camera stores.

4.9 Focal Reducers

The image sensors in webcams and their astrocamera derivatives are much smaller than those used in conventional digital cameras and thus have an inherently smaller field of view. Since the natural view for a webcam is similar to that through a 5mm eyepiece, some celestial targets are simply too big to fit into this limited field of view. For this reason, a focal reducer is often used to shorten the telescope's focal length, with the added benefit of making the field brighter and reducing the needed exposure time. Celestron and Meade offer focal reducers that change their normally *f*/10 SCT systems to either *f*/6.3 or *f*/3.3. These reducers are interchangeable among the two brands and install between the telescope's threaded rear cell and the visual back that accepts optical accessories. The *f*/3.3 reducer is not suitable for visual work because of distortion in the outer portion of the field, but the since the field of view of a webcam is naturally small, it views only the central area and is not adversely affected. Some imagers have had success in stacking the *f*/3.3 and *f*/6.3 reducers to achieve the equivalent of *f*/2 through an *f*/10 SCT.

In our look at Barlows we saw that varying the projection distance from the Barlow to the camera focal plane can alter its advertised magnifying power. The same applies to focal reducers; they will also vary the degree of their reduction of image scale with increased projection distance. And to complicate things further, telescopes that feature a moving mirror in order to focus the image have an even greater proportion of image scale change because their focal length is not constant.

For the webcam user, the large focal reducers offered by Celestron and Meade are essentially overkill for a webcam's small imaging surface. If the user already has one for other applications on an SCT-style telescope, it will also work fine with a webcam. But if a user is thinking of buying a reducer just for use with a webcam, a more economical option is

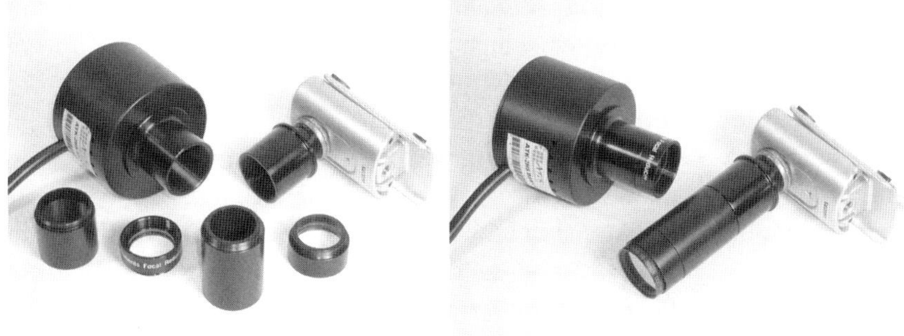

Fig. 4.15 *Focal reducers screw directly into the camera adapter barrel and are front-threaded to accept filters. Atik Instruments (camera on left) markets a 0.5 reducer that also fits other webcam-style cameras. Adding a Scopetronix 25mm eyepiece extension tube further increases the reduction factor. Similarly, installing the Mogg 0.6 focal reducer on the Mogg 0.3 reducer extension tube yields an effective focal reduction of 0.3, making an f/10 telescope perform like and f/3 system.*

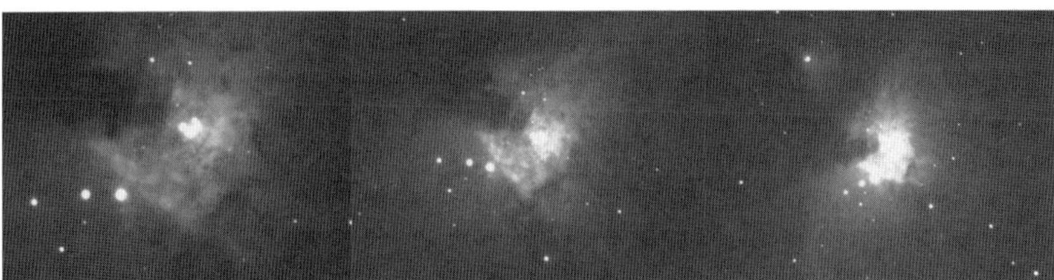

Fig. 4.16 *These three views of M42 are single exposures taken through a Celestron-8 telescope with an ATK-2HS camera. The left image shows the telescope's view at a normal f/10 focal length. The middle one utilizes an Atik 0.5 reducer, yielding f/5; while the right image adds a 25mm projection tube to the reducer, approximating an f/3 system. Photos by Robert Reeves.*

the Mogg focal reducer that screws directly into the nose of the 1¼-inch webcam adapter. Placing a small reducer in the camera adapter itself allows the device to be used with the camera on other telescopes as well as camera lenses. The Mogg focal reducer has a reduction factor of 0.6 and adequately illuminates the larger ⅓-inch sensor used on re-chipped, modified webcams and the SAC, and Atik commercial cameras when viewing through an 8-inch *f*/10 SCT. Atik offers its own 0.5 focal reducer which threads directly into the camera's 1¼-inch barrel.

The Mogg 0.6 focal reducer consists of a 125mm focal-length lens mounted in a cell that threads into the camera adapter. An optional 37mm extension tube increases the reduction factor to 0.3, but the user is warned that this configuration may not fully illuminate a webcam field.

As stated before, increasing the projection distance behind a focal reducer will increase the reduction factor. The standard 0.6 reduction factor is achieved with the Mogg reducer by threading the device into the front the 1¼-inch webcam adapter, which places the camera about 34mm from the reducer. Adding the extension tube increases this to 71mm and produces a 0.3 reduction factor. Inserting other accessories into the optical path, such as a SAC filter wheel, makes the projection distance 61.5mm and produces a reduction factor of 0.37. Stacking all three accessories—the focal reducer, the extension tube, and the SAC filter wheel—results in an unrealistic reduction factor of 0.11 and most telescopes will not achieve focus with this setup.

The focal reduction factor achieved with the above accessories is a straightforward number that we can apply to our telescope's basic focal length in order to do image scale calculations. But as we have seen elsewhere, depending upon the projection distance used, an *f*/10 SCT is not always the same focal length, and as the mirror is moved to maintain focus at different projection distances, the effective *f*-ratio also changes.

Craig Stark performed some tests with the Mogg focal reducer, extension tube, and SAC filter wheel to determine the exact focal length and *f*-ratio of his 8-inch *f*/10 LX-90 SCT while using combinations of these accessories. He found that his telescope, without a star diagonal, had a basic focal length of 1924mm and an *f*-ratio of *f*/9.6. Installing the 0.6 reducer on a webcam lowered the focal length to 1195mm and coincidentally achieved the expected *f*/6 even though the telescope's focal length was not normally a full 2000mm. Adding the extension tube would have theoretically achieved *f*/3, but in fact yielded a focal length of 766mm, at *f*/3.8. The reducer and SAC filter wheel should have ideally achieved a focal length of 882mm at *f*/3.7, but actually produced *f*/4.4.

The mechanical limit of mirror travel prevents achieving focus (at least with the LX-90) at greater projection distances, but another possibility allows achieving a theoretical *f*/2 (400mm) focal length with it: the stacking of two Mogg 0.6 reducers. This setup may seem to be an optically brutal way to achieve a super-fast optical system, but in fact it works surprisingly well with very little color fringing of stars near the edge of the field.

As we can see, the changing focal length caused by moving the mirror to achieve focus at differing projection distances slightly changes the overall focal ratio, but in practice it is of little concern that the focal ratio is *f*/3.8 instead of *f*/3. Mathematically there is a difference, and if science were being done with the images the difference would have to be accounted for. But for astrophotos taken for the fun of it, and that is what most of us do, the differences in focal length are of no real concern. After all, the beauty of webcam astrophotography, for both short and long exposures, is that the

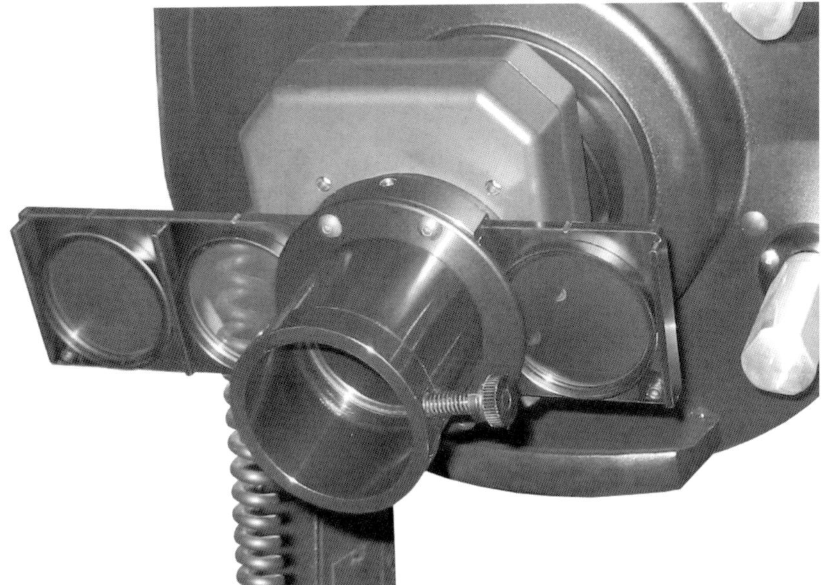

Fig. 4.17 *A filter slide such as the Lumicon Multiple Filter Selector allows the use of various filters with an attached camera. The 2-inch version shown here holds four filters while a fifth filter used in conjunction with others, such as an infrared blocking filter, can be installed on the front of the selector. Filter slides are more economical than filter wheels but have the disadvantage of exposing the filters not currently in use. Photo courtesy of Lumicon.*

results are instantaneous and any corrections needed because of a mis-exposure at varying focal lengths can be made on the spot.

4.10 Filter Trays and Wheels

The use of various filters increases the versatility of astronomical webcams. Often, multiple images of the same target taken through a succession of different filters are required, when as doing color planetary imaging with a monochrome camera. This process requires images taken through infrared blocking, infrared pass, ultraviolet pass, or various color Wratten filters, or RGB or narrowband SII, H-a, and OIII filters. Filters are traditionally threaded into the nose of the camera adapter, so switching them requires removing the camera to access the filter, then refocusing the camera after reinstallation—a cumbersome process. Fortunately, various devices are available that enable us to change filters without first having to dismount the camera, then reacquire an often invisible target.

The simplest is a Lumicon Multiple Filter Selector. This device inserts into the telescope focuser and the camera then inserts into the filter

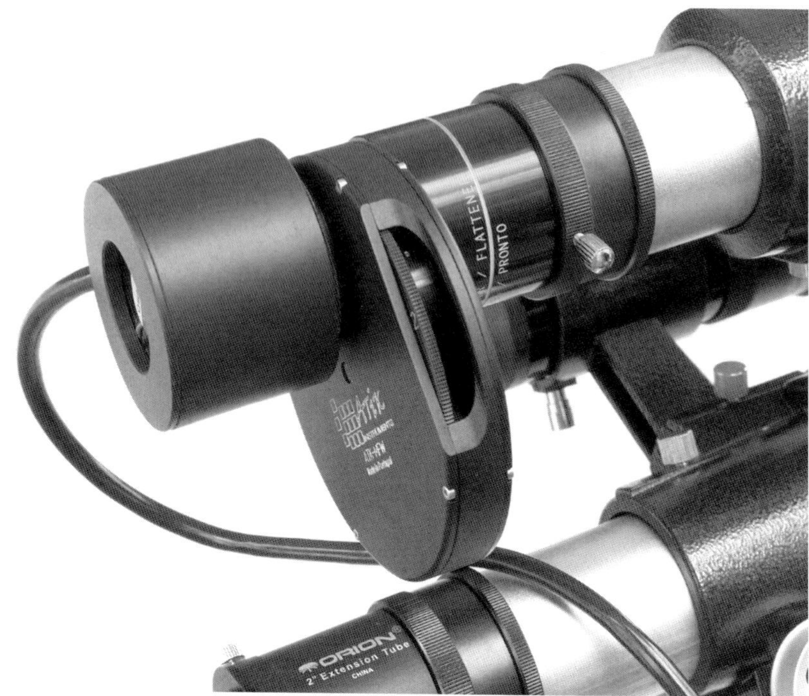

Fig. 4.18 *Atik Instruments makes a manual selection, five-position filter wheel that encloses and protects the filters. Although made for the Atik brand cameras, it attaches to any T-threaded camera. Photo by Jim Ferreira.*

selector. It is available in a 1¼-inch focuser version holding five standard eyepiece filters as well as a 2-inch focuser version that holds four larger filters. A sliding tray allows moving the chosen filter into the light path entering the camera while an additional filter can be stacked on the nosepiece of the selector for use in combination with the selected filter. The Lumicon 1¼-inch model offers the simplest and most economical way to rapidly switch filters on a webcam. It is also one of the few filter selectors that allows the use of a webcam equipped with a standard 1¼-inch telescope adapter. However, it does have the disadvantage of exposing the filters to the elements and possible fingerprints in the dark.

SAC imaging offers a four-position filter wheel that protects standard eyepiece filters within the body of the filter wheel holder, but the device is attached to a telescope by first removing the nosepiece from the SAC camera and attaching it to the filter wheel. The camera then attaches to the wheel and the entire assembly inserts into the telescope focuser. A fifth stacked filter can be installed on the outside of the SAC filter wheel.

Atik Instruments offers two ways to use standard eyepiece filters with

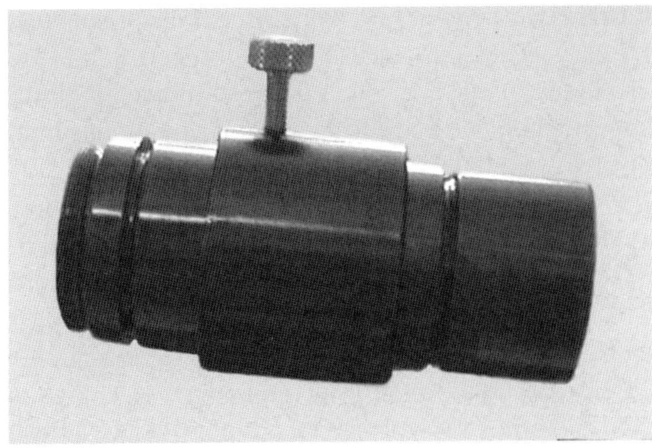

Fig. 4.19 *Solar system imagers will find the Planetary Atmospheric Dispersion Corrector (PADC), marketed by Adirondack Video Astronomy, to be an invaluable aid in achieving the highest planetary resolution possible. The inferior planets Venus and Mercury are usually imaged at low elevations where the effects of atmospheric dispersion are greatest. The PADC will realign the red and blue portions of the spectrum and reduce image smearing caused by dispersion. Photo courtesy of Adirondack Video Astronomy.*

their series of modified long-exposure webcam-style imaging devices. A filter support kit allows the use of interchangeable single filters on a camera without having to first dismount it. This is done by unscrewing the T-threaded 1¼-barrel from the camera body and replacing it with the filter support. The barrel is then reinstalled in the front of the support so the combined assembly can be installed in a telescope focuser. The filter is threaded into a holder that then slides into a slot in the support. Extra holders are available to mount additional filters. Any imaging device using T-threads can be adapted to the filter support.

Atik also offers a five-position click-stop filter wheel that installs with T-threads like their single filter holder. The 1¼-inch camera nose piece barrel installs on the filter wheel and a sixth stacked filter can be installed in the adapter. Its body fully encloses and protects the filters. The filter wheel is a slender 19.5 millimeters thick which minimizes back-focus problems with the limited focuser travel on Newtonian telescopes.

4.11 Wedge Prisms for Atmospheric Dispersion

As we saw in Section 3.12, the effects of atmospheric dispersion can be very detrimental to high-resolution imaging at low altitudes above the horizon. The Adirondack Planetary Atmospheric Dispersion Corrector (PADC) can control these bad effects to a large degree. There are two versions of the PADC, one uses a single four-degree wedge prism and another uses two separate two-degree wedge prisms. All the prisms are made from Schott BK-7 glass. This glass has refractive properties that are similar to the refractive index of air, and when properly aligned with the incoming light rays, can counteract the spectrum-broadening effects of atmospheric dispersion.

In practice, the wedge prism is rotated to control the direction of the refracted light passing through the prism. In the twin prism design, each can be rotated independently over an opposite 360-degree range to allow even more dispersion control and to prevent ghost images. It is recommended that the PADC not be used in optical systems with less than *f*/12 focal ratio. Below *f*/10 the angle of incoming light is so high that spectral dispersion will be noticed even when using the PADC. Use a Barlow to achieve the required focal ratio.

To use the single prism PADC, insert it into a telescope focuser and install an eyepiece in it. Align the scribe mark on the PADC barrel perpendicular to the horizon so it is at the 6 o'clock position when viewing through the eyepiece. Focus the telescope normally and note the red and blue fringing caused by dispersion. Rotate the entire PADC unit in either direction. The planetary target will roll around the field of view, but when rotated to the proper angle, the effects of dispersion will disappear and planetary detail will be enhanced. Center the target in the field, remove the eyepiece and install the camera with its infrared filter. If the target needs to be rotated in the camera field, turn only the camera, *not* the PADC. Refine the focus of the camera and expose the image. The dual prism PADC is used in a similar fashion, but first the two separate prism sections must be assembled so their scribe marks are aligned with each other. The entire unit is then installed in the focuser and used like the single prism unit. The advantage of the latter is that if needed, the separate prisms can be rotated up to 90 degrees to refine the unit to a specific amount of dispersion correction.

4.12 Camera Lens Adapters

Webcams are not limited to use on a telescope. Mogg and several other makers offer adapters that couple a webcam to a standard camera lens. A webcam-to-camera-lens adapter will attach directly to the back of the lens and has a ¼ x 20 threaded tripod hole to facilitate piggyback mounting of the lens and webcam on a telescope. The adapter threads into the webcam body like a regular 1¼-inch eyepiece adapter. The small size of a webcam's imaging sensor turns a standard camera's 200mm telephoto lens into a surprisingly narrow-field instrument. Though image resolution is low be-

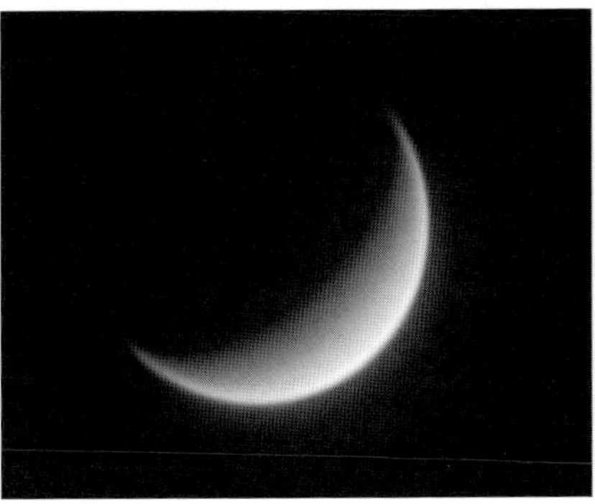
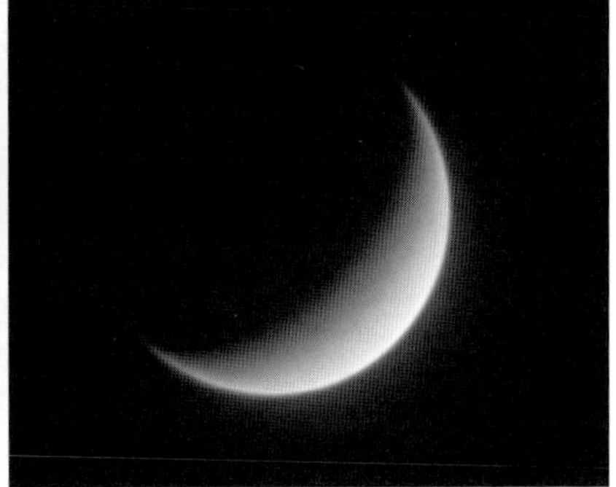

Fig. 4.20 *The Venus image at left was taken with a standard camera and telescope combination and exhibits the image-smearing effects of atmospheric dispersion. That at right was taken with the same telescopic setup with a PADC to reduce atmospheric dispersion. The result is a sharper, more detailed image. Photos by Russell Croman.*

cause of the camera's 640 x 480 pixel array, a webcam field of view through a 200mm telephoto lens is similar to that of a 35mm camera used at the prime focus on an 8-inch *f*/10 telescope. Objects like the full Moon or large nebulae fill the field. A webcam camera lens adapter, though, will not automatically focus an infinity focused lens. The target must be focused, as through a telescope.

4.13 Spectrographic Star Analyzers

The Star Analyzer is a high efficiency 100 lines/mm blazed diffraction grating mounted in a 1¼-inch cell that can be screwed into any webcam adapter threaded to accept filters. This accessory was developed by Robin Leadbeater, a British webcam astrophotography pioneer, and is commercially available from Paton Hawksley Education, Ltd, in the United Kingdom. It turns any webcam-style imaging device into a small, low-cost spectrograph capable of performing fascinating science. The delicate diffraction grating is protected by a coated glass cover that can be cleaned. The spectrum direction is marked on the filter cell, which can be secured in a given position with a locking ring.

Typically, in order to create a strong spectrum, the target object needs to be about five times brighter than required for normal imaging. Sensitive imaging devices such as the Philips CCD-based webcams, Meade LPI, or Celestron NexImage can record the spectrum of objects down to 4th magnitude through an 8-inch telescope. Long-exposure-capable monochrome imaging devices such as SC3-modified webcams, Atik IIhs, SAC8, or Meade DSI Pro can record spectra from objects down to 13th magnitude.

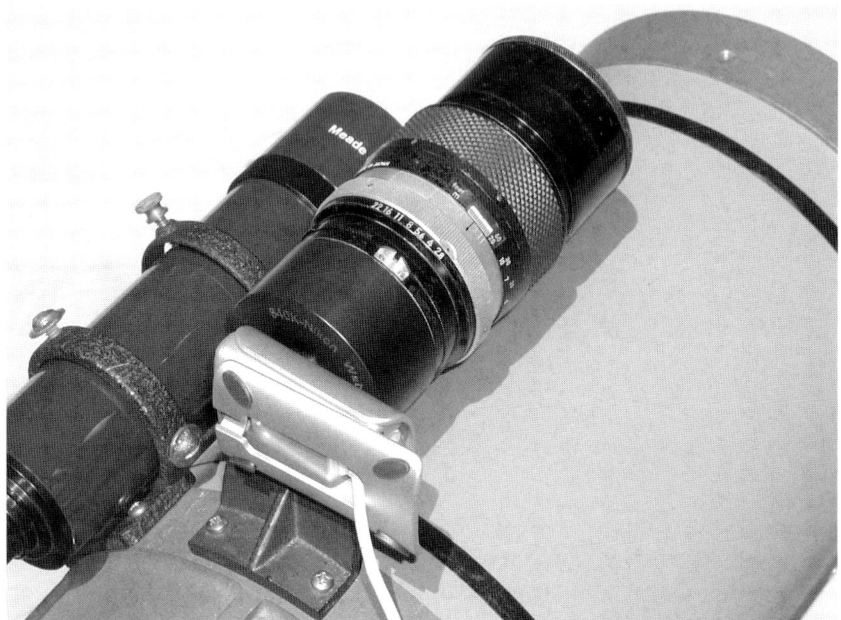

Fig. 4.21 *Conventional SLR and DSLR camera lenses can be adapted to a webcam with a Mogg camera lens adapter. Photo by Robert Reeves.*

The spectral range attained depends on the camera, but typically an unfiltered CCD is capable of recording from just under 400 nanometers in the violet to beyond 800 nanometers in the infrared.

The spectral resolution of the Star Analyzer is optimized for small CCD sensors like those found in webcams. Both the target and its spectrum can be recorded on the same image and the dispersion, or size of the resulting spectrum can be adjusted through the use of spacer rings to place the camera a certain distance from the grating. The greater the distance between the imaging sensor and the Star Analyzer, the larger will be the resulting spectrum. But it should not be so large as to exceed the size of the imaging sensor. The recommended minimum grating-to-sensor in millimeters is four times the pixel size in microns. For instance, a Philips SPC900NC with 5.6 micron pixels would require $(4 \times 5.6 = 22.4\text{mm})$ just under an inch. The maximum grating-to sensor distance still allowing the recording of spectra 700 nanometers wide across the diagonal of the sensor is 12 times the sensor's diagonal measure. For an SPC900NC, that works out to $(12 \times 4.5 = 54\text{mm})$ just over two inches. The maximum grating-to-sensor distance for modified cameras like the Atik IIhs, DSI, LPI, or SAC8 that use a larger ⅓-inch 7.4-micron CCD will be three

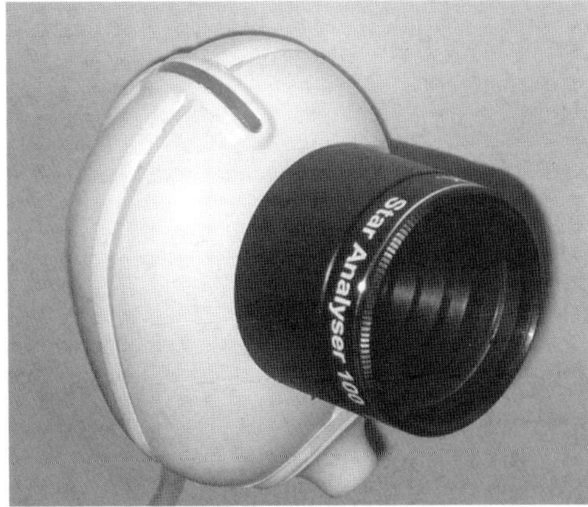

Fig. 4.22 *The Star Analyzer diffraction grating threads into the barrel of any 1¼-inch webcam adapter or webcam-style camera and turns the instrument into a small spectrograph. The direction of the spectrum is marked on the grating cell and can be locked in any orientation. Photo by Robin Leadbeater.*

inches.

The spectrum produced by the Star Analyzer will look similar to that shown in Figure 4.23. To enlarge the thin spectrum into the familiar wide bar as shown in Figure 4.24, use any image processing program capable of rotating images to align the spectrum so it is horizontal. Next, crop its vertical height so it is a strip only a few pixels high. Resize the strip to one pixel high, and then resize it again to 30 pixels high. Now the spectral lines are easily visible.

Analysis and the creation of spectral graphs such as Figure 4.26 can be done with the freeware Visual Spec program available on the web[1]. The AIP4Win image-processing program created by Richard Berry and James Burnell, and available from Willmann-Bell, also has features for spectroscopy.

4.14 T-Mounts

Many aftermarket camera telephoto lenses as well as some types of telescope eyepieces and accessories, such as various models of Tele Vue eyepieces and the Powermate, have T-adapter threads making the T-mount adapter a very useful webcam accessory. A T-mount is a universal camera lens adapter that was standardized by the photo industry after its introduction in 1957 by the lens maker Tamron. The concept of the adapter is that aftermarket camera lenses would have a standard rear threaded flange and

[1] http://astrosurf.com/vdesnoux/

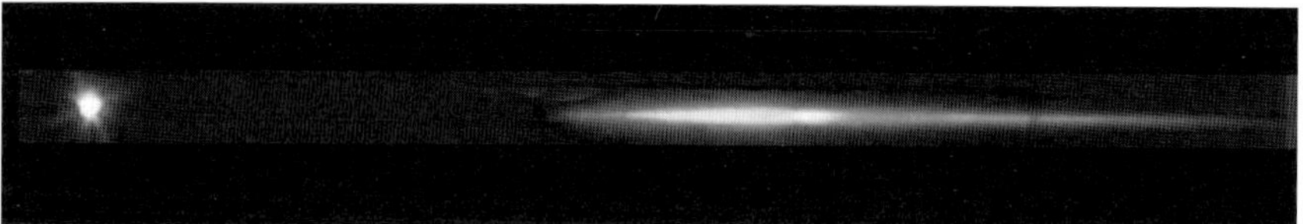

Fig. 4.23 *The spectrum of the star Vega and the star itself (left) were taken through an 8-inch f/9 reflector and a Philips ToUcam Pro using a Star Analyzer. Some dark absorption lines are visible in the raw spectrum. Photo by Robin Leadbeater.*

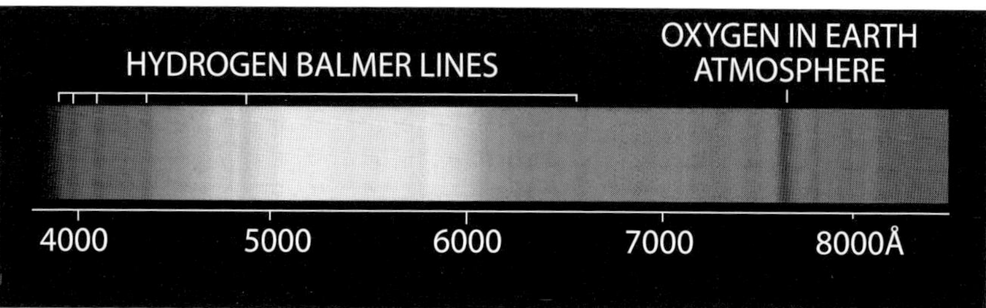

Fig. 4.24 *Using common image processing steps in Photoshop, a single line of pixels was cropped out of the Vega spectrum and vertically expanded. The enhanced spectrum now shows hydrogen Balmer lines in the atmosphere of Vega and oxygen lines from the Earth's atmosphere. Photo by Robin Leadbeater.*

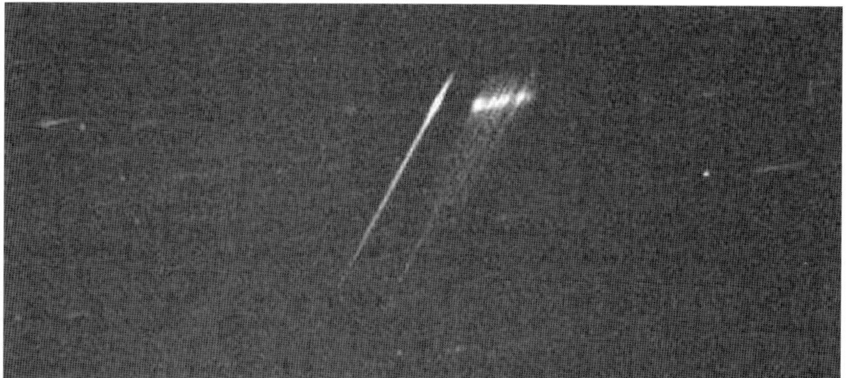

Fig. 4.25 *A Perseid meteor from the 2005 shower was captured above Cassiopeia using a long-exposure modified ToUcam utilizing a monochrome sensor and the camera's standard wide-angle lens. The Star Analyzer was placed in front of the ToUcam lens. Multiple emission lines are seen in the spectrum of the meteor's flare. Photo by Robin Leadbeater.*

would be attached to a camera brand-specific adapter, the T-mount, at the time of purchase. In its original concept, the T-mount consists of two parts: a T-adapter that threads onto the aftermarket lens, and a T-ring that mates to the specific brand of camera body the lens is being adapted to. The T-adapter and T-ring are designed to mate together with several setscrews. With this system, a particular aftermarket lens could be adapted to many

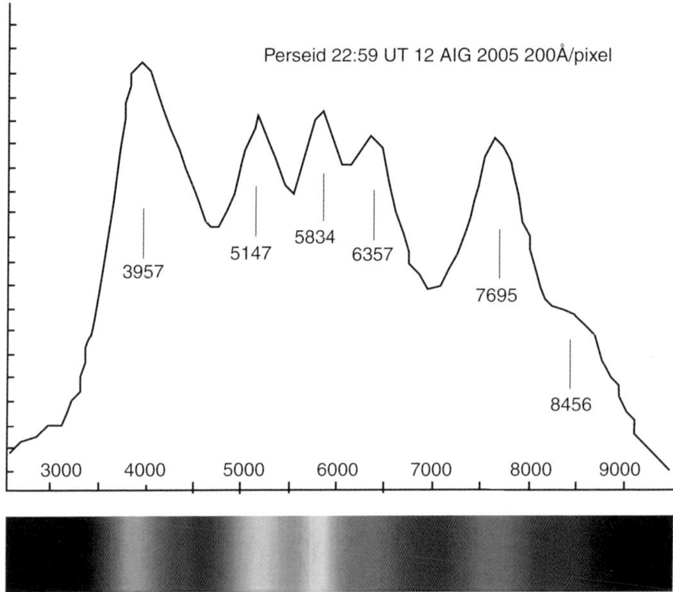

Fig. 4.26 *Visual Spec software was used to analyze the Perseid meteor spectrum and generate this graph showing the presence of calcium, magnesium, sodium, and silicon in it. Photo by Robin Leadbeater.*

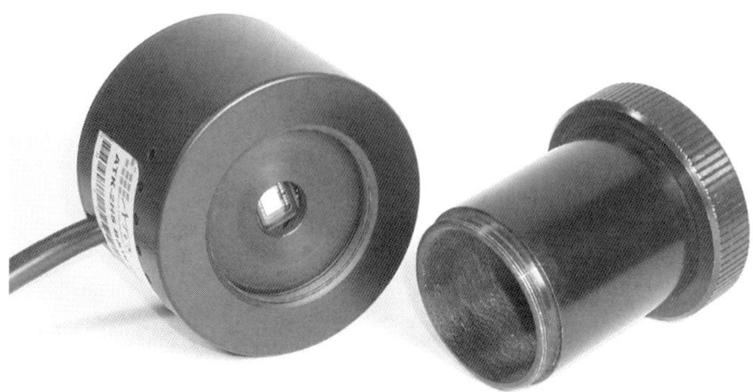

Fig. 4.27 *Cameras equipped with T-threads, such as the Atik models, can be adapted to a variety of astrophoto accessories and telescopes using T-adapters originally designed for conventional cameras. Photo by Robert Reeves.*

different makes of camera body by simply changing the T-ring to match the proprietary lens mount on the camera body. The internal thread size of the T-adapter is a metric thread 42 millimeters in diameter with a 0.75-millimeter thread pitch, referred to as M42 x 75 threads. In non-metric equivalent measure, this is 1.654 inches internal diameter with 33.866 threads per inch. In some cases, the concept has been modified so a given adapter is a solid one-piece item instead of two rings locked together with set-screws. The lens barrel or camera-to-telescope adapter must thread into the T-adapter and have sufficient wall thickness to be structurally sound. This restricts the internal diameter of the optical opening to about 1.5 inches. This same T-mount system has been adapted by major telescope manufacturers to allow a camera body to be attached directly at the focal plane of a telescope.

Chapter 5
Telescopes and Accessories

5.1 Types of Telescopes

Amateur telescopes for astronomy are a compromise between desired aperture, focal length, portability, and cost. A short focal length renders a bigger field but at the cost of resolution. A long focal length means higher resolution but will require longer exposure times and a heavier-duty mount. In the end, the best telescope is the one you have right now because it is the one you will be using tonight. Save for your future dream scope, one that is ideal for your long-term imaging goals, but adapt to what you have right now and enjoy the evening.

The best optics are always desired, especially for solar system photography. But for deep-sky objects, average optics on today's mass-produced telescopes work well. This is because even with ¼-wave optics, today's telescopes ranging from *f*/5.6 to *f*/16 produce a diffraction disk smaller than the photosites on the camera. At the same time, the "seeing" disk and tracking errors will always be larger than the diffraction disk. Bad optics will, of course, not work.

Telescopes can be divided into three general categories, each with strengths and weaknesses for astrophotography: refractors, reflectors, and compound telescopes like Schmidt-Cassegrain and Maksutov instruments.

A refracting telescope uses a lens to gather light and focus it at the eyepiece or camera focal plane. It is far more difficult, and thus expensive, to produce optical glass for, and manufacture, large objective lenses for refractors than it is to manufacture large mirrors for reflecting telescopes. For this reason, it is rare to see amateur-owned refractors larger than 6-inches in aperture. Refractors in the 4- to 6-inch range are considered "big" while a 6-inch reflecting telescope would be considered "small.

Refractors have the advantage of being fairly portable, even in the 6-inch aperture range. The inherent stiffness of a refractor's tube assembly allows it to hold its collimation, the alignment of its optical components, well. The smaller size of amateur refractors' primary optics, relative to those of reflectors, allows them to more rapidly achieve thermal equilib-

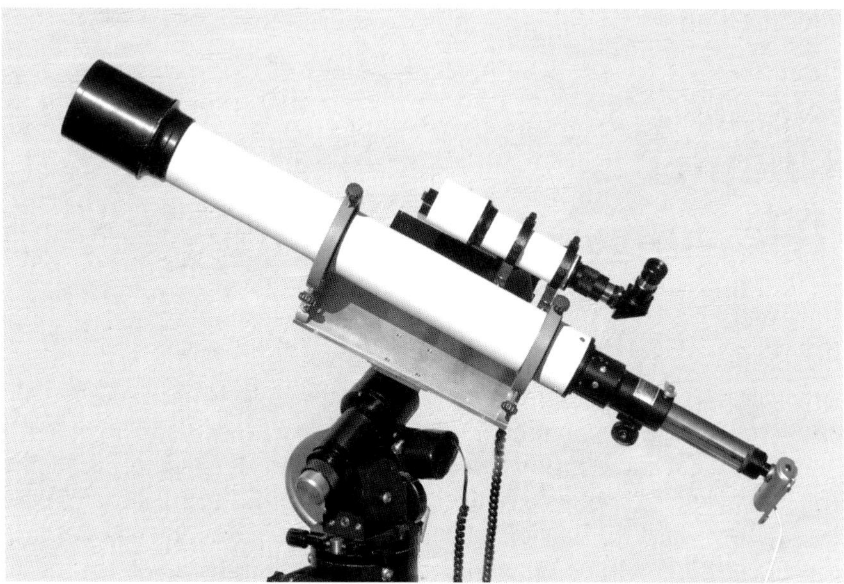

Fig. 5.1 *Refracting telescopes, such as this 80-mm model the author uses as a guide scope, use a lens to gather and focus light. Refractors have the advantage of a rigid tube that holds collimation well and do not have a central obstruction like the secondary mirror present with reflecting telescopes. Most refractors have the disadvantage of long tubes and high f-ratios. Photo by Robert Reeves.*

rium with their surroundings. A favorite feature of refractors is their unobstructed light path. Since they don't require the secondary mirror that must be present with reflecting telescopes, and which causes diffraction, and loss of light, refractors generally produce the highest-contrast images obtainable.

The primary disadvantages of refractors are small aperture, generally high *f*-ratio and concomitant need for heavier mount. But lately the trend in refractors has turned to what are called "short-tube" models with significantly lower *f*-ratios. This trend has enabled astrophotographers to use them to make stunning images of dim objects that were once regarded as the domain of systems with faster *f*-ratios.

Unless the objective lens has special glasses and coatings so that all colors focus at the same point (in which case it is called an apochromat), refractors generally suffer from color fringing around bright objects. This effect—called chromatic aberration—can be controlled to some extent through the use of proper filters.

Reflecting telescopes come in a number of different styles. The classical Newtonian has been popular with amateur astronomers for decades.

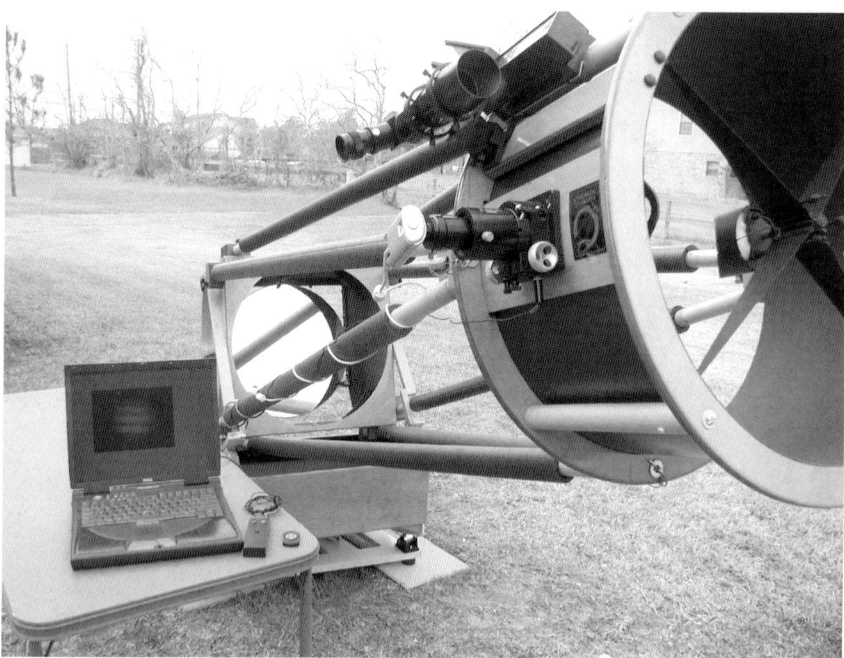

Fig. 5.2 *Reflecting style telescopes, such as this Obsession 20 Dobsonian, have the advantage of a simple optical system, but the structure of the telescope can be quite massive. Such telescopes are transportable, but cannot be classified as "portable." Photo by Glenn Schaeffer.*

The Newtonian primary mirror reflects light to a small 45-degree diagonal mirror mounted near the front of the tube, which relays the light to an eyepiece mounted at the side of the tube. Today the trend is away from large Newtonians on massive equatorial mounts toward large-aperture altazimuth-mounted telescopes known generically as Dobsonians, which takes their name from the design popularized by John Dobson in the late 1970s.

The classical Cassegrain is another reflector design. This optical configuration reflects the image from the primary mirror off the secondary and back through a hole in the center of the primary to an eyepiece mounted behind the telescope. Cassegrains have the advantage of folded optics that can achieve long focal lengths with relatively short tube assemblies.

Either reflector design becomes quite bulky with larger apertures and portability suffers. Also, both styles suffer from collimation problems and must have their optical alignment frequently checked for optimum performance. But reflectors have several good features that make them popular with astrophotographers; they are much more affordable per inch of aperture than refractors, are easier to manufacture, and have none of the chromatic aberration problems associated with refractors.

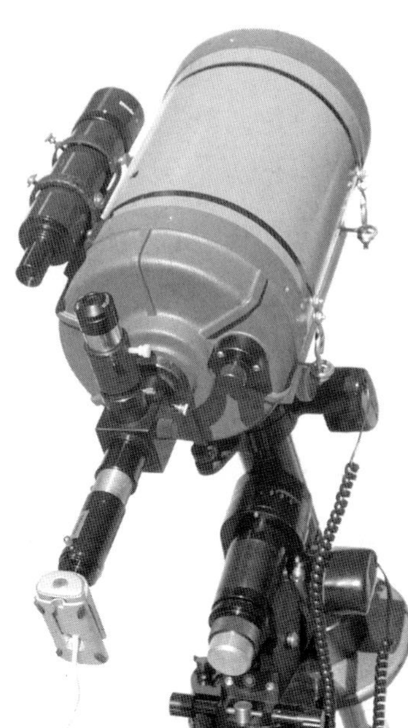

Fig. 5.3 *Schmidt-Cassegrain telescopes, and the similarly styled Maksutovs, are among the most popular telescope models in the world. The Schmidt-Cassegrain folded optic design has the advantage of a long focal length in a small, portable package. The exposed corrector plate is susceptible to dewing and the movable primary mirror can cause image shift while focusing. Photo by Robert Reeves.*

The Schmidt-Cassegrain (SCT) style of telescope is a compound design, that is, one that employs both mirrors and lenses to produce an image. In the SCT, light enters the optical system through a full aperture lens known as a corrector plate. The purpose of the corrector plate is to refract the incoming light rays just enough to exactly cancel the aberration of the short focal length spherical primary mirror, usually about *f*/2. From the primary the light rays reflect off a Cassegrain-style secondary, back through its central hole to the focus point behind the telescope.

The advantages of the SCT design are extreme compactness and portability for a long focal-length instrument of a relatively large aperture. Over the past three decades an entire industry has evolved around supplying accessories for SCTs making them one of the most versatile visual and photographic telescopes ever developed.

The Schmidt-Cassegrain design has some drawbacks that adversely affect its performance, but solutions to these problems have been engineered and are available as accessories for these instruments. Chief among the problems is the image shift inherent in focusing an SCT. The image is focused in this style telescope by moving the primary mirror forward or backward. This feature is the key to the versatility of this design and enables it to properly focus with a very broad range of optical accessories, but the moving mirror also causes the image to shift back and forth in the field as the mirror slightly tips while moving forward and back. With the small field of view that is typical with a webcam, this image shift can be frustrating. The solution is to use an auxiliary focuser that refines the focus after the moving mirror achieves approximate focus and is then "parked."

The large exposed corrector plate in front of the SCT design elevates dew from a nuisance to a real problem. The solution is a heated dew cap to prevent the large glass surface from cooling below the dew point. Every astronomical retailer now has dewcaps for all sizes of SCTs.

When the first Schimdt-Cassegrain instruments were released by Celestron and Meade, they were equipped with fork mounts to allow the telescope and mount to be carried as one compact unit. These mounts were very convenient and versatile, but suffered from stability problems. The basic fork mount is "springy" and flexes easily. The solution to the problem is to mount the SCT tube assembly on a standard German equatorial. Today, manufacturers offer this mounting option, as well as allowing the individual to purchase the tube assembly separately so as to equip it with an aftermarket mount like those available from Losmandy, Astro-Physics, and others.

Most SCT users, myself included, tend to call the position of the image formed at the focus of a Schmidt-Cassegrain telescope just outside the visual back adapter, the prime focus. This statement is technically incorrect. In a typical 8-inch SCT optical system, prime focus is where the image is formed by the roughly *f*/2 spherical primary mirror on the hyperbolic secondary mirror attached to the back of the corrector plate. The true terminology for the position of the image formed at the visual back is Cassegrain focus, which after reflecting off the secondary mirror is now closer to *f*/10.

5.2 Telescope Mounts

High-resolution images are difficult to achieve if the mount supporting the telescope is not steady and resistant to movement in a light breeze. High magnifications, and the webcam's small imaging sensor, and narrow field of view require a steady mount so the individual video frames are as sharp as possible. It takes little to jog the target out of the field of view using an 8-inch or larger telescope. Stacking webcam video frames, in a way, helps combat telescope shake because the stacking software allows you to eliminate substandard frames (though it is no substitute for an adequate mount).

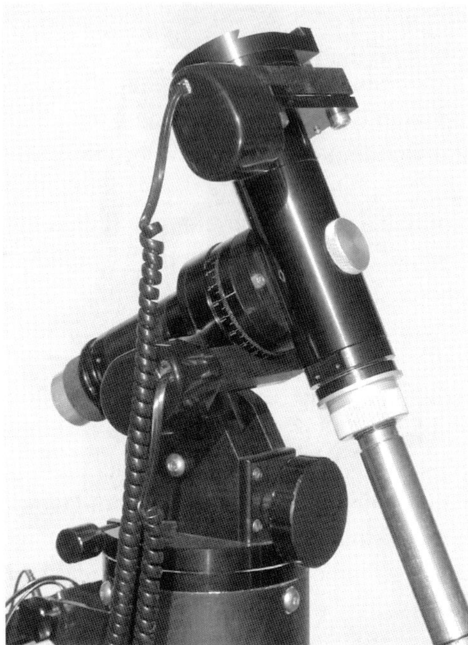

Fig. 5.4 *Mid-range aftermarket telescope mounts such as the Losmandy GM-8 offer affordable, high-quality electronic control capable of interfacing with autoguiding equipment. The GM-8 can carry up to 35 pounds. Photo by Robert Reeves.*

The higher the number of good frames, the higher the ultimate resolution and dynamic range of the finished image.

Use an equatorially mounted telescope if possible when doing high-resolution imaging. An equatorial mount, by the nature of its design, only needs to be motor driven on one axis in order to follow the apparent motion of the stars across the sky as the Earth rotates.

Altazimuth mounts use two motors to drive the telescope—one for movement in azimuth, and another for movement perpendicular to the horizon. Less expensive mounts use stepper motors, which due to their relatively large jumps between each step, may introduce vibration. Computer-controlled altazimuth telescope mounts can track objects across the sky for long periods of time, but photography through these instruments will only work for limited periods. This is because after a few minutes of tracking, a phenomenon called field rotation will complicate the stacking of multiple exposures or the creation of mosaics because successive images in the sequence will not be square with each other.

True equatorial mounts generally come in two varieties, the German and the fork. Users of fork mounts either love them or hate them. The good points are their compactness and versatility. Their bad points are flexure and a reputation for having inaccurate drive systems. Manufacturers have addressed the latter complaint, and in the past few decades have improved the accuracy of their mass-produced fork mounts.

German equatorial mounts, if well built and massive enough for the given application, have a reputation for good stability and accuracy. But they are another large component that must be transported and there is no such thing as a good lightweight German equatorial mount. Its design necessitates counterweights to balance the optical tube and accessories used with the telescope. The counterweights attached on a bar extending opposite the polar axis from the telescope tube increase the overall footprint of the telescope system and care must be used to avoid bumping the weights in the dark while the telescope is in use.

Some users of German equatorial mounts employ the lightest counterweight possible and place it at the end of the counterweight bar to balance the telescope. This may reduce the mass that has to be carried to an observing site, but it increases stability problems. A better system is to use a heavier counterweight that lies as close to the polar axis as practical. The reason is that the moment of inertia of a lighter weight at greater distance from the center of mass is larger than a heavier weight at closer distance. The closer the counterweight is confined to the center of mass, the quicker vibrations will dampen.

5.3 Optical Collimation

Camera and image processing techniques can be perfected with practice, however, the collimation of the telescope's optical train is an area often ignored by the novice imager. If you want a sharp image, your telescope's optical components must be collimated. No amount of processing can recover image sharpness that was never there to begin with. The highly regarded French astronomer Thierry Legault states that collimation "is as important as the tuning of a musical instrument: the images given by a misaligned telescope can be as awful as the sound given by an out-of-tune piano."

Collimation is a series of check, adjust, check again, and repeat-as-necessary steps. The actual mechanics are not difficult but the observer must exercise the patience and diligence needed to perform the procedure thoroughly.

Refractors typically are very robust in their collimation. The stiffness of their tube assemblies promotes good alignment of optical components. While there are some really good refractors on today's market, their repu-

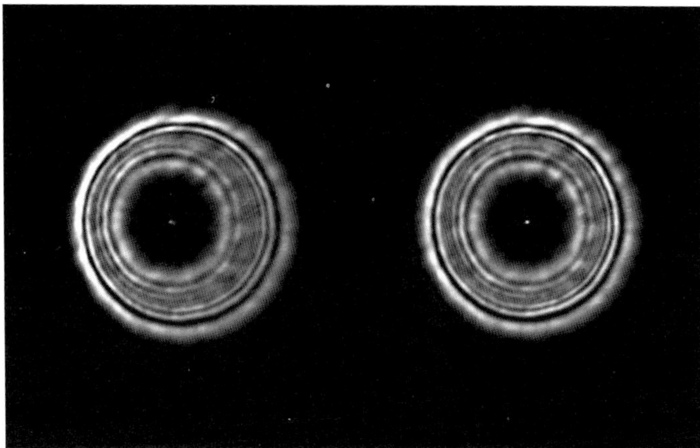

Fig. 5.5 *The out-of-focus star image on the left shows the need for secondary collimation in a Schmidt-Cassegrain telescope. The star's diffraction rings display an elongation in one direction, indicating that the secondary mirror is tilted off-axis. The right image shows the concentric diffraction rings typical of the view through a well-collimated scope in average seeing conditions. Photo by Robert Reeves.*

tation for image quality over other types of telescopes is more a statement about the less-than-optimum mechanical adjustment, or collimation, of the optics in reflectors SCTs. As a rule, if a refractor is out of collimation, it is usually because of physical damage to the telescope assembly.

Any kind of reflecting telescope, including SCTs, needs frequent tune-ups of optical collimation, especially if it is transported in a car. Newtonians are notoriously easy to get out of collimation. This is not a reflection (no pun intended) on their potential performance, but is, instead, a statement of fact because of the nature of the telescope's design. The mirror mounts of reflectors tend to shift slightly when they are transported. For best viewing and imaging results, it is almost mandatory that a collimation adjustment be a natural part of setting up a reflector. Casual users who view extended deep-sky objects at low power may not realize how out of collimation their optics may be. Vibration and minor bumping during transport are guaranteed to shift the mirror in a Newtonian configuration and can also slightly shift the secondary mirror in an SCT. Newtonian (and Cassegrain) telescopes are sensitive enough to structural sagging and shifting from gravity that some high-resolution image purists even tweak the collimation of their scopes, when moving from one target to the next, if it is in a significantly different part of the sky.

The increasingly popular Maksutov-style telescope is relatively immune to collimation problems because its secondary is a reflective spot on the back of the corrector plate and not a separately mounted mirror.

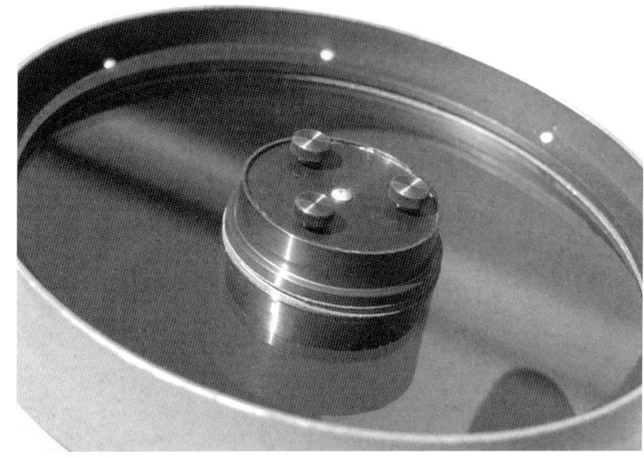

Fig. 5.6 *Aftermarket SCT collimation knobs such as those available from Scope Stuff in Austin, Texas, make it easy to quickly tweak SCT secondary collimation. Often, knobs such as these will protrude far enough to contact the telescope end cover, so care must be exercised to prevent damaging the corrector plate when installing the cover. Photo by Robert Reeves.*

SCT owners generally think their telescopes are not affected by shifting optics, but in fact, the adjustable secondary is quite sensitive to minute movement that can degrade image quality. For instance, when collimating to a high degree of precision, adjustments of the secondary screws of as little as 1⁄20-turn will produce noticeable differences in image quality, all other factors being optimal.

So, the bottom line is that if your photo techniques, seeing, mount, tracking, and image processing are good, yet your images do not compare with those taken with similar instruments, serious attention should be given to the accurate collimation of your telescope's optics.

Generally, a telescope should be collimated in the optical configuration it will be used with; that is, if it is to be used visually it should be collimated with an eyepiece, and if it is to be used photographically, it should be collimated with the camera attached. Either way, before you begin the collimating procedure, make sure the telescope is thermally stable. A hot mirror and associated tube currents can lead to erroneous adjustments.

5.4 Newtonian Collimation

There is more to collimating a Newtonian than simply placing the shadow of the secondary mirror in the center of an out-of-focus star. Stopping after this preliminary step can lead to a loss of up to 50 percent of image contrast. On a Newtonian telescope, the collimation procedure is performed in two steps, the geometric placement of the secondary mirror using a collimating eyepiece, then the precision alignment of the primary mirror.

This process can be accomplished fairly quickly by first using a laser collimation tool, followed by fine-tuning using a high-power eyepiece on a star. These laser devices are widely available and it is highly recommended that Newtonian users purchase one. However, a functional tune up can also be accomplished with very basic tools, if a laser unit is not available. A simple Newtonian collimation procedure is as follows:

Drill a small hole—about 2 millimeters in diameter—in the center of the bottom of a plastic 35mm film canister. Insert the canister into a 1¼-inch eyepiece focuser as deeply as possible. If the secondary mirror is out of adjustment, the reflection of the primary mirror will be offset or only partly visible. Three small screws on the back of the secondary mirror holder allow its adjustment. Slowly, in small increments, adjust one of the screws to see which way the primary mirror shifts as seen through the collimation tool. Adjust the screws to center the primary mirror's reflection within the telescope tube.

Next, place a center dot on the primary mirror. To do this, remove the mirror from the tube and cut a circle of paper with the same diameter as the mirror. Now, fold it in half two times. Where the two folds cross is the center of the circle. Punch a small hole at that point and place it over the mirror to use as a guide for marking the mirror's center.

Some users place a small ink spot at the center of the mirror while others place a donut-shaped adhesive notebook paper reinforcement ring. The theory is that the ring leaves a reflective surface at the exact center for later use with a laser collimation tool while the ink spot does not. In fact, it's best to do both, because the ink spot forms a "bulls eye" for the reinforcement ring.

With the mirror reinstalled, look through the collimation tool and center the mirror using the adjustment knobs on the back of its cell. Adjust them until the secondary spider vanes appear approximately equal in length. Further refine the mirror adjustment until the spot on the primary is in the center of the secondary mirror.

Of course, if a laser collimation tool is available, perform critical mirror adjustment with that.

Now it is time to view an individual star through a high-power eyepiece to tweak the alignment. Use one with a magnification of at least one per millimeter of aperture, that is, for an 8-inch telescope (200 millimeters), use about 200 power. Center and defocus a 1st magnitude star until it shows a donut pattern. The hole in the donut is the shadow of the secondary and must be perfectly centered. If it is slightly shifted, determine which collimation screw needs to be adjusted by placing your hand over the front of the tube and observing its shadow in the donut image. The shadow will correspond to the "clock angle" of your hand relative to the tube and indicate which screw needs adjusting. If a star diagonal is being used, be sure to account for the appropriate reversal of the image.

Once the donut hole is properly centered, switch to a higher power eyepiece to achieve 2x to 3x per millimeter of aperture and sight on a star of 2nd or 3rd magnitude, that is high above the horizon to minimize turbulence. This time, only slightly defocus the star from inside to outside focus. As the image comes into focus, the star's Airy disk, a small bright point surrounded by a complex series of rings, will be visible. The rings should expand and collapse concentrically. If seeing is poor, the Airy disk will not be resolved and final collimation will have to await better conditions. If the seeing is good and rings are not fully concentric, tiny adjustments of the collimation screws will be necessary. At this point, the adjustments are mere fractions of a turn, often little more than simply putting rotational pressure on the screws. Once the telescope is properly collimated, the Airy disk and rings will also present a uniform concentric appearance.

5.5 Schmidt-Cassegrain Collimation

While it is true that a Schmidt-Cassegrain will maintain collimation better than a Newtonian or classical Cassegrain, with my own 8-inch SCT, the bumps from rolling it over a wide crack in my driveway several times is sufficient to noticeably offset collimation, particularly the sensitive secondary mirror. When collimating to a high degree of precision, adjustments of the secondary screws of as little as 1/20 turn produce noticeable differences in image quality, assuming all other factors such as steady atmosphere and tracking being optimal.

A star diagonal can be used during collimation, but first it must be checked to insure that it does not shift the field of view off the optical axis, thereby making the process ineffective. Critically view a distant object both with and without the star diagonal and insure that there is no shift in the field of view.

An easy way to adjust collimation on an SCT is to orient the telescope so that one of the three adjustment screws on the face of the secondary holder is nearest the ground, when one is looking at the corrector plate. Orient the star diagonal and eyepiece so they point upward; i.e., in the opposite direction as the "lower" collimation screw. Insert a high-power eyepiece and center a moderately bright star, (no more than 3rd magnitude) in the field. Slightly defocus the star and carefully examine its image. See Figure 5.5.

If collimation is needed, refer to Figure 5.7. Note that the adjustment screws are designated A, B, and C. In the diagram, we see how the star's image elongation caused by poor alignment relates to how we initially positioned the telescope with one of the collimation screws toward the ground. Using the chart in Figure 5.7, we can determine which screw to turn, and which direction to turn it. Under no circumstances should the central screw that holds the secondary to the corrector plate be touched. Only the outer three screws are adjusted.

For example, if the star were elongated downward as seen in the eyepiece, the chart tells us to turn screw "A" in a counter-clockwise direction in order to center the brightest portion of the defocused star. Note that very little rotation of the screw is needed to make adjustments. A grossly out of collimation secondary would only require about a ¼ to ½ turn while a majority of "maintenance tweaks" would range between 1/20 of a turn to little more than just applying light pressure to the adjustment screw.

It is important not to over-tighten, or the screw can distort the star image by pinching or warping the secondary. I have found it easy to create triangular star images in an SCT in this manner. To prevent warping, for every turn of a screw in one direction, the other two screws must be turned in the opposite direction to relieve the strain on the secondary mirror. Equally important is to not turn any of the opposing screws to the point where the secondary becomes loose and can wobble. There must a slight tension on all three screws at the same time.

After each adjustment, the star will shift toward the edge of the field. The telescope must be moved to recenter the star, verify the adjustment, and continue with any additional adjustments, if needed.

The collimation screws that tilt an SCT secondary mirror are best described as tiny. They are traditionally adjusted using a very small Allen wrench. It is inconvenient to repeatedly insert the Allen wrench into successive adjustment screws. The first time you drop that tiny tool and find yourself on your knees searching for it on the ground in the dark, the wisdom of replacing the Allen screws with tool-free manual adjusters will become evident. Fortunately, such devices exist in the form of Bob's Knobs. These are replacement screws that have a wide knob instead of the small Allen head. These can be adjusted by hand, greatly simplifying the SCT collimation process.

For the convenience of fast collimation, Bob's Knobs are an incredible bargain, but they must be installed properly to prevent possible telescope damage. It is very important to first tilt the telescope downward to

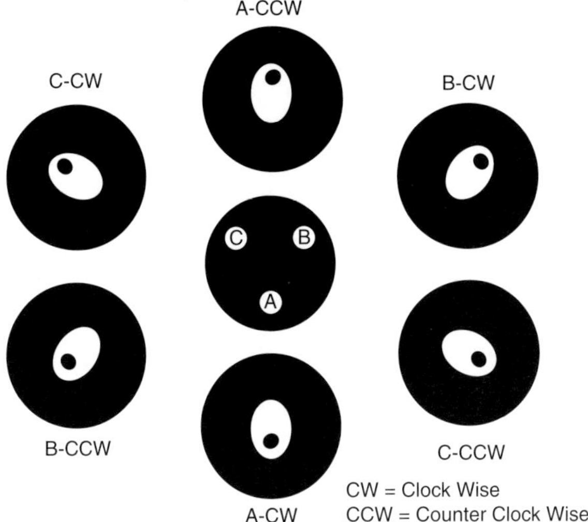

Fig. 5.7 *This depicts the secondary holder as seen from the front, with the collimation screw closest to the ground labeled "A." View a 2nd or 3rd magnitude star with the diagonal and eyepiece pointing up and note in which direction the star is elongated. Refer to the chart and turn the appropriate setscrew in the indicated direction to adjust the secondary mirror collimation. Always make very small adjustments. After a drawing by Steve Walters.*

prevent the secondary from falling onto the primary mirror if it should come loose. This is a cautionary step "just in case," because the secondary will not come loose if the knobs are installed one at a time. *Never* remove all three collimation screws at the same time.

Once the collimation screws are replaced with knobs, the secondary will be out of collimation. Coarse alignment can be achieved in daylight by looking into the front of the telescope from about ten mirror diameters away. Visually, the various reflections of the primary and secondary mirror should form concentric circles if the secondary is properly positioned. Once the secondary has been adjusted so that the reflections are concentric, the fine collimation must be done with a star.

5.6 Telescope Temperature Equilibrium

A stable telescope temperature is as important as good seeing. If the Sun has heated the telescope during the day, not only will unequal temperatures distort the mirror, the warm telescope assembly itself will affect the air nearby the telescope and create distortion in the incoming light rays. Essentially, a hot telescope creates its own poor seeing.

The larger the telescope, and consequently the larger and thicker the

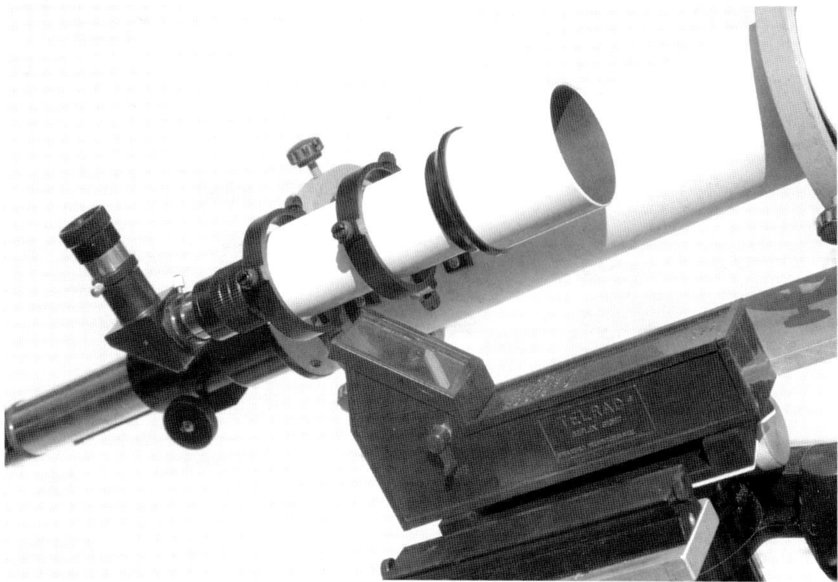

Fig. 5.8 *A good finder is essential for targeting planets and deep-sky objects with the small CCD in a webcam-style camera. The author uses a combination of a zero-power Telrad and a 9 x 50 right-angle finder to acquire objects. Photo by Robert Reeves.*

primary optic, the longer it will take the instrument to cool to thermal equilibrium with the surrounding air. It is not unusual for a 12- or 14-inch SCT to take two to four hours to reach thermal equilibrium with the nighttime air. In some cases, fans have been adapted to the telescope to assist in cooling the primary, but vibration can be a problem while they are running.

5.7 Aiming the Telescope

Telescope pointing errors of a half-degree are acceptable with visual astronomy because, with most amateur telescopes, wide-field eyepieces will still see the target. With webcam astrophotography, however, the field of view through a common 8-inch *f*/10 SCT is about one-fifth that through a low power eyepiece, thus half-degree pointing errors can be more than frustrating; they can render an imaging session useless if the target cannot be found. There are a number of ways to zero in on a dim deep-sky target or the small globe of a distant planet: well-aligned finders, substituting eyepieces for the camera, flip-mirror finders, setting circles, and GO-TO telescopes.

Shorter focal length telescopes like an 80mm short-tube refractor can be aimed at a planet with a well-aligned laser Red Dot finder or Telrad.

Zero-power finders such as these are good to within a quarter-degree pointing accuracy, but they only work with objects bright enough to be seen with the naked eye. Their pointing accuracy will place a target within the eyepiece field of a flip-mirror finder.

Moderate focal-length telescopes like an 8-inch SCT without a Barlow can be aimed with a well-aligned finderscope. Searching for planets with focal lengths above 2000mm will benefit greatly from using a flip-mirror assembly to allow viewing the telescope field at a lower magnification than the camera field. "Serious" deep-sky imaging of very dim targets may require a higher-power finder such as one of the popular 60mm or 80mm short tube refractors.

Finderscope alignment can be performed in daylight by aiming at targets at least a mile away to eliminate the parallax error between the finder and the main telescope's field of view.

With small telescopes on lightweight mounts the action of substituting a wider field "finder" eyepiece for the camera, then reinstalling the camera, will often throw the image out of the field of view. Such instruments will also benefit from using a flip-mirror assembly to allow using a wide-field eyepiece to find and center small planetary targets.

If "star-hopping" is the only option for finding a target, computer-based star charts such as MegaStar, will prove valuable. Zooming in on the desired star field on-screen is much easier than poring over charts on a dimly-lit observing table.

Manual setting circles on a polar-aligned telescope can get you close to a deep-sky target, but most circles are not accurate enough for the limited field of a webcam. For example, my own Losmandy GM-8 is a great mount, but the two-degree gradations on the setting circles are far too coarse. They are nearly useless for finding dim targets with a camera whose field of view can be 1⁄20 that of the circle gradations.

A sensitive camera will also help locate faint objects. In using an Atik IIhs model at *f*/5 on my 8-inch SCT, the Ring Nebula, M57, is visible with only one-second exposures, making an on-screen search of a target area a viable option. Similarly, sweeping for bright star clusters in light-polluted skies is possible using half-second exposures. For instance, the large globular cluster M22 is invisible in my 8 x 50 finder because of city lights near my backyard. However, slowly scanning back and forth through the sky area where the cluster should be quickly reveals the target on screen. This technique becomes tedious, however, with very dim objects that require 30-second to one-minute exposures to detect.

A flip-mirror finder is the best aiming system for targets that are bright enough to be seen in an eyepiece, such as planets and star clusters. But this device is less effective with dim galaxies and is of no help with faint nebulae. Some optical systems do not have enough back-focus to allow a camera to reach focus with a flip-mirror system unless the focuser mechanism is relocated. This is not a problem with SCT telescopes because they have generous focus travel.

Digital setting circles are a pricey option for a telescope mount, but today's models, when properly calibrated, are accurate enough to place dim objects within the field of a webcam at 1000mm focal length. Serious deep-sky imagers who target objects that are essentially invisible to the eye will find that a good set of digital setting circles will solve their telescope pointing problems.

GO-TO telescope mounts have become very popular because they eliminate the star-hop method of finding dim or unseen deep-sky targets. GO-TO mounts can be categorized into two classes: affordable units primarily designed for visual use, and high-end models designed for precision pointing. Most GO-TO mounts like the Celestron Nexstar and Meade LX series, when properly aligned, have performed satisfactorily for many users, but their pointing accuracy after cross-sky slews may call for a brief search, given the inherently narrow field of webcams. Higher priced GO-TO mounts like those offered by Losmandy, Astro-Physics, Mountain Instruments, and others, have the accuracy needed to place a target within the camera's field.

5.8 Star Diagonals

A star diagonal is primarily intended to aid viewing comfort at the telescope. With refractor, SCT, and Maksutov telescopes, it is impractical to view with a straight-through eyepiece when the telescope is aimed near zenith. Anyone who uses a straight-through finder on any kind of telescope will attest to that fact. But a star diagonal is not recommended for use with a webcam for several reasons. Since the image produced by a webcam is previewed on a computer screen, a webcam does not need to be mounted on a diagonal. It can be attached directly to the telescope visual back, or to auxiliary optics such as a focal reducer, Barlow, or flip mirror.

Sometimes it is necessary, because of telescope mount clearance limitations, to use a star diagonal with a camera, such as when imaging near the zenith through a wedge-mounted SCT using a flip-mirror assembly and Barlow. The Barlow's extension might cause the camera to strike the telescope mount if a diagonal were not used. But the need for physical clearance is an exception. The general rule for maximum image resolution is to avoid the use of a diagonal to eliminate as many glass and reflection surfaces as possible.

However, if one is absolutely necessary for some reason, try to use a mirror diagonal. They are both cheaper and pass more light than prism diagonals, resulting in a brighter image.

5.9 Schmidt-Cassegrain Mirror Lock

Maintaining good focus with a Schmidt-Cassegrain telescope will be almost impossible without a means of locking the primary mirror into a fixed position. An SCT mirror is designed to move on the central Cassegrain light baffle in order to focus the telescope, and thus some degree of mirror flop is normal. The natural mirror flop in the SCT design can alter the focus point when slewing an instrument from one part of the sky to another. Experienced Schmidt-Cassegrain users lock the mirror at a point of good initial focus then refine it further with an auxiliary rack-and-pinion focuser or electric focuser attached to the visual back of the telescope.

To secure the mirror after initial focus, some people install nylon screws through threaded holes drilled in the back of the telescope. These screws gently push on the mirror in three locations to prevent it from rocking as the scope is moved. (Needless to say, though, such modifications are done entirely at the owner's risk!)

My method for preventing mirror flop is to carefully secure the mirror edge using small plastic tabs that grip its rim. This technique, however, could potentially pinch the mirror, creating undesirable image artifacts. My SCT is used solely as a visual guide scope so perfect star image quality is not an issue, but if the mirror lock tabs are excessively tightened, the telescope produces triangular-shaped stars.

To perform the mirror lock modification on my 1970's vintage Celestron-8, I devised a system of three plastic tabs that gently contact the rim of the primary mirror at 120-degree intervals. Because the in-focus location of the primary can change with different combinations of camera, focal reducer, flip mirror, Barlow, or other extension tubes, a single location for mirror lock setscrews may not be wise. Thus the setscrews do not contact the mirror directly, but instead push on three plastic tabs, which in turn flex to contact the mirror anywhere along one inch of mirror focus travel.

To find the correct placement for the lock tabs, first focus to a typical infinity position. Next, remove the corrector plate to access the inside of the tube. On a classic SCT, this is accomplished by removing a retaining ring secured with eight screws. Be sure to mark the corrector plate so it can be aligned in the same position when reinstalled. Then measure the dis-

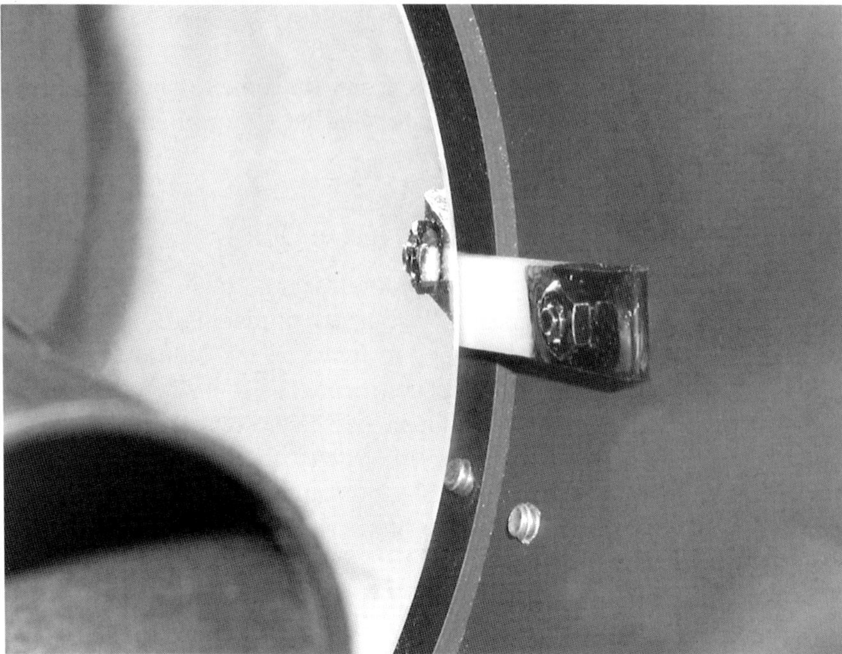

Fig. 5.9 *One of the three mirror-lock tabs that the author installed on his 8-inch Schmidt-Cassegrain telescope. The mirror has been retracted to show the plastic tab that is gently tightened against its edge with a setscrew from the outside of the tube. The forward part of the tab has been painted black to reduce reflections. Photo by Robert Reeves.*

tance from the front of the tube to the mirror and mark the outside of the tube at the same distance. Locate the desired position for the three setscrews at 120-degree intervals at the marked distance from the front of the tube. Now move the mirror to its lowest location with the focus control.

Stick masking tape as drilling chip traps to the inside of the tube where the holes for the screws and plastic tab anchors will be drilled. The setscrew holes should be tapped with ¼ x 28 thread to accept a fine thread ¼-inch bolt. Each stiff plastic tab should be 2 inches long, ½-inch wide, and anchored with a machine screw and nut one inch above each setscrew (see Figure 5.9).

Before an exposure begins, the telescope is focused and the mirror locks set in place. They should be tightened in sequence but just enough to still permit a slight movement of the star images: any tighter and the plastic tabs will distort the mirror and create triangular stars. If the focus is altered because of temperature change, the setscrews must be released before the refocusing. Be sure to release the mirror locks when the telescope is not in use.

5.10 Optical Problems

If high-resolution images never seem to be as good as would be expected for the observed seeing conditions, and mechanical checks of the camera alignment with the telescope appear to be in good order, then it is time to check the quality of the telescope's optics. Star tests as outlined in Harold Suiter's book *Star Testing Astronomical Telescopes*, published by Willmann-Bell, will reveal if there are problems with the optics or optical configuration of your instrument.

If it is still under manufacturer's warranty, star tests will reveal the nature of any defects so the telescope can be returned or exchanged for a better unit. A mass-produced model capable of at least ⅙-wave performance is sufficient for high-resolution imaging. Reasonable results have been achieved even with ¼-wave optics.

If the telescope is out of warranty and the optical components are in proper collimation but the performance is still not up to par, the user must accept the fact that this particular telescope will be inadequate for high-resolution imaging.

Before condemning a telescope as a poor performer, though, be sure to systematically eliminate all components that may contribute to poor focus. Is the focus still poor without accessories such as a Barlow, filter, or a star diagonal? If focus is good at prime focus with no other accessories, progressively add each component until the image noticeably degrades, then troubleshoot the component that contributed to the focus problem.

Regardless of what type of telescope is used, there are a number of design defects that can affect the quality of images taken with the instrument. These include hot spots, reflections, field flooding, and vignetting. All of these problems are easily diagnosed and cured, but left unchecked can rob an image of contrast or create colorcasts that lessen the quality of the image.

Hot spots or brighter areas on an image where there should be no additional exposure are most likely caused by cylinder reflection within the focuser (see Figure 5.10). To check for this, remove the camera or eyepiece and place your eye at about the same distance as the camera focal plane. The telescope objective should be bright and everything else should be dark. If light is reflecting off the shiny inside surface of the focuser barrel, it is also reflecting into the camera and creating hot spots. Black paint alone may also be reflective, but mixing sawdust with black paint creates a rough surface that breaks up and absorbs reflections.

Reflections from optical glass surfaces rarely cause any problem with webcam astrophotography, but it can occur. The user should be aware that

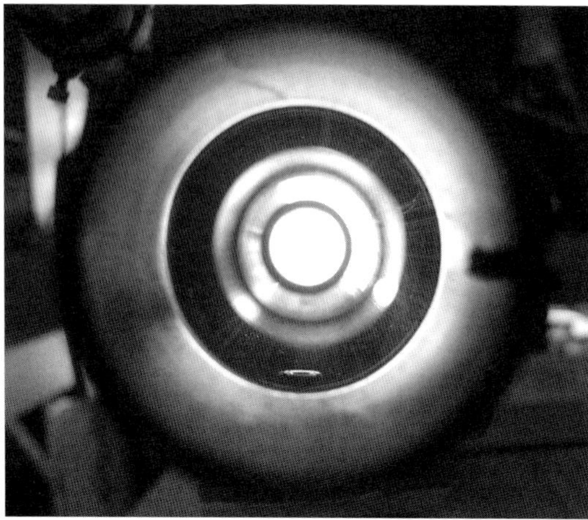

Fig. 5.10 *Contrast-robbing ghost reflections from bright objects just outside the field of view can occur if the inside of the focuser tube is reflective. A coat of dull black paint will reduce reflections and increase image contrast. Photo by Robert Reeves.*

reflections from the glass cover on the CDD sensor can bounce back from a nearby accessory optic like a focal reducer or eyepiece lens, and create an out-of-focus reflection of the telescope objective on the image. Changing the projection distance or switching to a different accessory lens may solve the problem.

Field flooding is unwanted light reaching the focal plane and reducing image contrast. This problem exists primarily in reflecting style telescopes. In Newtonians, light illuminating the tube wall opposite the secondary from the focuser can reflect into the focuser and wash out the desired image. Light entering the bottom of the tube around the cell of the primary mirror can also reflect off the secondary and contribute to contrast loss. These problems are minimal under very dark conditions but can significantly affect images taken under bright urban skies or daytime planetary imaging. Similarly, unwanted light passing an improperly designed Cassegrain or Schmidt-Cassegrain central baffle, or around the outside of the mirror, can also reach the focus of the telescope. Properly baffling the Cassegrain and blocking light paths around the back of the primary mirror will solve these problems. Newtonians benefit from an extended front tube and internal baffling forward of the secondary to reduce stray light reaching the tube wall opposite the focuser.

Vignetting is physical blockage of the light path from the primary optic to the camera focal plane. If present, vignetting prevents the use of the telescope's full light-gathering capability and will contribute to darkened edges around the field. The simplest check to see if vignetting is

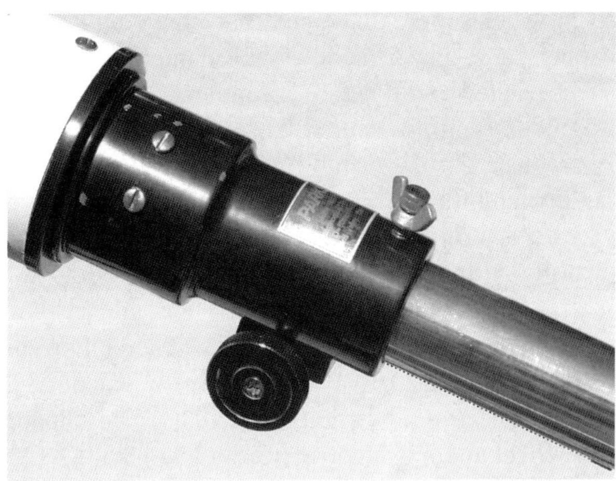

Fig. 5.11 *Inexpensive refractors make good autoguiding telescopes with a webcam. Many economy refractors have adequate optics, but lower quality mechanical components. Excess play in the focuser can seriously compromise the effectiveness of an autoguider, so here the author added a ¼-inch setscrew to eliminate drawtube "wag" while autoguiding. Photo by Robert Reeves.*

present is to view the camera sensor through the primary optic. If the full area of the sensor can be seen all the way to the edge of the primary lens or mirror, then all the light available will reach the camera. A less definitive method is to place the eye at the camera focal plane and insure that the entire primary optic is visible. The best check is to aim the telescope at a star pattern that covers the entire field of view and take two long exposures, one inside focus and the other outside. The stars should be well-defined disks with a Newtonian (or Cassegrain) secondary showing at the center. If no vignetting is present, the star disks will be round to the edge of the field. If vignetting is present, portions of the disk will be clipped off. It is not unusual to see some clipping in Newtonian telescopes designed primarily for visual use. These instruments use tall focusers and smaller diameter primary mirrors. Unless the vignetting is severe, its effects can usually be ignored. Severe clipping will require re-engineering the telescope.

5.11 12-V Power Supplies in the Field

As light pollution spreads and dark sky locations become a distant destination from home, remote power supplies become a critical accessory for any kind of astronomy. Today's electronic astrophotography is power hungry. No longer is there just a simple electric clock drive on the telescope like when I was a beginner. It is critical to have adequate remote power available because now telescopes have internal computers, stepper motors, electronic displays, anti-dew heaters, and even internal GPS receivers. Additional power is needed to run laptop computers controlling the telescope, webcam, autoguider, running separate astronomy programs, or a combination of all four. Music players or radios for late night entertainment at the telescope also demand their share of electrical power. If we are set up in our own backyard or driveway, this is no problem with a long extension cord, but powering all these devices 100 miles from home at a dark location calls for a remote power supply with considerable capacity.

There are a number of homemade and commercial products that can supply both 12-volt DC and 115-volt AC power for a modern astroimaging setup. Devices that need only 12-volts DC like a telescope drive can be hooked directly to a battery, but those needing 115-volts DC like a laptop computer require that a DC/AC inverter be used. Hardware stores now stock many models that will change 12-volt DC battery current into 115-volt AC current. If you decide to use batteries and an inverter as your power supply, be aware that some inverters are better than others and their output wattage alone should not be used as a buyer's guide. The 115-volt devices are designed to use sine-wave alternating current, that is the alternating current oscillates with a rise and fall that describes a wave-like pattern. Cheaper inverters using simplified circuitry may say they are outputting 115-volts AC, but in reality they can be producing pulse-wave current that instantly alternates in a square wave instead of a sine wave. Some electronic devices do not run properly with square wave AC current. Costco stores sell an excellent 1000-watt inverter that produces modified sine-wave current that is compatible with electronic devices. These inverters, sold under the Xantrex X-Power brand for $69.99, will power all telescope and electronic imaging needs in the field.

Another power-to-go option is a 140-watt 12-volt DC-to-120-volt AC inverter that plugs into an vehicle cigarette lighter, that is offered by Radio Shack as item #22-148. These inverters are actually manufactured by Xantrex but sold by Radio Shack. Be careful when powering your equipment with your car battery; a discharged car battery makes for a long walk home.

Some people use a spare car battery or deep-cycle marine battery to power their gear at remote locations. A separate battery will free you of the need to be right next to your car. Also, some major star parties do not allow vehicles on the observing field. High capacity, lead-acid batteries useful for remote power applications can be divided into 4 categories:

- **Car Batteries** are designed to provide a brief burst of high current to start an engine; they are not suitable for deep-discharge applications. Their thin plates are good for a quick release of energy, but deep discharging will cause rapid deterioration.

- **RV and Marine Batteries** are a compromise between car and deep-discharge batteries. Their plates are similar to engine starting batteries and will not withstand repeated deep discharging.
- **Motive Power Batteries** are deep-discharge batteries designed for electric vehicles such as golf carts and forklifts. They have thick plates that will withstand many deep-discharge cycles. These batteries are durable and have good storage capacity.
- **Stationary Batteries** are commonly identified as backup cells for telephone companies and computers. Most are not designed for deep cycling.

An understanding of how a battery is rated will help you determine the proper size for your needs. A car battery's "amp-hour" rating and its advertised "cold-cranking amps" are not the same thing. Cranking amps relates to a battery's ability to deliver large amounts of power in a short time while amp-hour relates to both the battery's overall power capacity and the rate at which it can be discharged.

As you use the battery, its output voltage slowly drops. A manufacturer's rated amp-hour capacity for a 12-volt battery actually refers to the point where the voltage drops to 10.5 volts. For this reason, amp-hour ratings alone should not be used as an absolute guide for voltage sensitive devices to run properly. Moreover, the older a battery is, the lower its amp-hour rating is. Old batteries, while functional, simply do not deliver the same amount of total power as new ones of similar rated capacity.

Some battery makers no longer print the amp-hour rating directly on the battery, so the user will have to test it to insure it has adequate capacity. A simple way to do this for a particular 12-volt battery is to discharge it under controlled circumstances and measure how long it takes to reach the minimum acceptable voltage. To perform this test, fully charge the battery then bridge the terminals with a wire-wound 12-ohm/20-watt resistor and time how long it takes the battery to discharge to the minimum acceptable voltage. For a 12-volt battery, the 12-ohm resistor will draw one amp. While it is true that the power draw will slowly change as the voltage drops, a fully charged 12-volt battery starts with a little more than 12-volts and the test will end at less than 12-volts. The average voltage during the test will be around 12 and the calculated amp-hour rating will be close enough for our needs. If the battery takes 30 hours to discharge to the minimum useful voltage during the test, you can count on 30 amp-hours of power from it when it is fully charged.

A note of caution: do not recharge lead-acid batteries with an amp-

Table 5.1
Deep Cycle Battery State of Charge

Battery Voltage	State of Charge (%)
12.7–13.0 V	100
12.5–12.6 V	80
12.3–12.4 V	60
12.1–12.2 V	40
11.9–12.0 V	20

hour recharge rate greater than one-tenth of the total amp-hour capacity. Higher charge rates will heat the battery and damage it. For example, recharge a 30-amp-hour battery with no more than a three-amp-hour charge rate. Most heavy duty battery chargers exceed this rate. These devices are designed for one-time emergency quick charging of car batteries by mechanics and their repeated use will eventually damage a battery by overheating it.

For safety's sake, carry a lead-acid battery in an enclosed plastic case. These can be purchased at boat stores. Do not add electrolyte to a battery (except to replace spills) and always recharge them within 24 hours of each use. For deep-cycle batteries, avoid discharges of less than 10 percent or more than 80 percent. A deep-cycle battery's service life will be extended if it is connected to a low-voltage disconnect device set at 80 percent depth-of-discharge.

Some points to remember. Heat kills batteries and cold weather reduces their available power. More deep-cycle and car batteries die from poor charging procedures than from old age. If the battery is often unused for long periods of time, it should be continuously connected to a "smart" or float charger that is matched to the battery type. These chargers will slowly recharge the battery when it drops below 80 percent state-of-charge. A cheap unregulated trickle charger can overcharge a battery and destroy it.

The state of charge for a deep-cycle battery can be determined with a digital voltmeter capable of reading within 0.1-volt DC. To check a deep-cycle battery in this manner, it needs to be at room temperature and not have been charged or discharged for several hours. Table 5.2 shows a typical voltage vs. charge state.

Remember that storage batteries hold a great deal of instantaneous energy. Never allow the positive and negative leads to come into contact with each other. Shorted terminals can overheat, burn you, or damage equipment.

5.11.1 Gel Cell Batteries

Rechargeable gel cells are popular portable 12-volt telescope and inverter power sources. Because they are leak-proof and spill-proof, they can be operated in virtually any position, although it is not recommended that they be used upside-down. Gel cells can also be recharged at any time during the discharge cycle without adverse effects, but they must be recharged slowly. A Sears DieHard wheelchair gel cell or a computer backup power supply gel cell will work well as a portable power source.

A gel cell is a high capacity lead-acid battery that is pressurized and sealed using special vents, and thus it should never be opened. They use a thixotropic gelled electrolyte that liquefies if shaken but returns to a solid if allowed to rest. Gel cells are known as "recombinant" batteries in that the oxygen produced by the positive plates combines with the hydrogen given off by the negative ones. The recombination of the oxygen and hydrogen produces water, which in turn replenishes the moisture in the electrolyte. This principle make gel cells sensitive to overcharging because that will produce a greater amount of oxygen and hydrogen which will vent away before it can recombine into water. Additionally, prolonged undercharging will shorten a gel cell's life by corroding the positive plates. They cannot be recharged using a high-amperage car battery charger or even an automobile alternator unless there is provision for regulating the voltage and current. If gel cells are recharged too quickly, there is a risk of swelling, which can cause them to burst their case, or possibly explode.

By their nature, these batteries need more care than standard maintenance-free lead-acid types. It is best to store them fully charged, and always recharge them immediately after use. Do not leave them discharged for prolonged periods or they will fail. Conversely, continuously overcharging a gel cell will dry out the electrolyte and shorten its life. Only use a temperature-sensitive trickle charger with a gel cell option.

Here are some tips for gel cell care and use:

1. A 12-volt cell has six 2-volt cells.
2. The maximum charging voltage for a 12-volt model is 2.4 volts per cell, or a maximum of 14.4 volts.
3. If the charging voltage is 2.4 volts per cell, the charger must be turned off when done.
4. If the charging voltage is 2.15 volts per cell, the charger can be left on indefinitely.
5. The maximum charge rate is the amp-hour rating divided by 5.
6. It takes about 40 percent of the total charging time to recharge the final 10 percent of a battery's capacity. The final charging current is about 1/

Table 5.2
Depth of Discharge vs. Gel Cell Charge Cycle Life

Average Discharge (%)	Battery Life in Charge Cycles
100	200
50	400
30	1300
10	2000

200 the amp-hour capacity of the gel cell.

7. Gel cells must be recharged every six months if not used.
8. Beware of flea market batteries that have been sitting around for a year or more.
9. There is a direct relationship between the depth of discharge and the life of a gel cell in terms of recharge cycles. This is summarized in Table 5.2.

5.12 Portable 120-volt Power in the Field

Digital imaging places a premium on the use of electrical power at the telescope. Sometimes electrical power is not available at dark locations but, telescope drives, laptop computers, and peripheral equipment still demand their usual voltage in order to operate. I solved the remote power problem by constructing a combination battery and inverter box with two built-in GFCI protected 120-volt outlets as well as twin 12-volt cigarette lighter outlets. The box was made from ½-inch plywood and all the electrical and cabinet hardware were found at the local Lowe's Home Improvement store. The only items purchased elsewhere were the two cigarette lighter outlets, which are available from automotive parts stores.

The power box was scaled around the size of the inverter used, in this instance a 1000-watt X-Power inverter by Xantrex available at Costco and other major chain stores. Xantrex (www.xantrex.com) makes other inverter models with built-in ground-fault circuit interrupt (GFCI) protection designed for use in RVs and marine applications and these models are available at RV and boating stores. If the user is not thoroughly familiar with electrical 120-volt wiring, I recommend using an inverter model that contains the built-in GFCI instead of wiring GFCI outlets to a non-GFCI inverter.

Under heavy use, the inverter needs to be exposed to open air for cooling purposes. For this reason, mine is mounted under the lid of the box on a removable platform above two 33-amp/hour 12-volt Power Sonic (www.power-sonic.com) gel cell batteries. The lid is opened to expose the

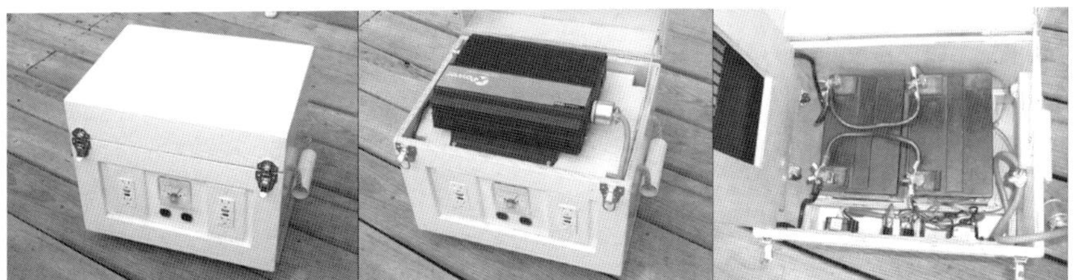

Fig. 5.12 *The author powers all 12-volt and 120-volt telescope and camera accessories with this battery box and inverter package made from ½-inch plywood. Two 33-amp gel cells power a Xantrex 1000-watt inverter and two 12-volt cigarette lighter outlets. Twin GFCI-protected outlets distribute power from the inverter. Photos by Robert Reeves.*

unit for cooling while it is operation.

The box dimensions were also scaled to allow the 12-volt input wiring and the 120-volt outlet plug leading to the two GFCI receptacles to clear the lid of the box when it is closed. To access the two batteries, the outlet plug is disconnected from the inverter and the inverter platform lifted out and swung aside. The 12-volt input wires are sufficiently long to allow the inverter to be removed without difficulty. It only needs to be removed to occasionally check the battery connections for corrosion. The batteries are held in place by a short plywood framework that cradles each one and prevents their movement during transportation. When in use, the batteries need ventilation, so the inverter platform has a large cutout under the inverter to allow air circulation.

It is not the purpose of this book to give instructions about cabinet making, but a point to remember about building a wooden case with a hinged lid is that it is easier to build an enclosed box, then cut it in half, than it is to build two open boxes that fit together at the hinge.

The box shown in Figure 5.12 is a basic design that can be enhanced with additional features or scaled down if the user desires. It has twin 120-volt GFCI wall receptacles so that more than one person can power their telescopes with it. Twin GFCIs are used so that if one user trips their GFCI for some reason, it will not shut down the other person's equipment. A six-outlet strip extension cord enables the connection of multiple items to the GFCI outlet.

With the two batteries installed, the power box weighs in excess of 60 pounds. User-friendly enhancements like recessed carrying handles can be added to the sides to aid in transporting it. Wooden skids can be attached to be bottom so the box can be placed on the ground out in the field and brass protectors can be attached to prevent damage from frequent transport. An on/off switch can be added to control the 12-volt cigarette lighter

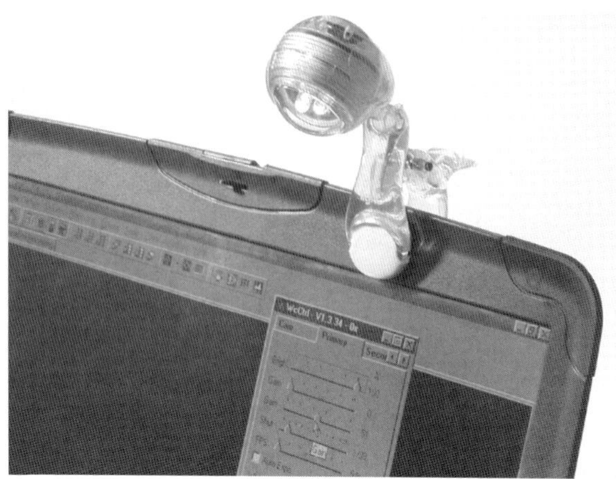

Fig. 5.13 *A clip-on LED "book light" makes it easier to operate a laptop keyboard in the dark. This "My Light" model was purchased at Costco and operates for months on two small mercury batteries. Photo by Robert Reeves.*

outlets. If heavy amperage use is expected on the 12-volt circuit, the on/off switch can control two standard 30-amp automotive relays, which in turn supply power to the outlets. When the inverter is not under an electrical load, its fan is off, making it difficult to tell if the inverter is operating. An LED status light can be added to indicate that the 120-volt inverter circuit is turned on.

5.13 Flashlights

There is an old astronomical proverb that says any flashlight used for astronomy should taste good, because its going to end up between your teeth at some time during the night. There are many models, both LED and bulb illuminated, that are advertised as "astronomy" flashlights. But when it comes down to universal availability of batteries, accessories, and replacement bulbs, few can beat the utility of an ordinary AA battery Mini Mag flashlight. These can be converted for astronomical use by inserting two layers of red cellophane behind the front lens.

The Mini Mag series also has a number of accessories that make it very useful at the telescope. A lanyard that clips to the base of the flashlight allows it to hang around your neck. As mentioned above, all astronomy flashlights will eventually end up in your teeth, as both hands are needed to perform a nighttime task at the telescope. The Bite-a-Lite accessory slips onto the end of a Mini Mag just for that purpose. The Mini Mag is available in various colors and I color-code mine to avoid accidentally turning on a white light instead of red one.

An astronomy flashlight should be relatively dim in order to preserve night vision and not incur the wrath of nearby astrophotographers. The Mini Mag can flare its light cone out into a diffuse wide pattern, or zoom into a tight spot when greater illumination is needed. A "red" flashlight should be truly red. An unofficial test to see one is really red is to shine it on the cover of a *Sky and Telescope* magazine in a dark room. If the magazine name is indistinguishable from the surrounding red border, the light is sufficiently red.

By necessity, we use a laptop computer extensively in webcam astrophotography. Operating the keyboard at night can be a challenge. The screen is seemingly bright enough to read by, but it also dazzles our eye to the point where we cannot see the keys. Constantly clicking on a small flashlight and holding it in your teeth to illuminate the keys is a less-than-elegant solution. My wife, Mary, found the solution at a Costco store in the form of a small clip-on book light. The MyLight unit shown in Figure 5.13 uses two blue-white LEDs that are powered by a replaceable button-sized battery. Similar lights are widely available. While astronomical logic would dictate that a red light would be preferable, the fact is that in webcam imaging we are working with computer monitors that, even when dimmed or covered with a filter, can still be so bright as to affect night vision. Also the white LED in the model shown can be swiveled to aim down at the keyboard, causing less problem with night vision.

5.14 Accessory Carrying Cases

A compartmentalized accessory carrying case will make an astrophoto session much easier. To keep all of my accessories organized, and to insure that nothing is left at home by accident, I constructed a wooden triple-decker carrying case for all my telescope equipment. The lid has a compartment for star charts and books while the body has two lift-out trays. Small optical items like eyepieces, Barlows, diagonals, and adapters go in the foam-lined top tray; while tools, flashlights and camera gear go in the lower one. Heavy or bulky items like counterweights and the dew blower go in the bottom compartment.

Camera gear and accessories are much easier to keep track of in the dark and protect from damage if they are in foam-lined carrying cases instead of a camera bag. The price of such cases purchased from a camera store can carry a bit of sticker shock. A less expensive option is a foam-lined plastic case from a sporting goods store. The cases I use for my photo gear were intended for holding competition pistols. The foam lining of these cases has to be carved out with an Exacto knife to fit the accessories

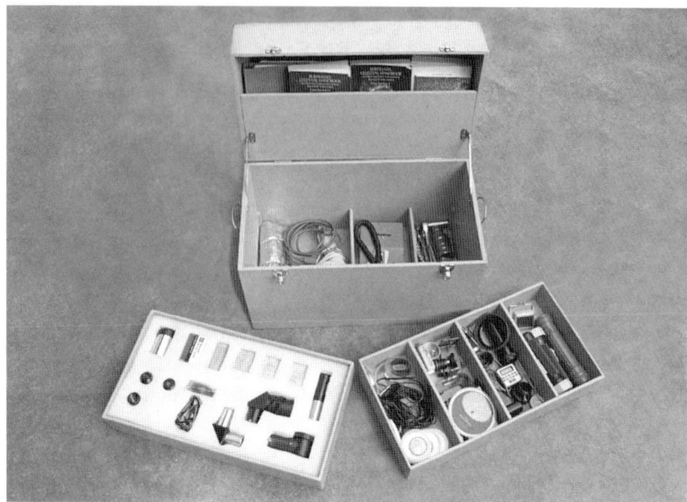

Fig. 5.14 *Finding that a critical item is missing after traveling to a remote dark site can end an astrophotography session. The author built a four-level plywood case to keep all accessories in one location. The foam-lined top tray contains delicate optics, the middle tray holds lightweight accessories, while heavy items are stored in the bottom. The compartment in the lid houses books and charts. Photo by Robert Reeves.*

you wish to store. Thick foam is so soft and spongy that cutting it with a sharp hobby knife often leaves a ragged or uneven cut because the foam has so much "give." To remedy this, soak the foam in water and squeeze most of it out so the foam is damp, but not dripping, then freeze it. It will now be stiff enough for accurate cutting with a sharp blade. Allow it to dry thoroughly after thawing.

Chapter 6
Using a Webcam for Astrophotography

6.1 Keys to Getting Good Images

Achieving good deep-sky images with a modified webcam has many of the same requirements we need for standard film-based astrophotography. These include:

- Good optics. While optics do not have to be perfect, the combined optical system does need to be at least ¼-wave and free from defects such as chromatic aberration, astigmatism and coma. With the exception of use under perfect seeing conditions, atmospheric turbulence will limit the performance of the best lens or mirror so imaging with ¼-wave or better optics will give satisfying results under typical multiple-arcsecond seeing. Good optics also need good mechanical support. Even the best optics can't perform well if poorly supported, or if carried on a flimsy mount.
- A good telescope mount and drive. The field of view of a webcam is small. This is a fact that we must accept and live with. A good telescope mount with accurate, fine-scale setting circles or a GOTO option will greatly ease the challenge of centering dim, even invisible, deep-sky targets. An accurate drive to adequately track the target is also necessary. A telescope with a poor drive system will soon be retired from astrophotography.

  A telescope drive with a lunar rate is very helpful for keeping lunar targets in the field of view. As the Moon orbits the Earth it will move by its own angular diameter every hour so being able to alter the tracking rate to compensate for the Moon's constant eastward march will simplify imaging lunar targets. But the lunar tracking rate is not perfect. The Moon travels around the Earth in an inclined orbit so it may also be traveling north or south in declination as well as eastward, something the lunar tracking rate does not compensate for.
- Accurate polar alignment. Being well-aligned on the celestial pole allows a good telescope drive to do its job. No matter how good the drive is, it will not properly track an object if its alignment is sloppy. This is a

critical step in successful deep-sky astrophotography.

- Adequate telescope cool-down. An often-omitted step is adequate telescope cool-down time. Imagers are anxious to get started as soon as it is dark, but often telescope optics are still warm from being in the Sun or in a hot car during the day. Allow at least an hour after sundown for optics to cool and relieve thermal strain in the glass and telescope structure. Built-in fans help speed cooling with larger and thicker telescope optics.
- Standardized exposure times for all video frames of the same object. Varying the exposure times within a video sequence will alter the noise characteristics of individual frames, making it more difficult to process the images later. If a very bright object is being imaged, where differing exposures are required to allow layering the images during processing to preserve detail in a bright core, use multiple video sequences, each with a different exposure. Ultimately, the user will find that for most deep-sky objects, image noise will be a limiting exposure factor and exposure length will be standardized for all faint objects.

6.2 Familiarization with Camera Controls

Before you venture out under the stars it is important to gain an understanding of the camera controls—what they can do and what they cannot do. You should familiarize yourself with the webcam software and controls in daylight. First, set up the webcam for "terrestrial imaging" like it was designed for. Take some movies of familiar objects and people. Work with the controls until the images look normal.

One way to develop your technique is to create an artificial night sky by wrapping aluminum foil over the end of a short cardboard tube and lightly pricking the foil with a needle to make pinholes. Aim your webcam into the tube and focus, (using the webcam's lens) on the artificial stars. To block stray light you can drape a dark cloth over the camera and the space where it mates with the rear of the tube. Using this setup you can explore how the camera's image capture, shutter, gain, brightness, and gamma controls affect the image. When you set up the webcam for its first astronomical imaging, things will run more smoothly than if you were learning these things completely in the dark.

Webcams like the Philips SPC900NC or those modified for long exposures are all able to do short-exposure astrophotography. If they have a long-exposure option it is simply not used. Those webcams modified for long-exposure imaging work the same way as a short-exposure model. The difference lies in the length of individual video frame exposures. For instance, each frame taken with a camera set at five frames per second represents the passage of ⅕ second as the video is played. A long-exposure

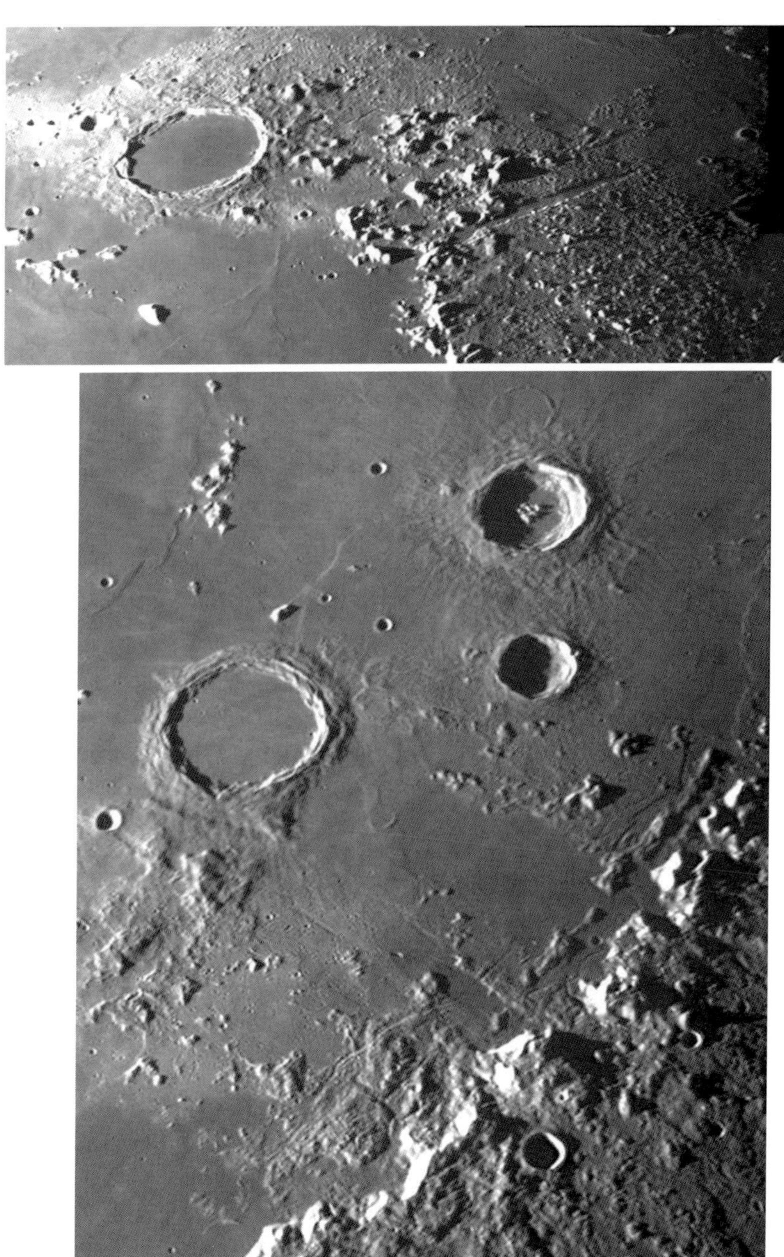

Fig. 6.1 *Good optics, polar alignment and tracking, seeing, and focus are essential for achieving good lunar and planetary images. Plato and the Alpine Valley (top), and the flat-bottomed crater Archimedes and the Apennine Mountains (bottom) are two- and three-frame mosaics captured with a ToUcam 840 on a Celestron-8 telescope operating at f/20. Photos by Robert Reeves.*

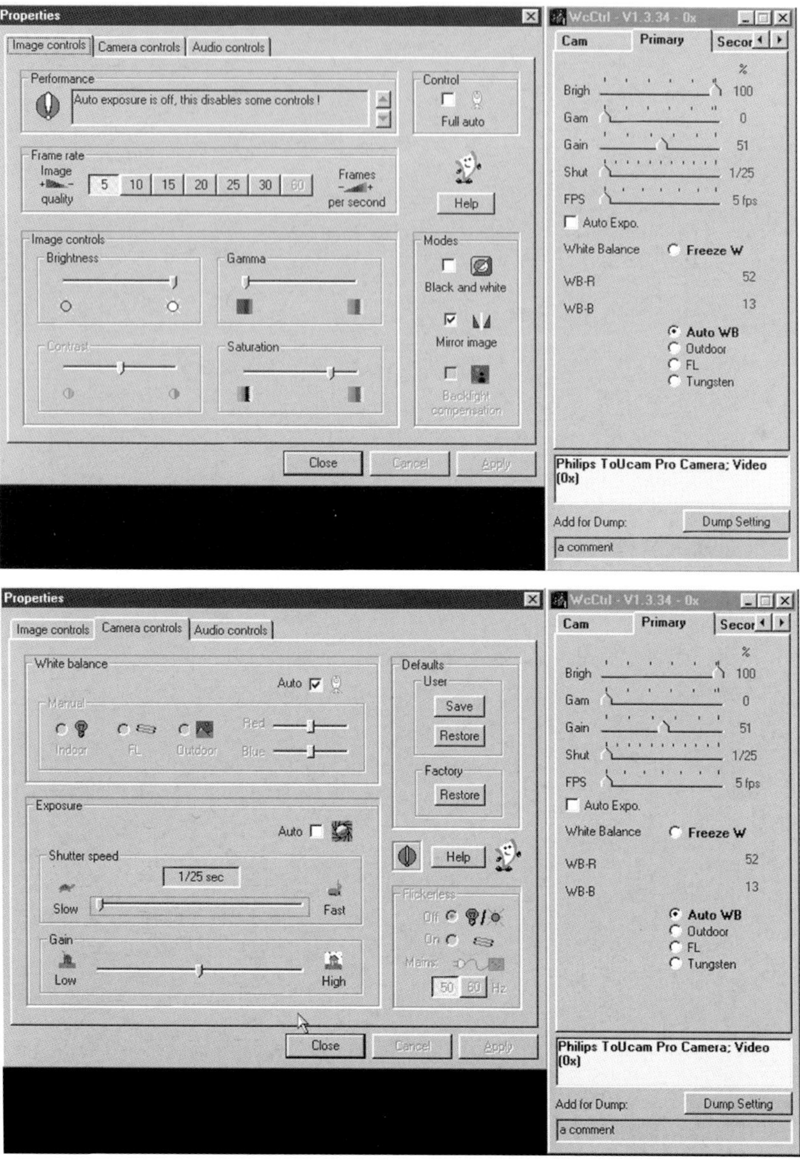

Fig. 6.2 *The two-screen Philips ToUcam control interface is shown above. The Atik 1 and 2 series cameras also use this interface. The frame rate, brightness, and gamma are controlled from the <Image controls> tab while shutter speed and gain are controlled from the <Camera controls> tab. The freeware program WcCtrl is shown next to the ToUcam controls and mirrors the settings from both ToUcam screens. The advantage of using WcCtrl in place of the Philips interface is that all camera settings are numerically displayed on one small screen that fits next to the K3CCDTools image capture program instead of two large screens that often overlay and block the view of the capture control screen. Screen shots by Robert Reeves.*

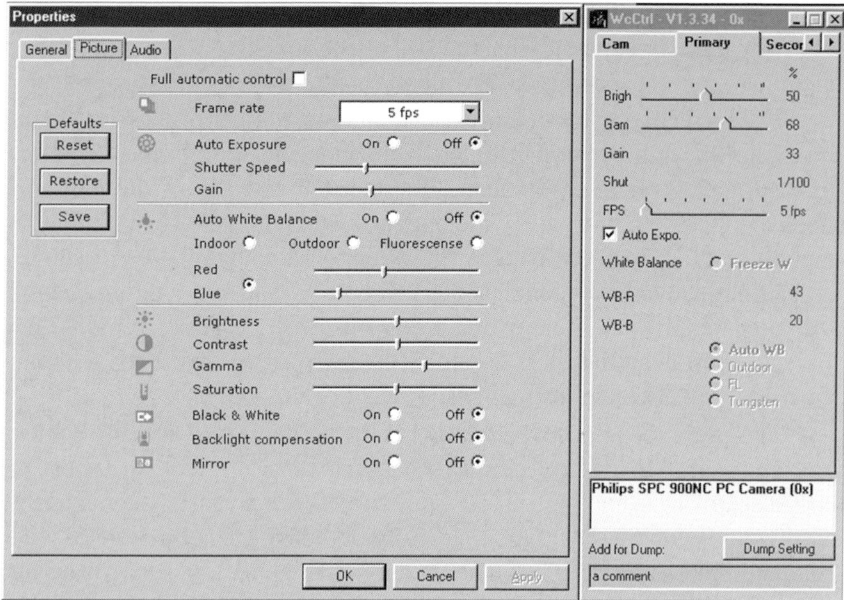

Fig. 6.3 *The Philips SPC900NC camera control interface places all controls on one screen, but lacks a numeric display, forcing the user to guess at actual settings. WcCtrl will numerically display the SPC900NC settings, but the current version of WcCtrl will not allow direct control of shutter speed and gain. WcCtrl does, however, allow numerical monitoring of those parameters when adjusted within the Philips interface. Screen shot by Robert Reeves.*

camera selectively slows down this rate. For example, a video taken with 30-second exposures for each frame, and then played at five frames per second, will show the passage of two and one-half minutes per second of video play time. However, the length of each exposure is irrelevant to the processing of the sequence of video frames into a single image. The software will stack a series of long exposures into a single image just like it does short exposures.

6.3 Focusing Aids

Of all the camera control activities the user will face in astrophotography, focusing a webcam will prove to be the most challenging. Since a live image is displayed on the computer screen it would seem a simple task to watch the image as the focuser is turned and quickly arrive at best focus. However, in the "real world" it is not that simple. The blocky star images on a computer screen do not look like the pinpoints seen in an eyepiece. There is a wide focus zone where the image does not look appreciably better on the computer. Fortunately, a webcam provides many frames per second while focusing on a bright star and focus analysis can be done relatively quickly compared to standard astronomical CCD cameras where it can take several seconds to download each new focusing exposure.

The focus of a telescope is dependent on a number of factors, including things like atmospheric turbulence, the mechanical stability and optical quality of the optical components, the thermal stability of the telescope tube, and the depth of focus of the optical system. It cannot be emphasized enough that focus should be checked often while imaging. Optics can sag and shift due to changes in telescope orientation, and temperature changes will alter the physical dimensions of an instrument's structure. Both factors can spoil even the most careful focusing. Occasional focus touchups should be considered a normal part of webcam imaging.

The faster the *f*-ratio of a telescope, the harder it is to actually achieve precise focus. Most people think fast instruments are easy to focus; the image seems to "snap" into focus quickly. However, fast systems have a narrow range where the image is actually in focus. A fast *f*/4 system, when compared to one of *f*/10, has a focus range that is 2.5 times narrower, and that range is measured in thousandths of an inch. So while the process may seem to be easy, finding the exact spot of best focus can be a challenge.

Shifting optics will also alter focus. If an SCT is being used, one should always assume that the primary mirror will shift slightly if the telescope is moved more than a few degrees across the sky. To see if mirror shift is a problem with a Newtonian telescope, take a 20- to 30-second image of a star cluster, and then repeat the exposure while aimed at a similar cluster on the opposite side of the sky. If both images are not equal in focus, the optics shifted when the instrument was moved.

Owners of most compound telescopes will find they have to deal not only with a shifting primary mirror but the thermal expansion and contraction of the metal telescope tube itself. As the temperature varies throughout the night, a metal telescope tube will cool and shrink measurably. Field experience has shown that the focus point for a metal tube, 8-inch *f*/10 Schmidt-Cassegrain, changes 0.035 inches with a 15-degree temperature drop. This is more than three times the acceptable depth of focus as shown in Table 6.1 and demonstrates that not only do we need to focus precisely; we need to refocus often as the telescope cools. For all of these reasons, SCT users will find the need for frequent refocusing in order to maintain highest resolution during an imaging session.

Schmidt-Cassegrain and Maksutov telescopes that move the primary mirror in order to focus the instrument have enough travel to accommodate almost all auxiliary optical devices, such as a camera and flip mirror or

Table 6.1 Depth of Focus (inches) vs. Focal Ratio

Focal Ratio	Depth of Focus	±Total Range
f/4	0.002	0.004
f/5	0.0025	0.005
f/6	0.003	0.006
f/7	0.0035	0.007
f/10	0.005	0.010

focal reducer. On Newtonians it may be necessary to move the focus point of the telescope in order for a webcam to achieve focus. This may involve moving the mirror forward inside the tube, installing a low-profile focuser, or both.

If you find a webcam will not achieve focus on your Newtonian telescope, the focal point is most likely too far inside the focuser to reach the camera when it is attached with a 1¼-inch eyepiece adapter. To find the exact focal point, cut the bottom off a plastic 35mm film canister and tape over the hole with frosted cellophane tape. While not optically flat, this will provide an adequate surface similar to the ground glass screen in the focuser of a 35mm camera. Slide the device in an out of the telescope focuser while aimed at a bright object like the Moon and watch for the point of best focus. When this point is reached, measure the difference between it and where the webcam's sensor lies when attached to the telescope. If the webcam's sensor lies beyond the range of the current focuser, sometimes installing a low-profile focuser will drop the camera to the proper point. The camera should be able to "overshoot" the focus point slightly in order to go back and forth to refine the exact focus.

If installing a low-profile focuser still leaves the required focus point a few millimeters short of reaching the camera sensor, slightly shimming the primary mirror upward may be all that is needed. Use a thermally stable material like stiff plastic to make the shims. If more than several millimeters of additional travel is needed to move the focal point onto the camera's sensor, it is time to bite the bullet and move the mirror cell a little further up the tube. If this step is taken, be sure to rebalance the telescope.

If a standard camera lens being used with a webcam does not achieve as good a focus as it does when used on a conventional camera, make sure that chromatic aberration is not the culprit. A webcam's built-in infrared filter will be removed when the camera is adapted to a camera lens. The CCD in a webcam is quite sensitive to the infrared portion of the spectrum while a camera lens is designed to properly focus only visible light. This can result in fuzzy images arising from either well-focused visible light but poorly-focused infrared wavelengths; or well-focused infrared but poorly-focused visible light. Either the visible light or the infrared will have to be filtered out to achieve good focus. Camera lens filters described in Chapter 4 will eliminate the infrared portion of the spectrum and sharpen the focus of star images.

Other components capable of degrading focus are a loose focuser or T-adapter. Either will introduce play in the optical system that shifts the position of the camera relative to the point of best focus. While aimed at a focused star, gently push on the camera fore and aft as well as side-to-side. Excess play in either a focuser or a T-adapter will reveal itself as a slight shifting of the image independent of telescope movement. Additional set-screws may help stabilize a focuser, but a loose T-adapter will have to be replaced.

There are a number of things the user can do to improve the focusing performance of their telescope when using a webcam:

- Insure that the optics are well collimated. Use of a laser collimator with Newtonian optics will simplify this task.
- Use an electric focuser to prevent shaking the telescope.
- Use an IR/UV blocking filter to eliminate wavelengths that do not focus in the same plane as visible light.
- Use a Hartmann mask or diffraction focusing aid to zero in on best focus.

If the focus never seems to be quite right in spite of best efforts with the above techniques, then an examination of optical adjustment and quality is in order. With good telescope optics under good seeing, an out-of-focus star should present an equally round star image at the same point on either side of best focus. If the out-of-focus star disk is asymmetrical and brighter at the edge of one side of the disk, then optical collimation should be suspect. If the star still is not round on either side of focus after collimation, then the physical quality of the optics may be the problem.

6.4 Manual Focusers

If you do not have an electric focuser for your telescope, one way to achieve finer manual control of the focus knob is to make it larger. This can be done by creating a “Jiffy focuser” like the one popularized by Bill Arnet. The name derives from the fact that Arnet modified the focus knob on his Schmidt-Cassegrain by attaching the lid from a wide-mouthed Jif peanut butter jar to it. This greatly enlarged knob makes it easier to move the focuser in smaller increments without shaking the telescope.

If the configuration of your instrument’s focus control does not allow the use of a Jiffy focus device (on my Classic C-8 it strikes the visual

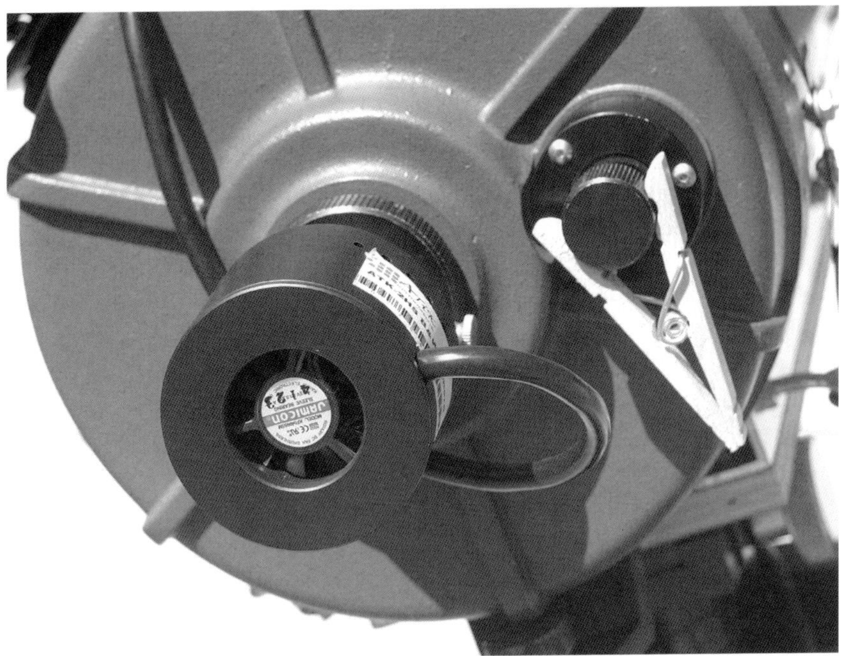

Fig. 6.4 *Grasping the focus knob with two fingers creates image shake, making critical focusing difficult. A simple solution is to clamp a clothespin to the knob, which can then be easily and gently turned with a fingertip. Photo by Robert Reeves.*

back), an alternative is to first achieve coarse focus by hand, then clamp a spring-loaded clothespin onto the focus knob. The clothespin extending radially away from the knob provides enough leverage to rotate it with gentle fingertip pressure. This is especially useful in cold weather when gloves make it harder to operate telescope controls. Removing the clothespin will move the knob enough to alter focus, so it is best to leave it in place once focus is achieved. Since the clothespin protrudes from the focuser, care must be taken not bump it when operating the telescope in the dark.

Because a webcam is focused by viewing its output on a computer screen, it stands to reason that the larger the image is on the screen, the easier it will be to monitor the focus. The output from a webcam is 640 x 480 pixels. Reducing screen resolution to about 800 x 600 allows the webcam image to fill the screen within the webcam's own operating program.

Alternatively, the K3CCD Tools camera control program allows up to 8x zoom for enlarging the image to help with focusing. Another image enlargement option is to use the "accessibility" options within the Microsoft Windows operating system. The "magnifier" option enlarges the view under the area around the cursor. The user can set the degree of magnification, reverse colors, and adjust the contrast of the magnified view.

Zeroing in on initial focus can be done quickly. Turn the focus knob so that the planet, star, or lunar detail progressively shrinks. Achieving fine critical focus is another matter; this will take a little while. The magic of webcam astrophotography lies in its ability to stack many individual frames to achieve a final image with very high resolution. This cannot be done successfully if the telescope is not properly focused to begin with, so webcam imagers soon learn to expect to take several minutes to achieve critical focus.

On some telescopes, particularly inexpensive refractors, the rack-and-pinion focuser is stiff and coarse, making fine focusing a challenge. To help achieve optimum focus, some inventive tinkerers have adapted the planetary reduction gear systems from a defunct electric screwdriver to create a slow-motion focuser drive. The adaptation of such devices to a particular focuser is unique on a case-by-case basis, but the principle remains the same for all applications. The gear set is attached to the focuser so that when the high-speed input is turned, the gears turn the focuser at a greatly reduced rate, often with a reduction of 7 to 1. The gear assembly itself must be anchored to the telescope with a bracket, or the planetary gear assembly will simply rotate around the telescope's focus knob when it is turned.

A good tip while focusing an SCT is to pre-load the focusing screw that moves the mirror. As the focus is racked back and forth in progressively smaller increments while zeroing in on the best focus, deliberately overshoot the best focus point so that it is approached while turning the focus knob in a counterclockwise direction. This is because the right-hand thread of the focus push bolt will raise the SCT mirror when turned counterclockwise. When the telescope is aimed upward, this preloads the play in the focus mechanism so it does not settle backward several thousandths of an inch as gravity pulls the mirror back against the mechanical play in the focuser threads. As shown in Table 6.1, the depth of focus can be very shallow and a shift of only several thousandths of an inch can cause noticeable degradation.

6.5 Electric Focusers

Manually turning the focusing knob will in itself shake the telescope. Considering the inherent high magnification when using a webcam, telescope wiggles during focusing can be frustrating while trying to watch a jiggling image on the computer monitor. An electric focuser will eliminate tele-

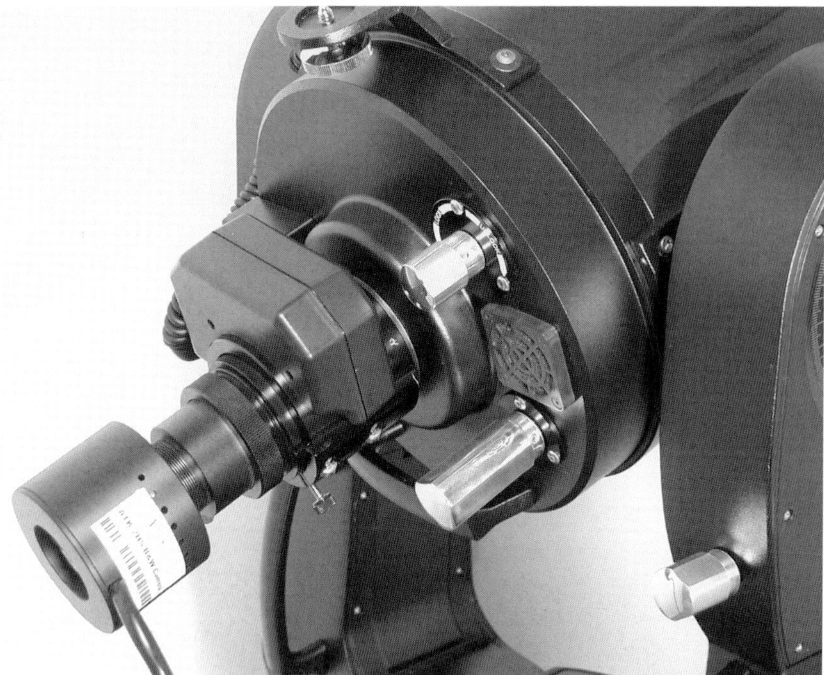

Fig. 6.5 *The Meade #1209 Zero Image Shift Microfocuser will couple to any SCT-style telescope. After achieving coarse focus with the telescope's normal focusing mechanism, the Microfocuser allows focus to be refined electronically, without the image shift common when focusing an SCT by moving the primary mirror. Photo by Robert Reeves.*

scope shake when focusing. Many imagers have adapted high-torque, low-speed 12-volt motors to remotely operate their telescope's focus mechanism. Commercially made electric focusers are also available, with those by JMI and Meade being very popular.

Reflector and refractor telescopes usually keep an image centered while focus is shifted back and forth. As we have seen, however, Schmidt-Cassegrain telescopes will display some degree of image shift while the focuser moves the primary mirror forward and backward. This movement allows the mirror to tip slightly during the process, and as a result the image shifts back and forth in the field. At high webcam magnifications, especially with a Barlow, this can move the image completely out of the field of view. An auxiliary focuser that installs between the SCT's visual back and the camera or flip-mirror assembly will eliminate the image shift problem. The mirror remains stationary while the auxiliary focuser moves the imaging accessories to achieve focus.

The Meade Zero Image Shift Model 1209 microfocuser is a popular

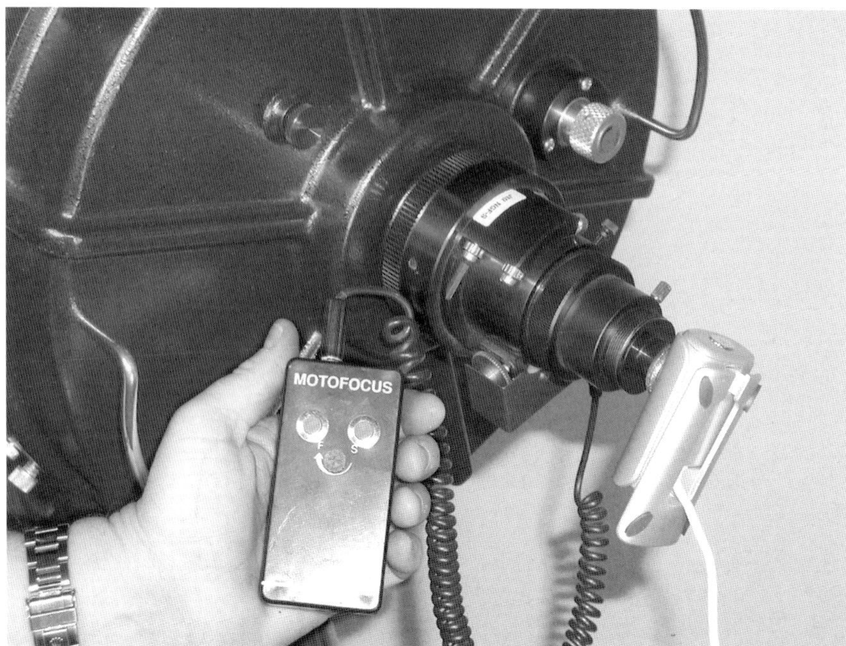

Fig. 6.6 *The JMI Model NGF-S Motofocus is shown coupled to a Celestron-14 telescope. After the telescope's normal mechanism achieves coarse focus by moving the primary mirror, fine focusing is done electronically with the Crayford-style JMI focuser. Photo by Robert Reeves.*

item with Schmidt-Cassegrain owners (see Figure 6.5). This unit has a critical slow-motion option that allows closing in on the true focus without motor overshoot or backlash. The Model 1209 is identical to the 07079 microfocuser that comes standard with the Meade LX200GPS telescopes except this version is controlled by a separate battery-operated hand controller instead of the telescope's control paddle. The unit is powered by eight AAA batteries, fits all SCT-style telescopes, and can accommodate both 2-inch and 1¼-inch accessories. It has four speeds, from fast to very slow. Since the unit fits directly into the telescope's visual back, coarse focusing can still be accomplished through the telescope's normal focusing knob, then fine focus achieved with the microfocuser.

Users of other types of telescopes have the option of aftermarket focusers such as Starlight Instruments' Feathertouch featuring a 10:1 reduction ratio for fine focusing, or the Tele Vue Focusmate with 6:1 reduction. Electric focusers such as the JMI Motofocus can be adapted to these reduction gear focusers for hands-off shake-free remote focusing (see Figure 6.6).

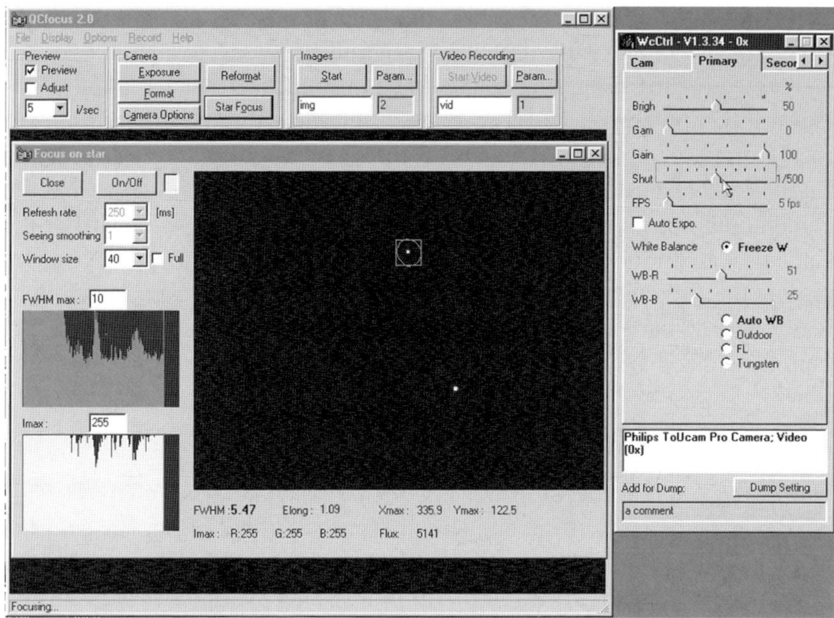

Fig. 6.7 *QCfocus is a freeware program that allows the exact determination of star focus by full-width half-maximum analysis of a star image. The selected star is at best focus when the upper graph displays the lowest possible reading or the FWHM numeric display shows the lowest possible number. The program works with WcCtrl to allow setting shorter exposure times if a bright star saturates the graph display with continuous maximum readings. Screen shot by Robert Reeves.*

6.6 Software for Focusing

K3CCD Tools has an option that displays the peak pixel values of a star. The brightness values of a star's pixels are at maximum when a well-focused star's light is concentrated into the smallest possible area. The point of best focus has been reached when pixel values are at maximum.

Although not advertised as a focusing aid, some users find the FFT tab in K3CCD Tools useful for finding best focus on star and planetary images. The FFT function calculates the quality of an image within a box at the center of the field of view, so the star or planet should be kept inside the box while focusing. The point of best focus can be determined by monitoring the FFT numerical display. When the FFT numbers are at their maximum, the focus is at its best. The FFT function reacts quickly to instantaneous changes in seeing, so the focus must be moved slowly and the numbers observed to make a mental average over several seconds.

Users of standard digital cameras have focusing-aid programs such as DSLRFocus that analyze star images and report statistics on the half- round full-width maximum of a given star image. Fortunately, a similar resource exists for webcams as the freeware program QCfocus, available over the web.[1] At first, the program appears to be in French. However, after installation it can be converted to an English interface by first clicking on the menu option labeled <Affichage>, then clicking <Langue>, selecting English and restarting the program. QCfocus presents a live view from the camera and allows selecting a star for focus analysis. Run WcCtrl at the same time to allow selection of a shutter speed high enough to prevent saturating the star image. Analysis of the star's focus is displayed both graphically and numerically; best focus is achieved when the "FWHM max" graph and the "FWHM" numeric display read the lowest possible numbers.

6.7 Hartmann Mask

One of the simplest focusing aids is a device called a Hartmann mask. This is an opaque cover with several sub-aperture sized holes that fits over the front of a telescope or camera lens. This device is also known as a Scheiner disk because it was perfected in 1619 by German astronomer and sunspot observer, Christoph Scheiner. In its basic form the mask has two holes, each about one third the diameter of the optic it is being used on, and separated by a distance equal to the holes' diameter. The smaller and closer to the edge the holes are, the better defined the image is, however there is a practical limit on how small the holes can be and still be useful. For visual focusing the holes have to be large enough to allow an image bright enough to see. For photographic focusing the holes can be smaller because the camera can use longer exposures and is more sensitive than the eye. In practice, each hole in the mask creates a separate image. As the focus is adjusted, the multiple images will either converge or diverge, depending which way the focuser is turned. When the multiple images merge into one, the telescope or lens is focused and the mask can be removed.

Hartmann masks can be made from any opaque material such as cardboard, Masonite, sheet metal, or stiff plastic. Foam core art board can be used to make inexpensive masks for large telescopes while any stiff cardboard or heavy construction paper can be used for camera lenses. For durability with repeated use, small cardboard masks can be mounted in filter step-up adapter rings. Check hobby stores for embroidery or needlepoint hoops that may fit over the end of your telescope tube to hold a mask in place. Hobby stores may also have artwork templates useful for cutting out various size circles and triangles in the mask. If a more durable mask is needed for heavy use on a large telescope, thin sheet metal can be cut with

[1] http://www.astrosurf.org/astropc/qcam/programme.html

a jigsaw and fabricated into a durable mask. Standard 5-gallon plastic bucket lids fit a 10-inch SCT if the inner circumference of the lid is lined with felt to act as a "gripper". These lids are available separately at Home Depot hardware stores for about a dollar.

The advantage of the Hartmann mask is that it can be used for extended objects like the Moon while knife-edge focusers can only be used on point sources like stars. A disadvantage of it is that it darkens the image, but this will not be a problem for those using an image capture program to display the image on a computer screen.

There are a variety of hole patterns that are popular for use with a Hartmann mask—two opposed round holes, three round holes in a triangular pattern, two round holes with a third triangular hole, or three triangular holes. Each pattern works and the user's preference determines the style that is best for their needs.

The standard round two- and three-hole patterns work by showing either two or three separate images when out of focus. The telescope is in focus when all images are merged together.

A two-hole mask with a third triangle works like the two-hole mask but the triangular hole will serve as a clue as to whether the telescope is inside or outside of focus. When out of focus, the two round holes are visible as well as the third triangular hole (see Figure 6.8). When the telescope passes beyond the focus point, the triangular hole will be pointing the other way. Place the triangle on the north end of the telescope tube and the two round holes on the east and west sides. The triangle will allow you to determine which way to move the focuser when viewing an object. Some experimentation is needed to see how the modified mask works with a particular telescope. For instance, if the triangle is seen on the northern side of the two round images, determine which way the focuser must bc turned in order to move the triangle toward the two round holes. If true focus is overshot, the triangle will appear on the southern side of the two round images. The direction the focuser needs to be turned, either in or out, can thus be determined by analyzing how the north-mounted triangle reacts through your optical system. Focusing on targets in the future will be easier because the north-south position of the triangle will indicate whether to turn the focuser in or out to improve it.

A pattern of three triangles works similar to three round holes, but each of the triangles creates three diffraction spikes around each image. Buy turning up the gain on the camera, these spikes are easily seen. When the spikes from each image merge into one another, the telescope is in focus.

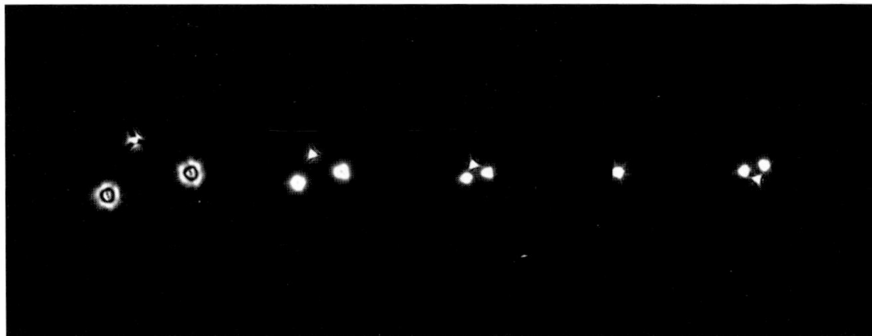

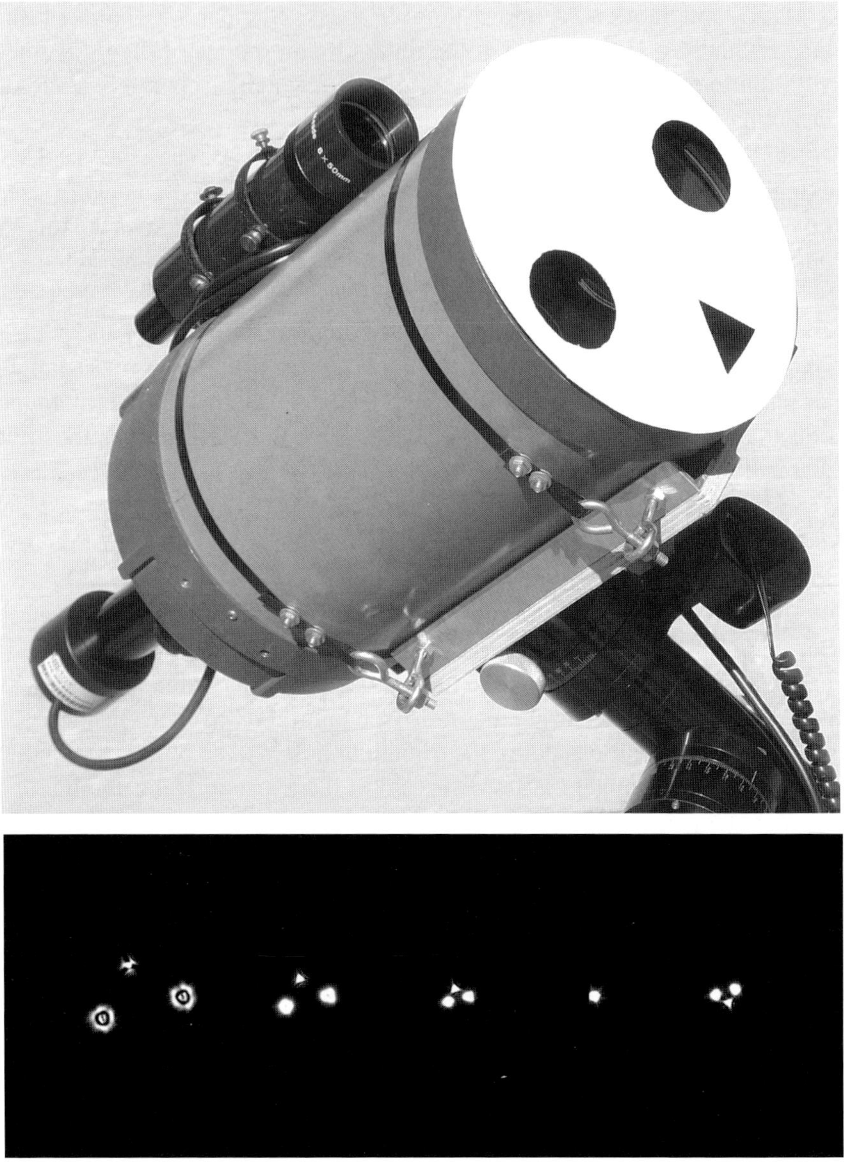

Fig. 6.8 *Although it bears a humorous resemblance to the popular logo of an American hamburger chain, a dual aperture Hartmann mask augmented with a triangular aperture will aid in rapidly focusing a star image. Such a mask will display three separate star images when the telescope is out of focus. The three images merge when good focus is achieved. Notice the image produced by the triangular aperture reverses position when passing from inside to outside focus allowing the user to determine which way to turn the knob to improve focus. Photos by Robert Reeves.*

If the seeing is not good, some prefer using only a two-hole Hartmann mask because when focus is nearly achieved, the resulting image is still elongated and the need for further focus refinement is indicated. A three-hole mask can produce some confusion as the point of good focus is approached, because the three independent star images begin to merge into a larger triangular or square pattern that resembles a focused round star image. However, if the seeing is exceptional, the three-hole mask will produce better focusing performance.

6.8 Diffraction Focusing Aid

A popular means for finding exact stellar focus when seeing is poor is to use the diffraction spike method. In poor seeing, a star can look blob-like and blurry over a wide range of focus. By providing a spider-like artificial obstruction like thin bungee cords stretched across the front of the telescope, thin crossed wooden dowels, or coat-hangar wire, an out-of-focus bright star image will display a dual diffraction spike pattern. When the diffraction pattern merges and its spikes are brightest and most contrasty, the image is at best focus. The spider can be attached to an embroidery hoop that mounts to the front of the telescope. After focusing, the spider assembly can be removed for imaging.

Most users feel the diffraction pattern method produces a crisper focus than that achieved with a Hartmann mask. To see if this is the case with your set-up, focus as best you can with a mask, then without disturbing the aim of the telescope, substitute the diffraction device of choice across the aperture and see if the focus can be further improved.

6.9 Using Webcam Characteristics to Achieve Good Focus

Focusing a bright star is difficult with a webcam because the star image blooms on the computer screen. Reduce the exposure, sometimes to as little as a thousandth of a second, and reduce the gain to increase the contrast between the star and the sky background. Progressively reduce the exposure time to where it is short enough that the star will completely disappear if it is out of focus, then pop back into view when it is close to best focus. By working only in the narrow zone of best focus achieved when the star is visible, much of the guesswork about the point of best focus will be eliminated.

A similar webcam focusing technique for the Moon is to use very short exposures of the lunar mountain peaks protruding into sunlight just beyond the lunar terminator. Progressively shorten the exposure time until

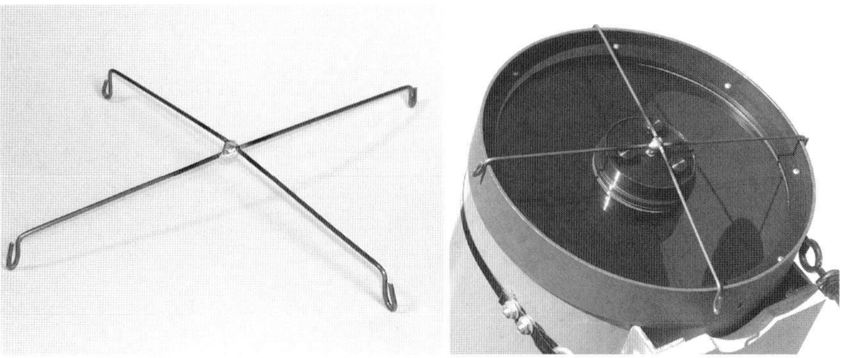

Fig. 6.9 *A simple spider diffraction focusing tool can be made by brazing or soldering two stiff wires into an X pattern. The author used coat hangar wire to create this tool that clips to the front of his Celestron-8 telescope. Photo by Robert Reeves.*

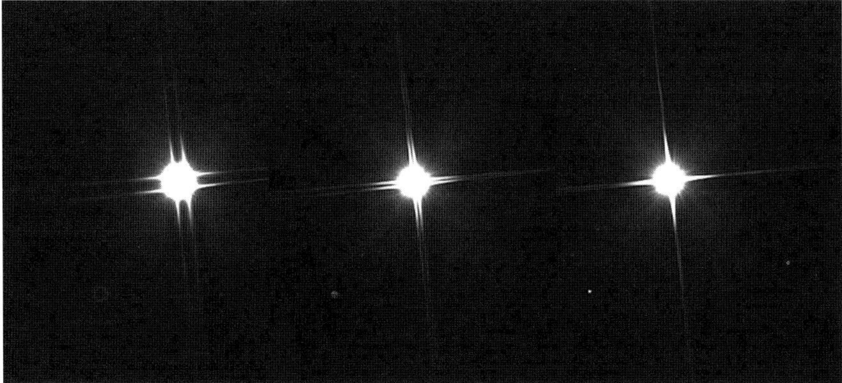

Fig. 6.10 *A spider in front of an out-of-focus telescope will produce a dual diffraction pattern. As the telescope is focused, the diffraction pattern merges. Notice that the dim star below the target star is invisible at the left but distinct when the proper focus is achieved on the right. Photo by Robert Reeves.*

the mountain peaks appear as pinpoints of light similar to stars. If the exposure is short enough, the peaks will disappear completely if the image is just out of focus, then pop back into view when the focus is best (see Figure 6.11).

Long-exposure webcams can use a variant of this technique where the greatest numbers of faint stars are visible if the image is properly focused. If the image is slightly out of focus, a certain number of faint stars will disappear because their light is spread out and too diffuse for the CCD to register. A series of long exposures at differing focus positions will show more and more stars until the point of optimum focus is passed, then the numbers of faint stars will rapidly decrease.

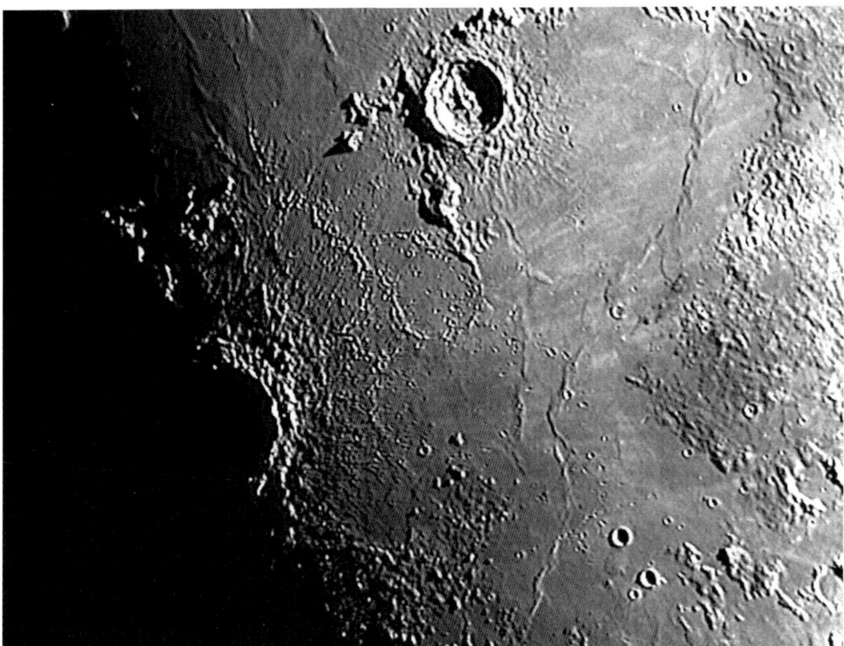

Fig. 6.11 *The star-like pinpoints of mountain peaks protruding into sunlight from the dark side of the lunar terminator can be used as focusing aids. To achieve good focus, raise the shutter speed until the peaks are invisible unless they are in perfect focus. Photo by Robert Reeves.*

Strange as it sounds, if the atmosphere is turbulent and the image is hard to focus, the turbulence can actually help improve the final stacked image. This is because the turbulent cells of atmosphere that sometimes defocus the image can also, at times, dramatically improve the focus of a telescope that is close to, but not precisely on, the exact focus point. By using programs such as AVI2BMP or RegiStax to select the very best-focused frames, the resulting image stack will be created using those images "fine-focused" by atmospheric turbulence.

6.10 Determining the Field of View

For the most part we are not concerned with the exact focal length to the nearest millimeter of an imaging system. When shooting through an 8-inch *f*/8 reflector we assume a focal length of 64 inches, or about 1600mm. With an 8-inch *f*/10 Schmidt-Cassegrain telescope we assume a 2000mm focal length. But when imaging through a compound telescope such as an SCT or Maksutov, the focal length and focal ratio may not be what they seem.

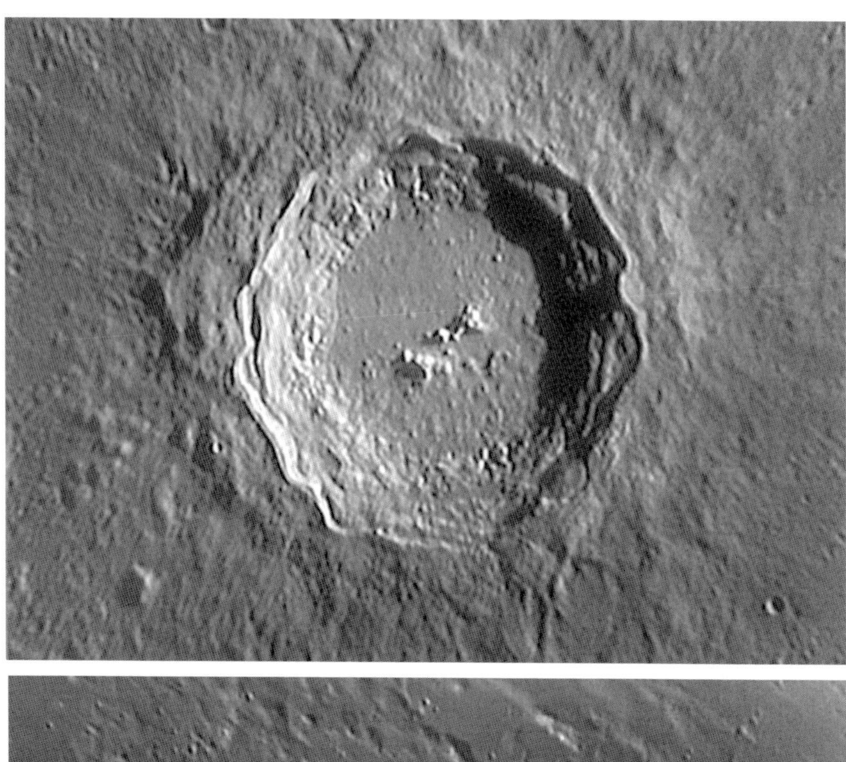

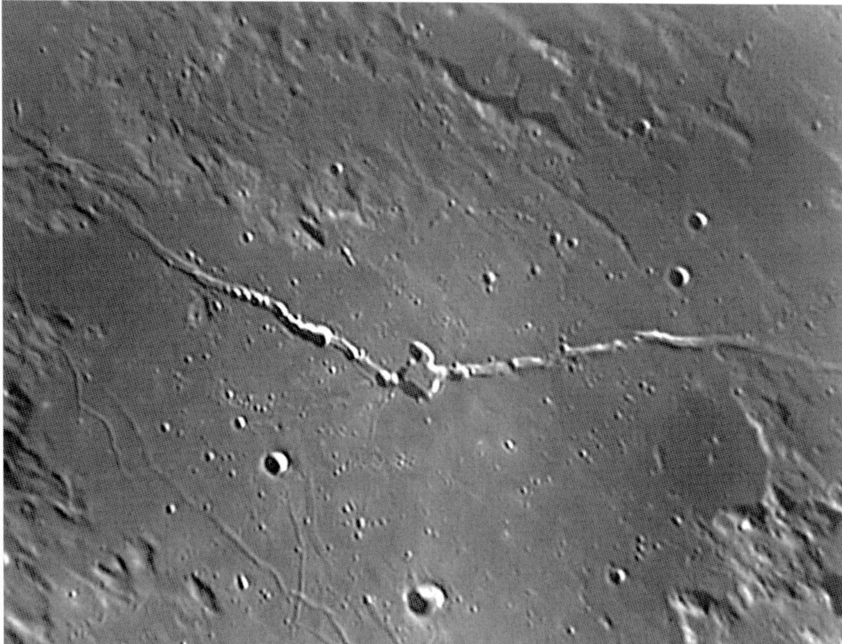

Fig. 6.12 *Using good focusing techniques with webcam astrophotography will allow modest telescopes to reveal a wealth of lunar detail. The crater Copernicus (top) and the Hyginus Rille (bottom) were imaged with a ToUcam 840 and a 12-inch reflecting telescope operating at f/29. Photos by Zac Pujic.*

The classic formula for determining the field of view of a CCD-type detector on a telescope is

$$\text{FOV} = \arctan\frac{\text{Pixel Size}}{\text{Focal Length}}.$$

If a scientific calculator is not available, a simple arithmetic way to calculate the field of view is with the formula

$$F = \frac{3437 \times W}{F_1}$$

where

F is the field of view in arcminutes,
W is the sensor dimension in millimeters, and
F_1 is the focal length in millimeters.

These mathematical models work well with refractors, Newtonians, and classical Cassegrain systems where the main optic remains fixed during focusing and only the eyepiece or camera is moved to achieve focus. However, they break down when applied to designs in which the primary mirror is moved in focusing. While moving-mirror optical systems upset the mathematics of this formula, the capability to move the mirror in such systems is a great advantage in that it allows us sufficient focus travel to accommodate accessories such as focal reducers, extension tubes, camera adapters, and flip mirrors.

If achieving focus were totally dependent on the movement of a rack-and-pinion focuser, many optical accessories could not be used without modifying the position of the telescope's focuser or primary optic. However, the flexibility of the Schmidt-Cassegrain and Maksutov designs allows the focal plane to move through a very large zone of travel. This allows the optics to still achieve focus even if attachments move the imaging plane of the camera further back. However, the actual effective focal length and focal ratio will be different than the telescope's "advertised" focal length and focal ratio. This can be confirmed next time you view the Moon through a Schmidt-Cassegrain. First, observe without a diagonal by viewing through an eyepiece inserted directly into the visual back on the telescope. Note the size of the Moon compared to the eyepiece field of view. Next, install a diagonal and, using the same eyepiece, you will note an increase in image size, proof of longer effective focal length."

For casual imaging a change of focal length and focal ratio due to moving optics is of no consequence. If we achieve a good image, we are happy. However, if science is the goal, such as astrometery of comets,

Table 6.2
Changing Focal Length in an 8-inch *f*/10 SCT with a 5.6 Micron Pixel Size Image Sensor

Optical Configuration	Resolution (arc-sec/pixel)	Focal length (mm)	Focal Ratio
Prime focus on visual back	0.635	1819	9.1
Prime focus w/extension	0.564	2048	10.24
Focal reducer under visual back	0.731	1580	7.9
Focal reducer with extension tube	0.899	1285	6.45

asteroid tracking, or double-star measurement, knowing the exact focal length is critical. The change in apparent focal length by adding accessories to an SCT can be significant.

The placement of a focal reducer can have large effects. The calculated focal length using, for instance, a 0.6 focal reducer is entirely dependent on the camera focal plane lying at an exact location relative to the rest of the optics. If the adapter used to mount the camera is a different dimension, the focal length of the system will change. For instance, if the camera is closer to the focal reducer the effective focal length will be longer, and thus the magnification higher. Conversely, if the camera is farther away, the focal length will be shorter than that advertised for the focal reducer, and the magnification will be less. As an example, adding a 68mm-long extension tube to an 8-inch *f*/10 SCT will change the focal length by an appreciable amount. Matthias Meijer measured the effective focal length changes to a typical 8-inch *f*/10 SCT with various optical accessories. His results are summarized in Table 6.2.

Why is this change in apparent focal length important? As the above table illustrates, it changes both the system's resolution in arcseconds per pixel and the image plate scale.

The field of view and resolution for a webcam with a particular telescope combination is easily determined using software such as K3CCD Tools. After entering the telescope and camera parameters, the program displays the camera's field of view in arcminutes. Tables 6.3 and 6.4 show the resulting calculations for various optical configurations.

6.11 Targets for Webcam Astrophotography

Modified webcams are quite sensitive to starlight. Commercially modified models, such as those by ATIK or other webcam-style imaging devices like the SAC Imaging cameras, are amazingly sensitive, being able to record stars dimmer than 9th magnitude with less than 10 seconds exposure. Such good sensitivity opens up an enormous range of deep-sky ob-

Table 6.3
Field of View in Arcminutes with Various Imaging Systems with 640 x 480 Pixel Array

Telescope Type	Focal Length (mm)	5.6-micron pixels	7.4-micron pixels
ETX 70	350	35.2 x 26.4	46.5 x 34.9
ETX 90	1250	9.9 x 7.4	13.0 x 9.8
ETX 105	1470	8.4 x 6.3	11.1 x 8.3
ETX 125	1900	6.5 x 4.9	8.6 x 6.4
Short Tube 80	400	30.8 x 23.1	40.7 x 30.5
5-inch SCT	1250	9.9 x 7.4	13.2 x 9.8
8-inch SCT	2030	6.1 x 4.6	8.0 x 6.0
9.25-inch SCT	2350	5.2 x 3.9	6.9 x 5.2
11-inch SCT	2800	4.4 x 3.3	5.8 x 4.4
14-inch SCT	3556	3.2 x 2.4	4.2 x 3.1
Tak FS 102	820	15.0 x 11.3	19.9 x 14.9
Tak FS 128	1040	11.9 x 8.9	15.7 x 11.7
Tak FS 152	1216	10.1 x 7.6	13.4 x 10.0

Table 6.4
Resolution of Various Imaging Systems

Telescope	Focal Length	Visual	Resolution in Arcsec per Pixel for	
Type	(mm)	(Arcsec)	5.6 Micron Pixels	7.4 Micron Pixels
ETX 70	350	1.63	3.30	4.36
ETX 90	1250	1.27	0.92	1.22
ETX 105	1470	1.09	0.79	1.04
ETX 125	1900	0.91	0.6	0.80
Short Tube 80	400	1.43	2.89	3.82
5-inch SCT	1250	0.91	0.92	1.22
8-inch SCT	2030	0.56	0.57	0.75
9.25-inch SCT	2350	0.49	0.49	0.65
11-inch SCT	2800	0.41	0.41	0.55
14-inch SCT	3556	0.32	0.30	0.39
Tak FS 102	820	1.12	1.41	1.86
Tak FS 128	1040	0.89	1.11	1.47
Tak FS 152	1216	0.75	0.95	1.26

jects as potential targets. All the Messier and NGC objects were first discovered visually, thus a webcam is capable of recording virtually all of these thousands of targets.

Many of the brighter Messier objects are too large for the small field of view of a telescope-mounted webcam, but such targets can be imaged

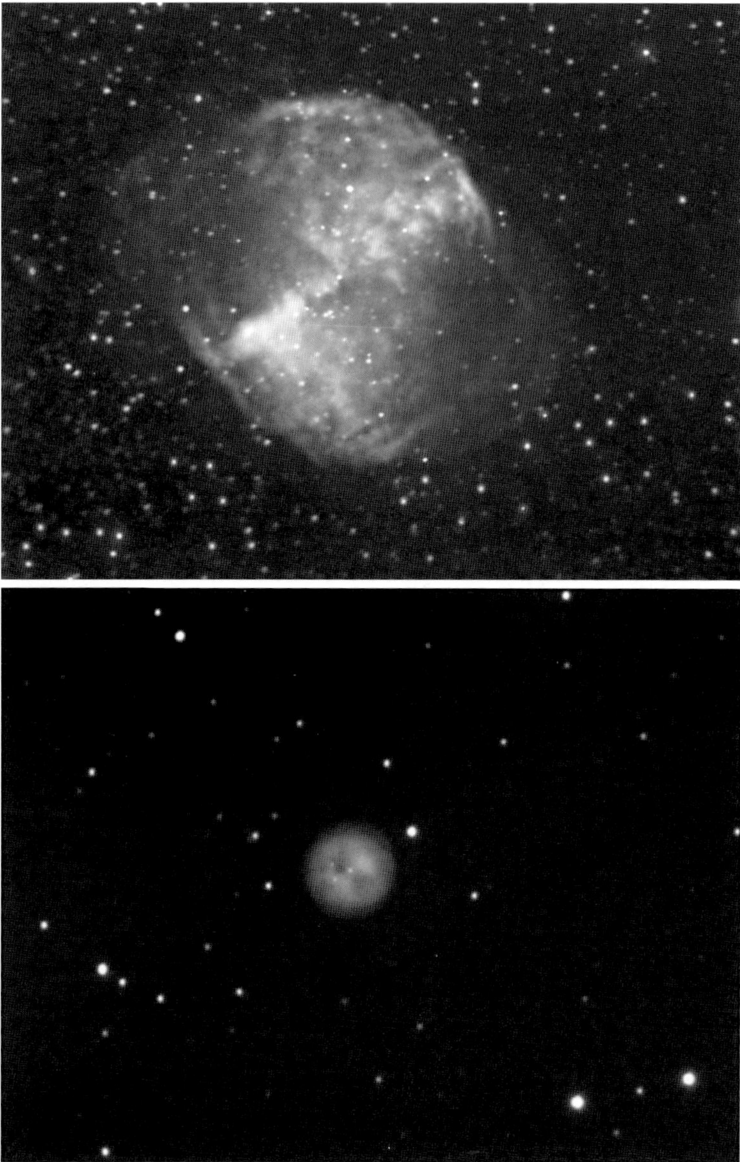

Fig. 6.13 *Planetary nebulae are a class of objects in which no two are alike. The Dumbbell Nebula, M27 (top), was imaged with a Meade DSI camera and 10-inch LX200GPS telescope. The Owl Nebula, M97 (bottom), was recorded using an ATK-2HS camera on an Orion 120ST refractor. Photos by Matt Taylor and Jim Ferreira.*

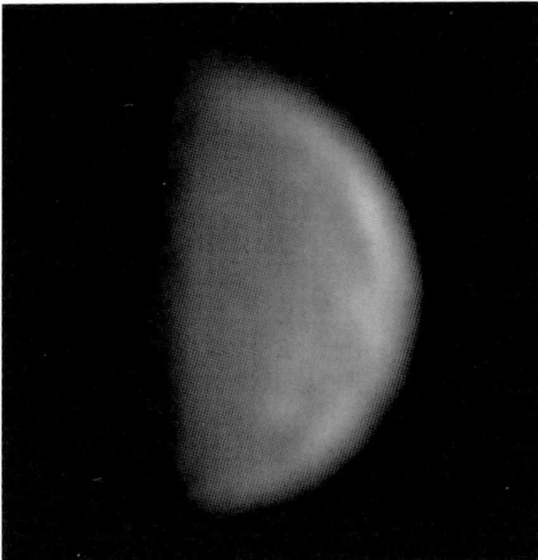

Fig. 6.14 *Mercury, an elusive target for the astrophotographer, was imaged using a 12-inch reflector at f/29 and a ToUcam Pro with a #25A red filter. The AVI sequence was captured at 10 frames per second with 1⁄100 second exposure. Photographer Zac Pujic states, "I consider my greatest achievement to be this image of Mercury. It shows a region called the Skinakas Basin and some possible crater rays (as bright spots)."*

using a moderately long telephoto lens. There are thousands of star clusters and galaxies that do fit the field of such cameras and offer long hours of rewarding astrophotography.

Here is a look at some of the targets available for the astrophotographer with a modified webcam:

- **Globular Clusters.** About 160 of these compact balls of stars populate our galaxy. Though many lie in the direction of the galactic center, others are scattered throughout the sky. Depending on their proximity to our side of the galaxy, globulars range from eyepiece-filling apparitions like Omega Centauri and 47 Tucanae to miniscule specks on the other side of the Milky Way. Some globulars are close enough that moderate telescopes will resolve individual stars within the cluster.
- **Open Clusters.** There are thousands of these groups of loosely-bound stars throughout the galaxy, although most are concentrated near the galactic plane. These gravitationally associated stars are less concentrated than globular clusters and are thus less intrinsically bright. The range of concentration of their component stars is what makes them interesting. Open clusters can be widely spaced and spectacular like the Pleiades (M45) or concentrated and granular, like M35. Often an open cluster resides within the emission nebula from which its stars were created such as NGC 661 and the Eagle Nebula (M16), and NGC 2244 and the Rossette Nebula.

- **Planetary Nebulae.** These are the gas shells ejected by dying stars. Although they are typically small compared to other objects, they usually have a high surface brightness. This makes them a good target for the small field of view of a telescopic webcam. Popular planetary nebulae include the Ring Nebula (M57) and the Dumbell Nebula (M27).
- **H II Regions.** Some of the major showpiece deep-sky objects are H II emission regions, which glow from the emission of ionized hydrogen and include targets such as the Orion Nebula (M42) and the Lagoon Nebula (M8). Although many are too large to fit within the field of a webcam, they make good targets when using a telephoto camera lens with one.
- **Galaxies.** The most plentiful deep-sky targets are galaxies. Over 50,000 are brighter than magnitude 15 and larger than 90 arcseconds. Galaxies come in numerous varieties and no two are alike. Besides large nearby examples like M31 and M33, distant and small galaxies span the range from "armless" elliptical galaxies, to shapeless irregular galaxies, to familiar disk-shaped spiral galaxies. The latter can display anything from an edge-on appearance to full face-on disks. Searching for supernovae is a good application for webcam galaxy imaging. Comparing a new image of a galaxy to one from last month will reveal any new bright apparition.

Both standard and modified webcams are among the best imaging tools available for solar system photography. The Sun, Moon, and bright planets are targets for standard cameras while modified long-exposure cameras will image the dim outer planets, asteroids, and comets. Some hints for planetary photography are:

- **Mercury.** The innermost planet may be fairly bright, averaging about 1st magnitude, but its proximity to the Sun means it is never more than 18 degrees away from the brightest object in the sky. The planet quickly sets in the evening or is rapidly lost in sunrise glare. Being inside Earth's orbit, Mercury displays phases like the Moon. The planet is fully illuminated only when it is on the opposite side of the Sun and thus at its smallest. When closest to Earth, the shadow side faces us. Mercury's inherent small size, averaging about six to eight arcseconds at quarter phase, makes it a difficult target. Compounding the problem is the large air mass at low elevations. Fortunately, Mercury is bright enough that it can be seen in daylight if the telescope is pointed properly. Use a red filter to help reduce sky brightness and image the planet when it is high in the sky to minimize atmospheric turbulence. When imaging Mercury during the day, take care to avoid accidentally pointing the telescope at the Sun. Even a brief flash of direct telescopic Sun will damage the camera.
- **Venus.** The brightest planet in the sky is also the most featureless. Venus is completely covered with bland atmospheric clouds so no surface markings can be seen. Like Mercury, it orbits the Sun inside Earth's orbit and

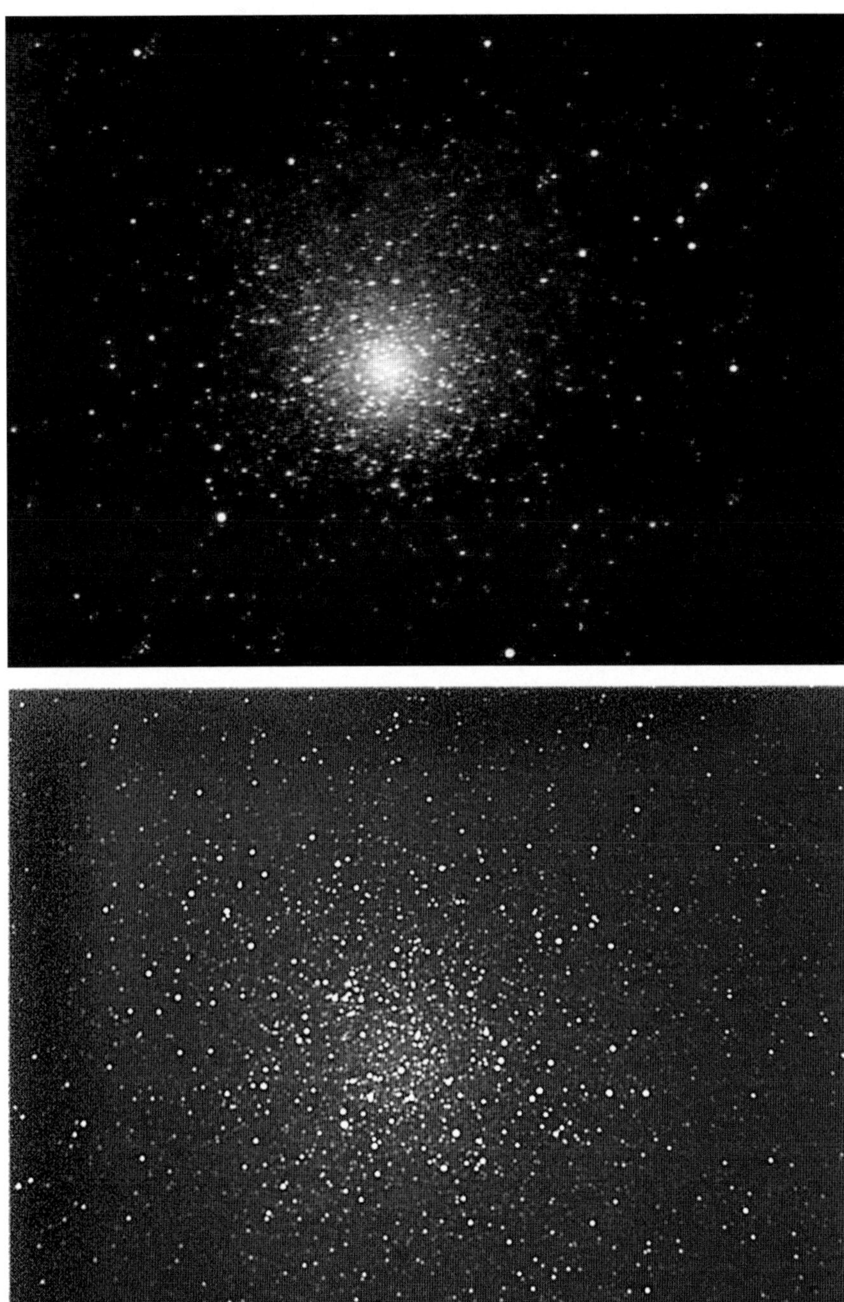

Fig. 6.15 *Globular clusters and open star clusters are bright enough to be imaged from urban areas. M3 (top) was recorded with a Meade DSI camera and 10-inch LX200GPS telescope, while M22 (bottom) was imaged with an ATK-2HS camera and Celestron-8 telescope operating at f/5. Photos by Dave Street and Robert Reeves.*

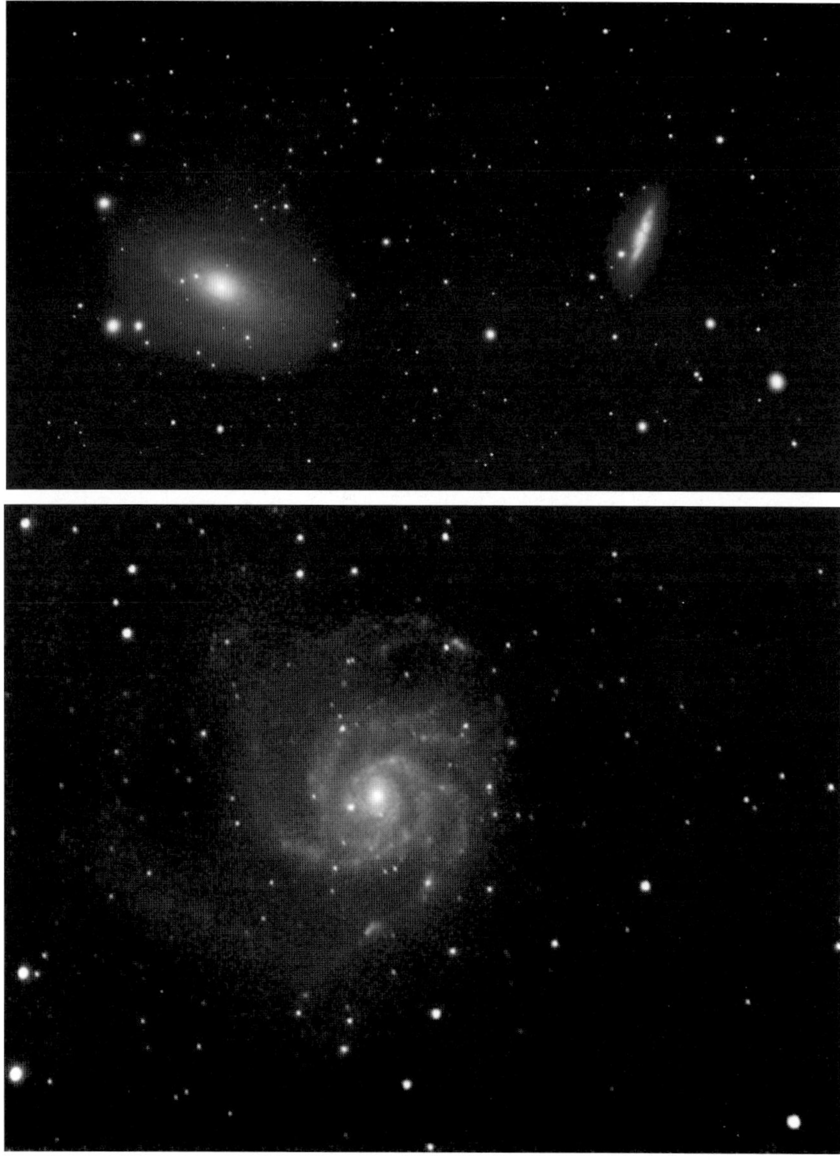

Fig. 6.16 *Galaxies supply a never-ending list of varied targets for the astrophotographer. These well-known spirals, M81 and M82 (top) and M101 (bottom), were all imaged with an ATK-2HS camera on an Orion 120ST refractor. Photos by Jim Ferreira.*

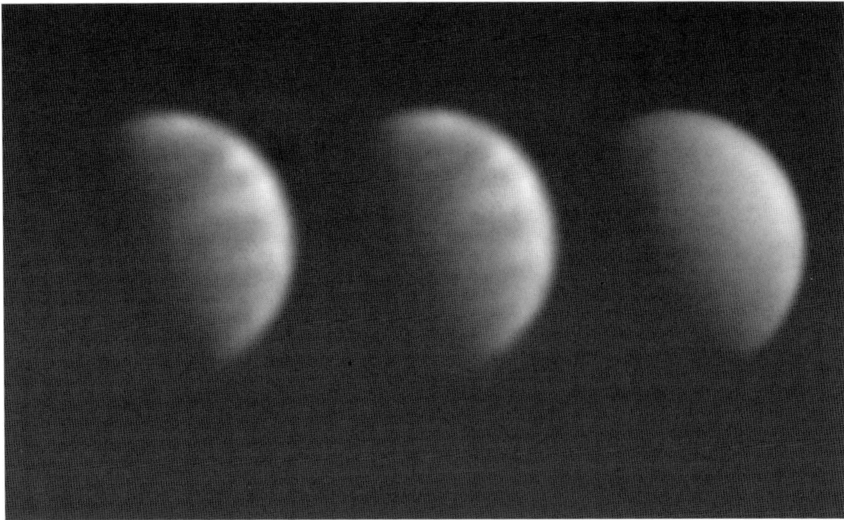

Fig. 6.17 *The planet Venus is visually featureless, showing only its ever-present cloud cover. Webcams are sensitive to the ultraviolet portion of the spectrum, thus using a UV pass filter that blocks visible light will reveal the cloud features of Venus (left), while the planet is featureless when imaging at longer wavelengths (right). Photo by Damian Peach.*

presents phases. At quarter phase the planet is a respectable 25- to 30-arcseconds in diameter. An ultraviolet pass filter such as the Venus-U or a violet filter such as Wratten #47 will accentuate any apparent patterns in the dynamic cloud structure around Venus. Like Mercury, it, too, can be observed in the daytime at high elevations. Take care not to overexpose its dazzling disk.

- **Mars.** It is a fact that more surface detail can be seen on Mars using a webcam than can be seen with the naked eye. I like to muse that if webcams had existed 100 years ago, Percival Lowell would have seen Mars in an entirely different light and science fiction and the thrust of our space program would have been an entirely different. Because Mars travels outside Earth's orbit, about every two years the two planets pass close to each other. During this time Mars can reach a diameter ranging between 14- and 25-arcseconds while being fully illuminated by the Sun. It possesses a thin atmosphere that, depending on the wavelength of light used, can affect how much surface detail is visible. Views can be quite pleasing without a filter, but a red or infrared pass filter will make surface features stand out more clearly. Because Mars rotates slightly slower than Earth, it presents only a slightly different face toward us at the same time each night, so about a month is required to view the entire Martian surface in this manner.
- **Jupiter.** The largest planet in the solar system presents a constantly changing target at least 40-arcseconds in diameter. Jupiter's rapid rota-

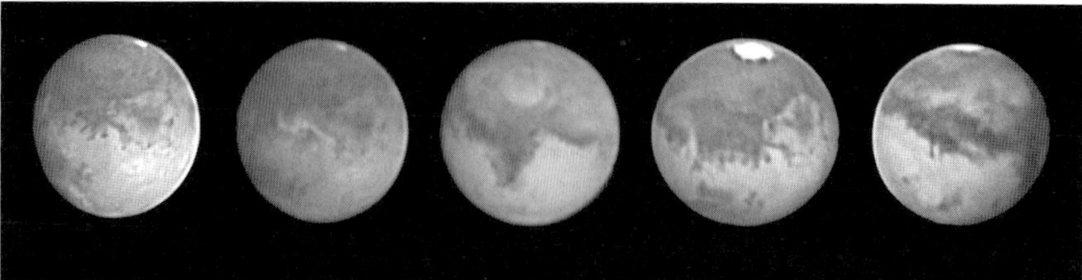

Fig. 6.18 *Approximately every two years, Mars reaches opposition and becomes a popular target for astrophotographers. The two left and center images were taken during the 2005 opposition with a ToUcam 840 and a 20-inch Obsession at f/20. The far left two images show the dramatic evolution of the 2005 Mars dust storm as it filled Vallis Marinaris. The center image shows the prominent feature Syrtis Major. The right two images were taken with a ToUcam 740 and a 12.5-inch reflector at f/25 during the 2003 opposition. Left three photos by Glenn Schaeffer, right two photos by Steve Foster and Frank Hinson.*

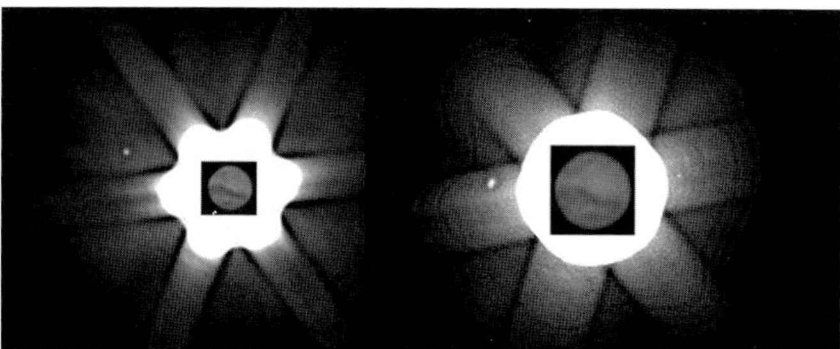

Fig. 6.19 *The Martian moons Phobos and Deimos were imaged using an Atik 1HS camera and a Takahashi Newton 250 reflector. Deimos (left) imaged with a 0.5 focal reducer yielding f/7.1 while Phobos (right) was imaged at prime focus at f/11.9. Both images are a stack of 400 separate ½-second exposures. A short exposure of the planet was superimposed on the overexposed image of Mars to show the actual distance scale between Mars and each moon. Photos by Paul Maxson.*

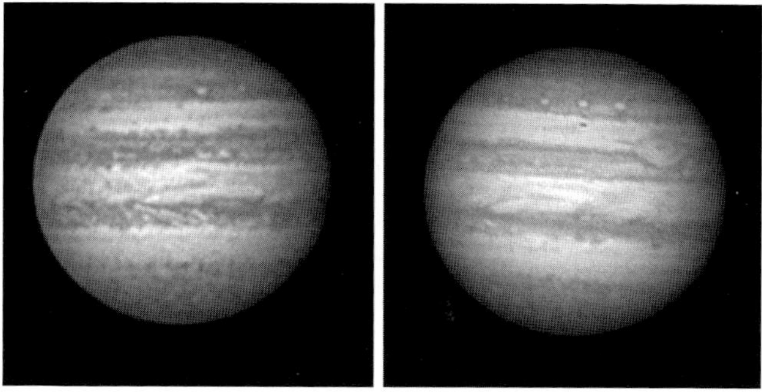

Fig. 6.20 *Even the "dull" side of Jupiter (left) opposite the Great Red Spot is an ever-changing mass of cloud belts. Jupiter's rapid rotation brings the Great Red Spot (right) into view almost every night. Photos (left) by Steve Foster and Frank Hinson, (right) by Glenn Schaeffer.*

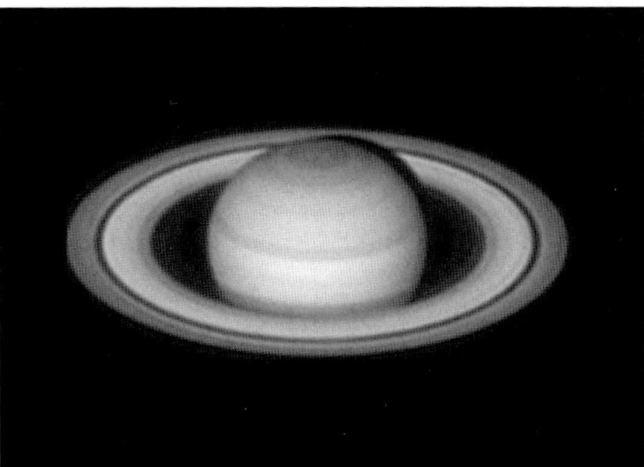

Fig. 6.21 *The rings of Saturn exhibit delicate detail that more than make up for the blandness of the planet itself. No matter what telescope and camera are used, Saturn's rings are always a delight for astrophotographers. Photo by Damian Peach.*

tion incessantly changes our view of the alternating light and dark cloud belts along with the Great Red Spot and other atmospheric storm features. In addition to a dynamic atmosphere, Jupiter also has its four bright Galilean satellites, which are good webcam targets in their own right. The satellites have an albedo half that of Jupiter itself, but they are easily imaged as their alignment with the planet constantly changes. Telescopes with long focal lengths can successfully capture detail on some of the moons in spite of their approximately one arcsecond size.

- **Saturn.** The magnificent 40-arcsecond diameter ring system around Saturn makes this planet an inviting webcam target. The atmosphere of Saturn is less colorful and changeable than Jupiter's but the ring system more than compensates. Even small telescopes, under good seeing, have little trouble imaging details such as the 0.4 arcsecond wide Cassini Division in the ring system. Saturn's largest moon, Titan, is also a small telescope webcam target while larger instruments have no trouble imaging half a dozen of the brighter moons with standard webcams.
- **Uranus, Neptune, and Pluto.** The outer planets, ranging from four arcseconds for Uranus to starlike for Pluto, are the realm of modified webcams. Shining at sixth magnitude and fainter, these targets need high magnification and more exposure than possible with unmodified models. With sufficient magnification, modified webcams can produce pleasing views showing the featureless globes of Uranus and Neptune, but no amount of magnification will show Pluto as anything other than a pinpoint among the background stars. Still, it is a "doable" target for modified webcams and imaging the distant planet will give the observer "bragging rights" for having recorded an object most people have never seen.

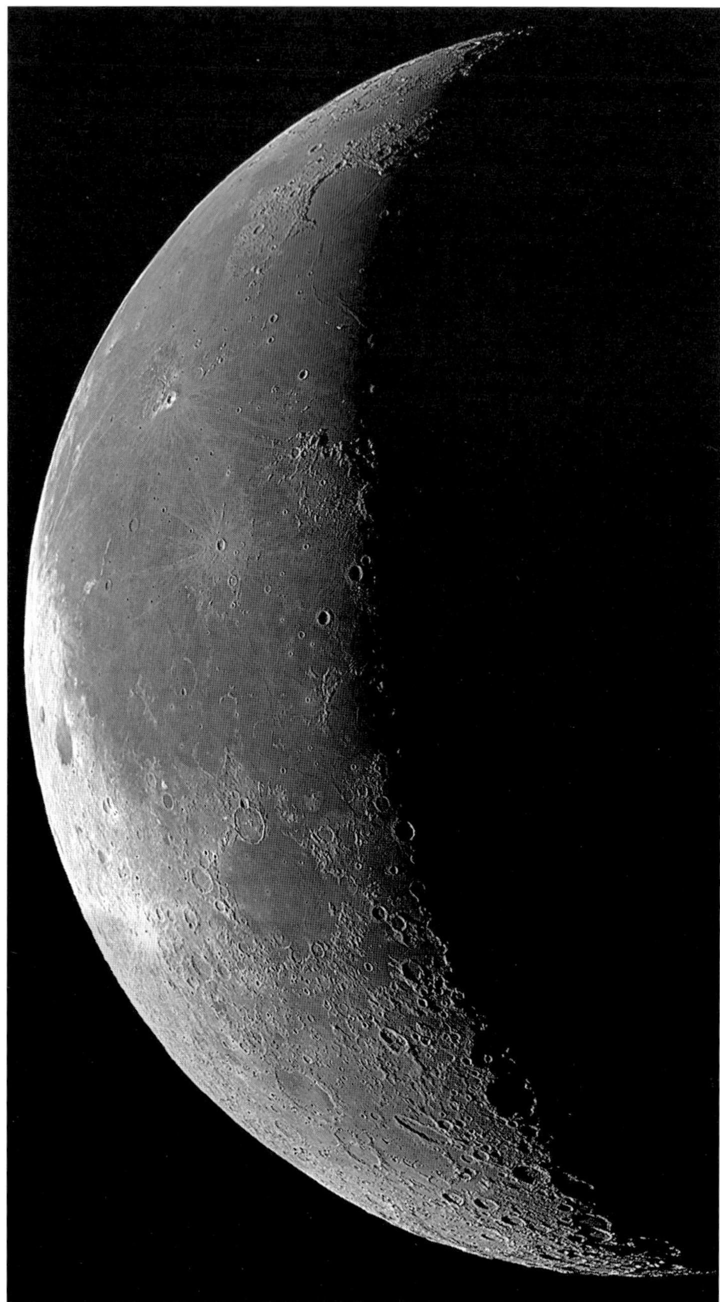

Fig. 6.22 *The Moon is a target astrophotographers can enjoy all year. How to create high-resolution wide-field mosaics of the Moon using many small, but highly detailed webcam images is explained in Appendix D. Photo by Robert Reeves.*

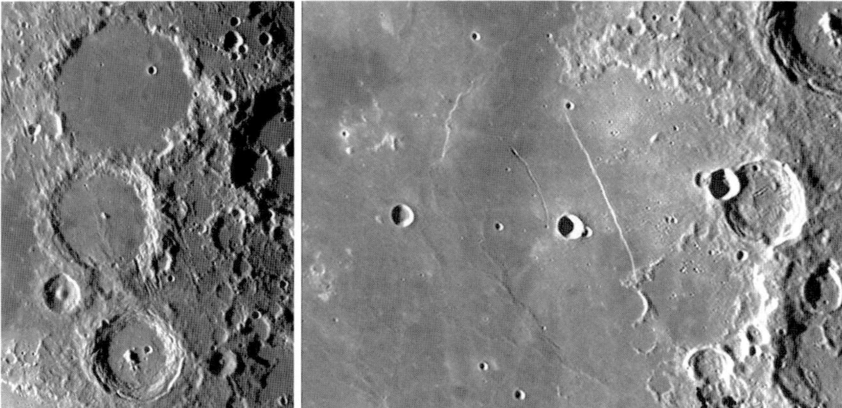

Fig. 6.23 *Just as deep-sky photography offers different classes of objects to enjoy, the Moon offers different types of geology for the astrophotographer. The above images were taken with an ATK-2HS camera and Celestron-8 telescope at f/20. At the left, a two-image mosaic shows heavily cratered terrain featuring the giant craters Ptolemaeus, Alphonsus, and Arzachel. At the right smooth lava plains are punctuated by the Straight Wall. Photos by Robert Reeves.*

6.12 Lunar Photography

The Moon is an easily found, frequently available target that can be imaged from bright urban locations. In recent decades, as their telescopes have become more powerful and deep-sky targets have become easily accessible, amateur astronomers have come to ignore the Moon and consider it merely a source of celestial light pollution blocking deep-sky views. However, the Moon presents the astrophotographer with an amazing display of topography that can keep him or her busy forever.

The Moon is essentially colorless, so images taken with a color ToUcam and a black-and-white Atik IIhs appear nearly the same shade. Most photographers simply convert color webcam output to grayscale in order to save image disk storage space. However, the Moon does possess subtle coloration in many areas and working with extreme color contrast stretching in an image processing program can reveal the difference in color shading between older and newer lava flows on large mare, or the differing mineralogy of deep areas exposed by cratering.

The brightness of the Moon varies with phase. The landscape of a fully illuminated Moon is about 20 times brighter than the same lunar area at a crescent phase. This is because at low Sun elevations, the hilly and cratered surface is rich in shadows. This presents the photographer with the interesting challenge of correctly exposing a celestial body than can be simultaneously harshly illuminated along the limb while also displaying extremely dim slanted illumination along the terminator between the Sunlit and dark hemispheres. High-resolution narrow-field imaging solves this problem by concentrating on such a small portion of the Moon that the exposure is uniform. However low-power images showing the entire lunar disk, as well as high-resolution multi-image mosaics showing the entire disk, need to be uniformly exposed to present an evenly illuminated body. The solution is to expose to reveal detail in the brightest portion of the Moon and let the dim terminator areas fade out naturally. Some detail along the terminator can be recovered through image processing, but if the brighter limb is overexposed, detail in that area is lost.

6.13 Solar Photography

WARNING: Solar observing is one of the few activities in amateur astronomy that is dangerous. If a solar filter falls off your telescope, your eye can suffer immediate irreparable damage. Unfiltered sunlight can also seriously damage optical and photographic equipment. Always insure that a solar filter is attached to the telescope in a fail-safe manner. The finderscope should also be filtered, capped, or removed. **Take extra care if children are present.**

The Sun is only about the same angular diameter as the Moon, but it is about a million times brighter. In addition, the solar spectrum contains harmful amounts of infrared and ultraviolet radiation that must be filtered out to prevent risk of personal injury or equipment damage. But solar safety is not expensive. Good full-aperture white-light filters are available for amateur telescopes from many manufacturers. Baader Planetarium Astro-Solar Safety Film, and Tuthill Solar-Skreen are both inexpensive and very effective Mylar-like filters for white-light viewing of the whole solar disk, or close-ups of sunspot regions. Orion and Thousand Oaks market full-aperture glass solar filters with a photographic density of 5.0x, which pass only 1/100,000 of incoming light. These filters offer views ranging from a pale yellow to a silvery white solar disk, but the Sun's features are essentially colorless and the choice of viewing appearance in a filter is a matter of personal preference.

Viewing the Sun in hydrogen-alpha light is a relatively new field for amateur imaging. Solar details such as prominences and flares that are invisible in white light are easily seen in the very narrow hydrogen-alpha wavelength centered at 656.28 nanometers. *Do not*, however, confuse nighttime light-pollution hydrogen-alpha filters with solar hydrogen-alpha filters. They are not the same. Solar hydrogen-alpha filters typically pass only an ultra-narrow 0.7 Angstrom of the spectrum while nighttime hydrogen-alpha filters pass everything from 640 nanometers to the infrared.

Coronado Instruments and DayStar both manufacture affordable solar hydrogen-alpha filters for amateur use. These instruments allow backyard astronomers to take incredible solar images that not too long ago were only the province of professionals. Webcams can be adapted to these devices and produce highly detailed views of eruptive events.

6.14 Verifying Fine Detail

The high planetary image resolution possible with webcam photography through amateur telescopes has allowed the recording of solar system features that were unknown to professional astronomers just a generation ago. Planetary imagers working under excellent seeing conditions with well-tuned telescopic and photographic systems now detect details on Mars, and even the Galilean moons of Jupiter, that were unseen until NASA spacecraft charted these far worlds.

For instance, Ganymede at opposition reaches a diameter of 1.7 arcseconds, well within the ability of amateur instruments to detect differences in surface albedo. A 12-inch *f*/10 telescope working at about *f*/35 will result in an image scale of 0.13 arcseconds per pixel (using 5.6 micron pixels) and produce an image of Ganymede about 13 pixels across. This is sufficient to reveal large dark areas on it such as Galileo and Nicholson Regio, and the bright ray pattern Osiris. Io is a smaller moon, never exceeding about an arcsecond in size, but the dark volcanic area known as Pele can be detected. Unfortunately Europa and Callisto do not have albedo features of sufficient magnitude to allow detection with small Earth-based telescopes.

Comparison with widely published planetary maps will reveal the nature of details in Mars photos, but the ability to discriminate detail on Galilean moons is new territory for amateur imagers. How does one prove the existence of detail on a world that never exceeds two arcseconds in angular diameter? Even without the assistance of professional sources such as the on-line JPL Planetary Simulator or NASA maps of the Galilean moons, it is possible to verify the existence of surface detail on Jovian moons. This is done by comparison of successive images taken one orbit later when a moon is at the same point in its orbit around Jupiter. The four Galilean moons are all locked in synchronous rotation around Jupiter, that is, they all have a rotation period equal to their orbital period and thus keep the same face toward their parent planet. Images taken one orbit later will be viewing the same area of the moon and should, under good conditions, show a similar albedo pattern. For instance, Ganymede orbits Jupiter every 7days, 3 hours, and 45 minutes. Images taken just over a week apart should

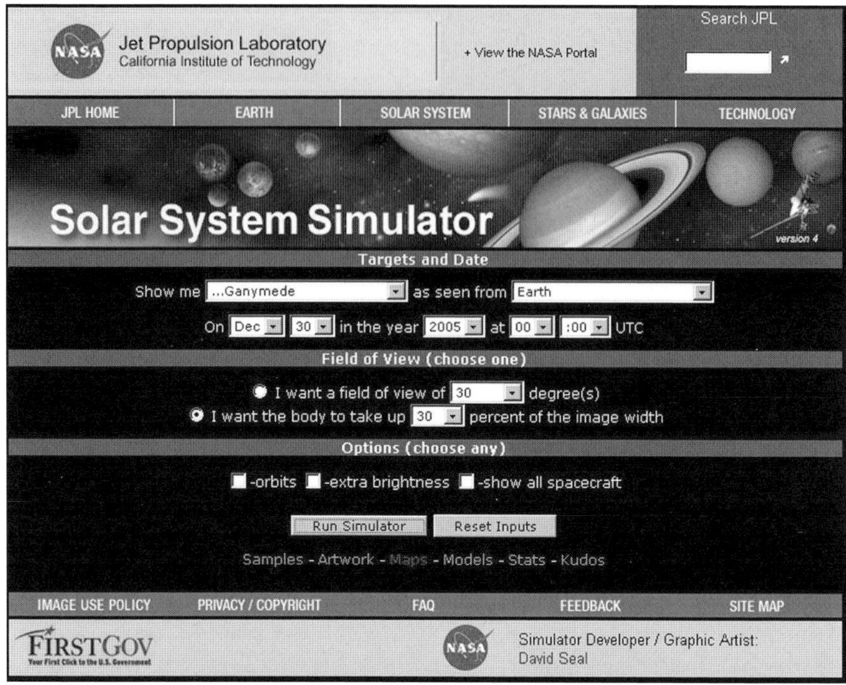

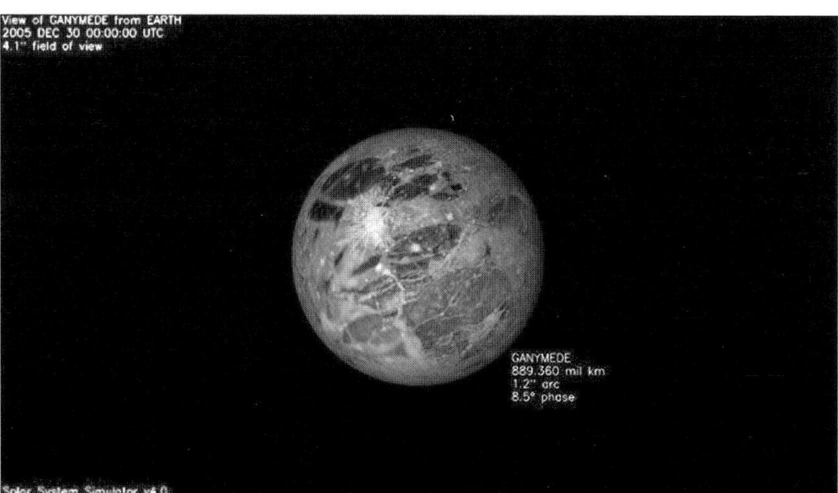

Fig. 6.24 *If your optical system is capable of resolving detail on tiny objects like the Galilean moons of Jupiter, you can test it by comparing your images with a simulated view generated by the JPL Solar System Simulator. Views of any solar system object, as seen from any location in the solar system, at a designated time and image scale, can be called up by entering the correct parameters. Screen shots by Robert Reeves.*

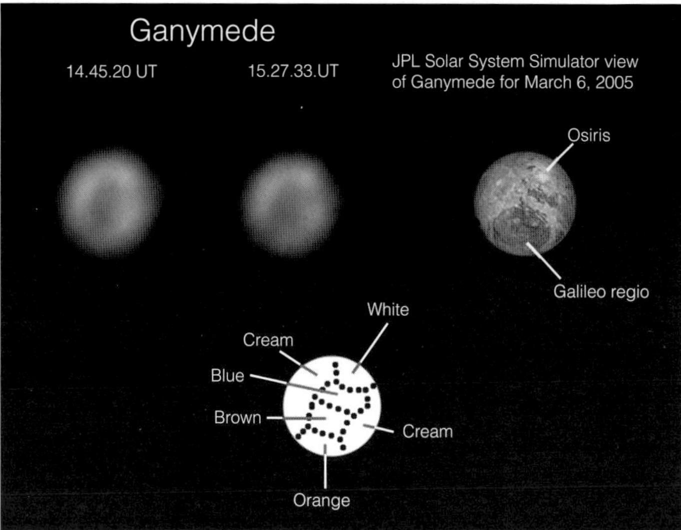

Fig. 6.25 *Zac Pujic from Brisbane, Australia, captured two successive images of Ganymede using a 12-inch reflector and a ToUcam Pro. The two images were taken almost an hour apart and show similar detail. Careful analysis of the coloration on the Galilean moon was compared to a view generated by the JPL Solar System Simulator. Without doubt, the images show the dark Galileo Regio and bright Osiris regions of Ganymede.*

thus show the same albedo pattern as the moon returns to the same point in its orbit. Repeatable albedo patterns have a great likelihood of representing actual surface detail.

When imaging tiny distant moons, be careful not to overly sharpen the images in programs such as RegiStax or Photoshop. Sharpening filters make details more visible by emphasizing differences in brightness values between pixels, but they also stretch noise into realistic-appearing planetary "detail." When imaging targets as tiny as Galilean moons, any noise can easily appear as spurious detail.

Once you have decided you have detected details on a Galilean moon, a good arbitrator for judging whether the detail is real is the JPL Solar System Simulator.[2] This is an on-line service that shows the appearance of any planetary body at a given time. Comparison of the predicted image with the actual image will show the reality of any detail. However, be aware that the simulated views are not created using visible light images so any colors may not match those of your view; but the areas of light and dark should match if the detail is real.

[2] http://maps.jpl.nasa.gov

Chapter 7
Webcam-Style Astrocamera Variations

7.1 Long-Exposure Modified Webcams

Modifying a webcam for long-exposure astrophotography, that is, turning an ordinary webcam into an entry-level CCD astrocamera, can be a rewarding challenge for those familiar with basic electronics and the techniques of working with circuits and soldering small components. Figure 7.1 shows some of the results that can be achieved. Such alterations are traditionally called "SC" modifications, so named for Steve Chambers, the person who pioneered the modification of webcams for long-exposure use. Chambers offers the following designations for the various types of modifications performed on webcams:

SC1	Basic long-exposure modification
SC1.5	Basic long-exposure modification with amp-off switch
SC2	Basic long-exposure modification with amp switch and separate interlaced frame download
SC3.1	Replacement of CCD with a Sony ICX424
SC 3.2	Replacement of CCD with a Sony ICX424 and basic long-exposure modification
SC 3.3	Replacement of CCD with a Sony ICX424 and basic long-exposure modification with 2x1 on-chip binning
SC 4.1	Replacement of CCD with a Sony ICX414
SC 4.2	Replacement of CCD with a Sony ICX414 and basic long-exposure modification
SC 4.3	Replacement of CCD with a Sony ICX414 and basic long-exposure modification with 2x1 on-chip binning.

SC-modified cameras and commercially-modified cameras such as those available from Atik require additional hookups to the computer through serial or parallel ports to control the long shutter times as well as the USB interface to download image data. Make sure the computer destined for use with such a camera can provide the necessary interfaces. Some newer laptops do not have such hardware. USB hardware that emulates a parallel port will be required in such a case.

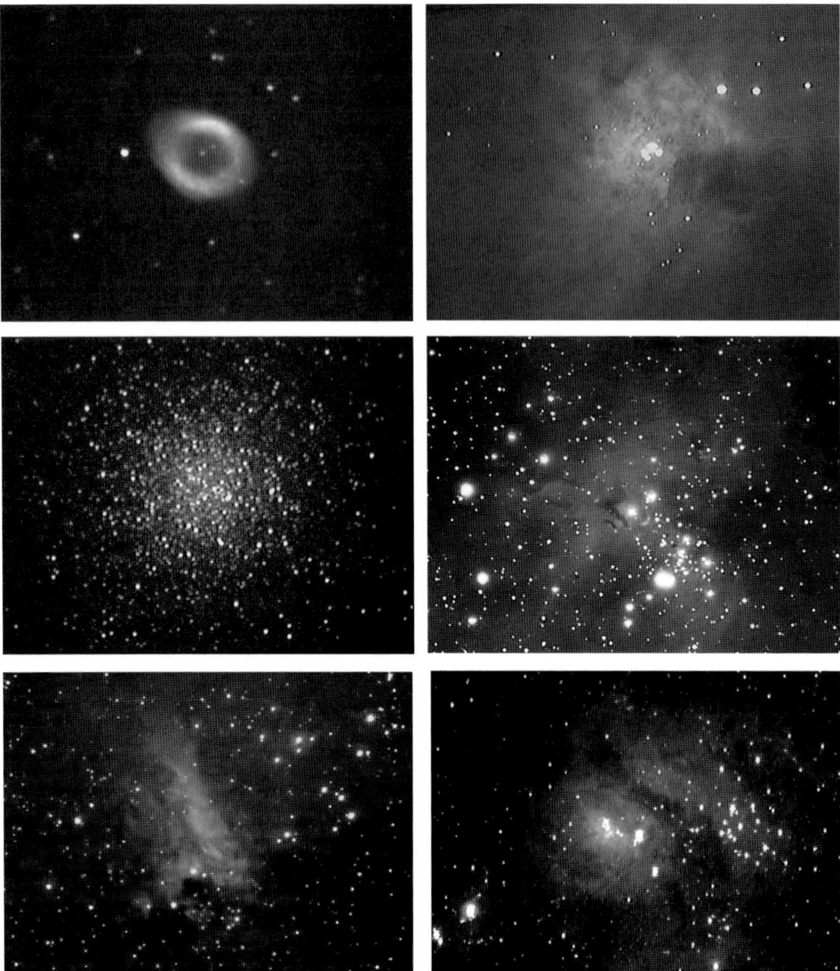

Fig. 7.1 *These images were taken with several versions of "SC"-modified long-exposure capable Philips ToUcams. The M57, M42, and M13 images were captured in color using a basic long-exposure modified camera mounted on a 10-inch LX200 telescope. M16 and M17 were captured using an SC3.2 camera incorporating a Sony ICX424 black-and-white CCD shooting through a Burgess Optics 127 f/8 refractor. M8 was taken with the same SC3.2 camera through a StellarVue AT1010 refractor. M16, M17, and M8 views were originally color images created by combining black-and-white images taken through red, green, and blue filters. Photos by Ashley Roecklein.*

Modified cameras are traditionally cooled in some fashion to reduce thermal noise in the sensor and thus increase the signal-to-noise ratio. This is important when imaging deep-sky objects where dark frame subtraction is being applied. Because of the cooling devices, webcams that have been modified often look radically different than their original commercial version. There is a difference between air-cooling and Peltier-cooling a camera. Air cooling uses a fan to constantly circulate air through the camera to remove heat generated by the camera's electronics. Peltier-cooling directly cools the CCD sensor itself, preferably 20 to 30° F below ambient, to reduce thermal noise in the sensor. Peltier cooling should only be applied to long-exposure capable cameras.

Cameras capable of long-exposure modification using the Steve Chambers technique are:

- PCA645VC
- Philips Vesta PVCK 675K, 680K and 690K
- ToUcam Pro PCVC 740K and 840K
- SPC900NC
- Logitech QuickCam 3000Pro, 4000Pro and VC parallel port version.

Which camera is best is a subjective concept because few people have access to all types of webcams for a direct comparison. Examination of websites that feature work done with various models will help, but any decision should be tempered by close examination of the telescope used to produce the resulting image.

The procedure for performing SC modifications to webcams is beyond the scope of this book. Those who wish to perform such electronic tinkering will find detailed instructions and illustrations for modifying most popular cameras through a quick Internet search. If you do pursue an SC webcam modification, some basic electronics "work rules" to observe during the process:

- When working with exposed electronic components such as a printed circuit board (PCB) full of voltage-sensitive components, always use an appropriately grounded wrist strap. This is very important on cool dry days when static buildup is a certainty. If you don't think this can happen to you, remember how you were "zapped" when you touched you car door handle on a very dry day?
- The soldered connections needed to modify the electronics on a webcam PCB are very small, measured in fractions of an inch. Practice the soldering techniques on a scrap PCB before trying to work directly on "live" camera circuits.
- Always verify a soldered joint using a volt-ohm meter (VOM). Never rely on visual inspection to insure that the soldered connection is making clean contact. A "cold solder" joint, one that has a frosty, dull surface instead of a bright silvery surface, is a sure candidate for a poor connection. But even if the solder joint is bright and silvery, verify it by testing its continuity with a VOM. The tiny connections needed on a webcam PCB

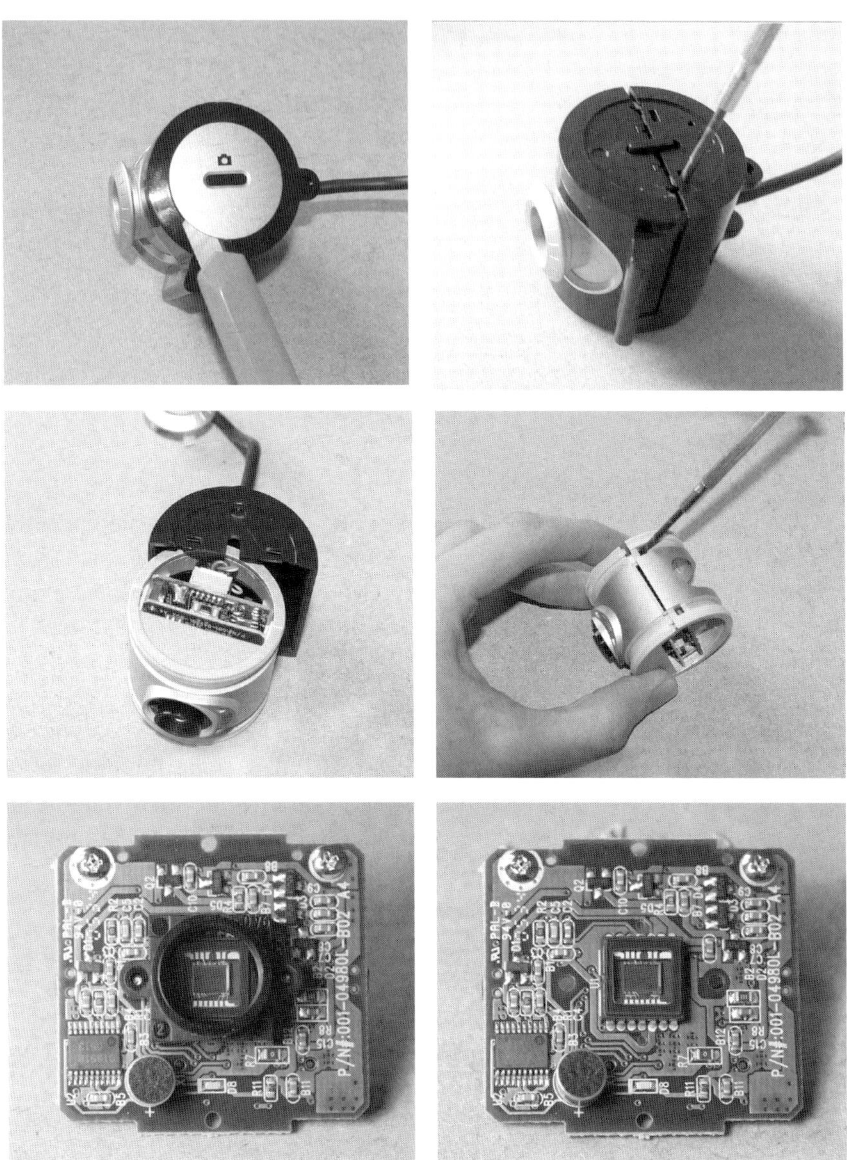

Fig. 7.2 *Opening the case on a Philips SPC900NC to perform "SC"-long-exposure modifications is much more complex compared to other lightweight webcams. Shown here are the steps to access the camera's electronics. The side covers are pried off with a flat tool and the outer case is pried apart. The inner case is extracted, revealing the two printed circuit boards of the camera. This case is then pried apart and the circuit boards extracted. The front board contains the threaded lens-mounting collar (lower left). With the lens collar removed, the Sony ICX098BQ CCD sensor can be seen in the middle of the circuit board. Photos by Matthias Meijer.*

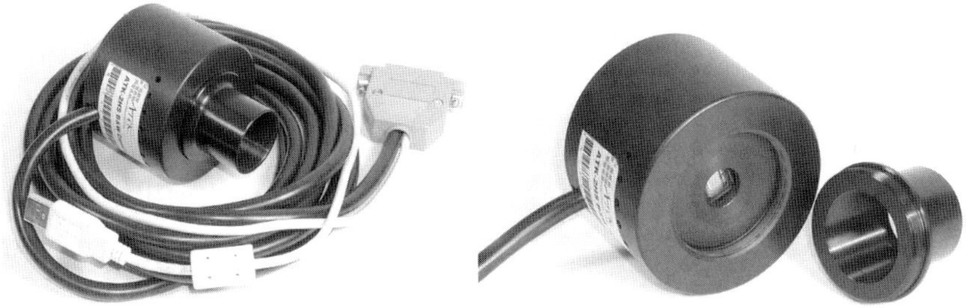

Fig. 7.3 *The Atik 1 and 2 series cameras are based on a commercially modified Philips ToUcam and have a small fan to provide forced air cooling to lower image noise by reducing the CCD temperature. A 12-foot cable allows convenient working distance from the telescope. Camera operation and image download are through the USB cable while long-exposure shutter control is via the parallel cable. Removing the telescope adapter nosepiece from an Atik 1 or 2 model (right) reveals T-threads, allowing any T-thread accessory to be attached to the camera. Photos by Robert Reeves.*

may require sharpening the VOM probes to a fine point in order to make contact with the trace on either side of the connection. Lead probes can be sharpened using a bench grinder or electric Dremel-style grinding tool.

Not everyone has an understanding of electronic circuits or is skilled in electronics fabrication. Fortunately, for those of us who lack such knowledges, there are a number of long-exposure-capable webcams or webcam-style astrocameras available from companies like Atik, Celestron, Meade, Orion, and SAC. And best of all, the cost of these products is not prohibitive.

7.2 Atik

The Atik ATK-1C "one-shot" color camera is a modified ToUcam Pro II 840K. Before it was replaced by the similar-performing SPC900NC, the stock out-of-the-box ToUcam was perhaps the most sensitive and widely used webcam for high-resolution lunar and planetary imaging. As reincarnated in ATK-1C, the camera is electronically modified to "SC 1.5" specifications to allow long exposures, has the "amp-off" feature (which limits noise buildup during long exposures) and has a built-in cooling fan to limit thermal noise, a feature very useful on warm summer nights. This model also comes with a basic version of the widely praised K3CCD Tools camera-control and image-processing program. The electronics assembly is remounted into an astronomically friendly housing that fits into a standard

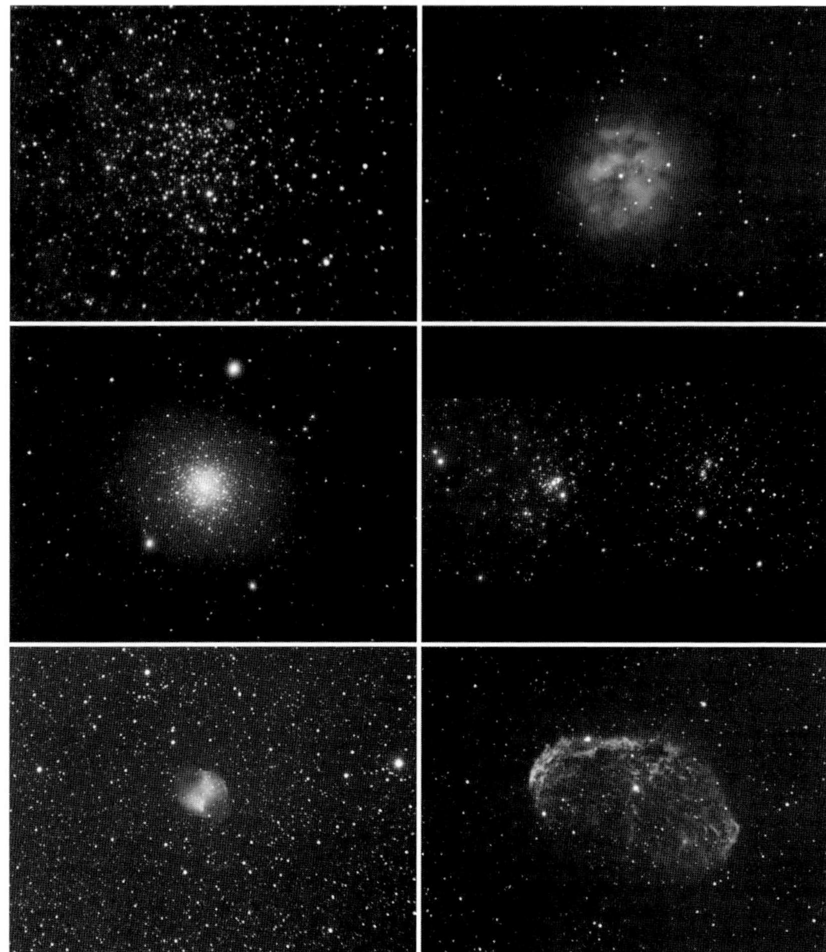

Fig. 7.4 *These images were captured using a black-and-white ATK-2HS camera. The M46 star cluster with the planetary nebula NGC 2438 (upper left) was imaged using 20 x 35 seconds exposure through an Orion 120ST refractor operating at f/2.5. IC-5146, the Cocoon Nebula (upper right) was taken using an Orion 80ED refractor and 20 x 5 minutes exposure. The globular cluster M3 (middle left) was captured using an Orion 120ST refractor operating at f/3.5 and 20 x 30 seconds exposure. The Double Cluster (middle right) was imaged using the same Orion 120mm refractor. Exposure data were not recorded. The Dumbbell Nebula, M27, (lower left) was taken using 20 x 150 second exposures with an Orion 80ED refractor operating at f/3.75. The Crescent Nebula, NGC 6888, (lower right) was imaged using the Orion 80ED at f/6 and an H-alpha filter with 20 x 5 minute exposures. Photos by Jim Ferreira.*

1¼-inch eyepiece holder. The ATK-1C costs about twice as much as the standard ToUcam, but it allows the "electronically-challenged" to not only immediately begin imaging deep-sky objects with a very capable CCD camera, but also to use the same device for short-exposure webcam photography of solar system objects.

The Atik ATK-1HS II is based on the same design as the ATK-1C, but the color sensor has been replaced with a Sony ICX098BL monochrome CCD. The 640 x 480 array, 5.6-micron pixel size monochrome CCD provides the same field size and resolution as the one-shot color camera, but with significantly increased sensitivity. Since the CCD does not have a Bayer mosaic RGB filter, each pixel receives all wavelengths of light instead of a single color. The ATK-1HS II also utilizes the same 12-foot parallel cable interface used with the color camera. Its USB interface piggybacks from the final 18 inches of the cable, allowing a single cable to extend from the camera head. The cable remains very flexible even in cold weather. Also like the color camera, the monochrome camera's 1¼-inch nosepiece can be unthreaded to either reveal female T-threads.

The Atik ATK-2C follows the "SC 3" modification architecture and provides a 55 percent larger field of view by using a Sony "one-shot" color ICX424AQ CCD. The larger sensor has the same 640 x 480 pixel array, but the larger 7.4-micron sensors allow higher sensitivity. The camera's construction and software controls remain the same as with previous Atik cameras. The Atik ATK-2HS is similar, but utilizes the monochrome Sony ICX424AL CCD with 7.4-micron pixels.

As someone who has done film astrophotography for over 40 years, I am impressed with the performance of the ATK-2HS. I qualify this endorsement by saying that my personal experience is only with the ATK-2HS, so I cannot comment on other cameras from first-hand use. I would expect the performance of models from other makers that use similar imaging sensors to be equally impressive, however.

7.3 Celestron NexImage

Celestron's official website states, "Celestron's NexImage Solar System Imager is remarkably similar to the Philips ToUcam Pro, at least on the inside." While not admitting that the Celestron NexImage Solar System Imager is actually a ToUcam 840 in a different housing, the specifications for the camera seems to bear out this assumption and operationally it is the same as a ToUcam. This is good for the astrophotographer who desires an unmodified short-exposure webcam for lunar and planetary work because the venerable ToUcam 840 is now out of production. The NexImage camera is thus a viable replacement and has the advantage of already possessing a 1¼-inch telescope adapter nosepiece, an accessory that must be purchased separately with the ToUcam. Celestron also offers the standard webcam accessories needed for solar system imaging, such as a focal reducer and UV/IR filter, both of which thread directly into the barrel of the eyepiece adapter, and a 2x Barlow for increased focal length.

Unlike the Meade Lunar and Planetary Imager which needs USB 2.0 and a hefty amount of computer horsepower to make the camera run, the NexImage runs nicely on a 333Mhz Pentium II processor and Windows 98SE and above. It comes bundled with AmCap video capture software that has a "look and feel" very similar to the traditional ToUcam software. The popular video processing program RegiStax is also bundled with the camera, and the NexImage instruction manual contains an excellent tutorial explains its operation.

7.4 Meade Lunar and Planetary Imager

After the ToUcam 840, perhaps the most popular webcam-style imaging devices are the Meade Lunar and Planetary Imager (LPI) and Deep Sky Imager (DSI). The entry-level LPI is not a modification of an existing webcam, but is instead a new 640 x 480 pixel camera built from the ground up to be user-friendly and easy for a novice to operate. Experienced webcam imagers who have also used the LPI generally agree that an "SC" modified webcam will outperform it. One of the primary reasons for the webcam's better performance is the fact that it uses a CCD sensor while the LPI uses a CMOS sensor. While it is true that the CMOS sensor used in conventional digital cameras has evolved into a sensor that is excellent for celestial photography, the cheaper CMOS sensors used in webcams are inferior to CCDs for astrophotography. A standing joke in the astronomical webcam community is that CMOS does *not* stand for "Catch More Outer Space."

Side-by-side comparison between the LPI and a basic ToUcam 840 shows that the Meade camera has about half the sensitivity of the latter. However, unlike an unmodified ToUcam, the LPI can expose from anywhere between 1/1000 to 16 seconds, offsetting the inherent lack of sensitivity. This allows the LPI to perform better as an autoguider. Meade claims that stars as dim as 10th magnitude can used, but does not state what kind of telescope allows this level of performance. Users report that with an 8-inch telescope, autoguiding is possible with at least magnitude 7.5 stars.

The LPI lacks a lot of the control flexibility that is available with some webcams, but the fact is that many novices can use it to quickly achieve good pictures. The Meade instrument also has the advantage of

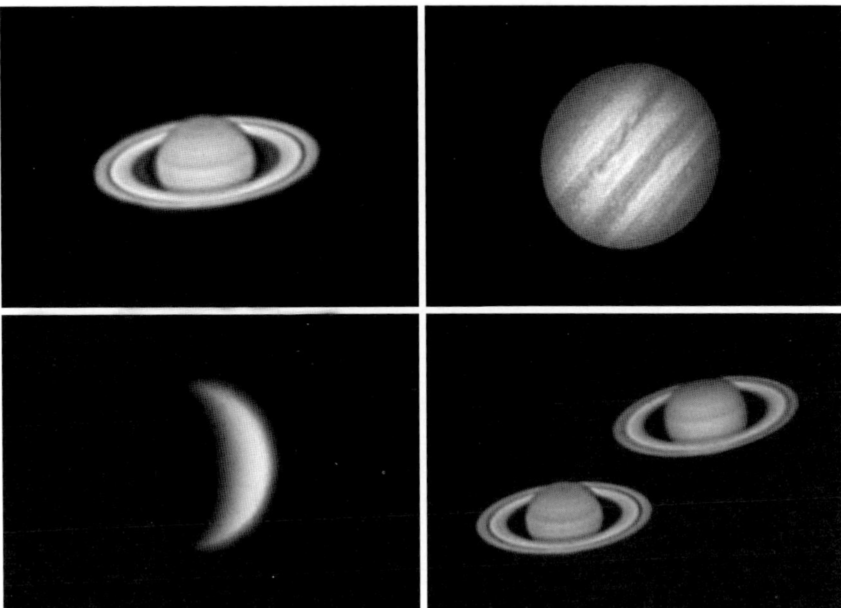

Fig. 7.5 *These images were taken through a 10-inch LX200GPS telescope using the Meade Lunar and Planetary Imager. At the upper left, Saturn was imaged by stacking 254 frames taken at f/10. Jupiter was photographed at f/20 using a 110-frame image stack. Venus shows a crescent phase in a 16-frame stack imaged at f/10. The dual Saturn images at the lower right compare the Meade LPI and DSI cameras. Both images were taken at f/18 and each combines 60 frames. Photos by Dave Street.*

being ready to use right out of the box (after software installation) while a webcam requires a telescope adapter. Users will however want to replace the restrictively short three-foot USB cable with a longer one.

The LPI will work with non-Meade software and many users prefer to operate the camera with the popular K3CCD Tools software. The Meade Autostar Suite software will produce results, but its operation is geared for entry-level beginners. Experienced users will find that the versatility and image data displayed with programs such as K3CCD Tools allow much more control of the process than that provided by what has been called "black box" processing. The Meade software does have a helpful feature in its "Magic Eye" focus indicator. Twin orange triangles on a graphical display move together as focus improves, with red lines indicating maximum sharpness. Another plus is that the software incrementally names image files so none are overwritten by later image sequences. Most users prefer to process the camera output in RegiStax, but the Autostar Suite image-processing options allow "track and combine" so the software will

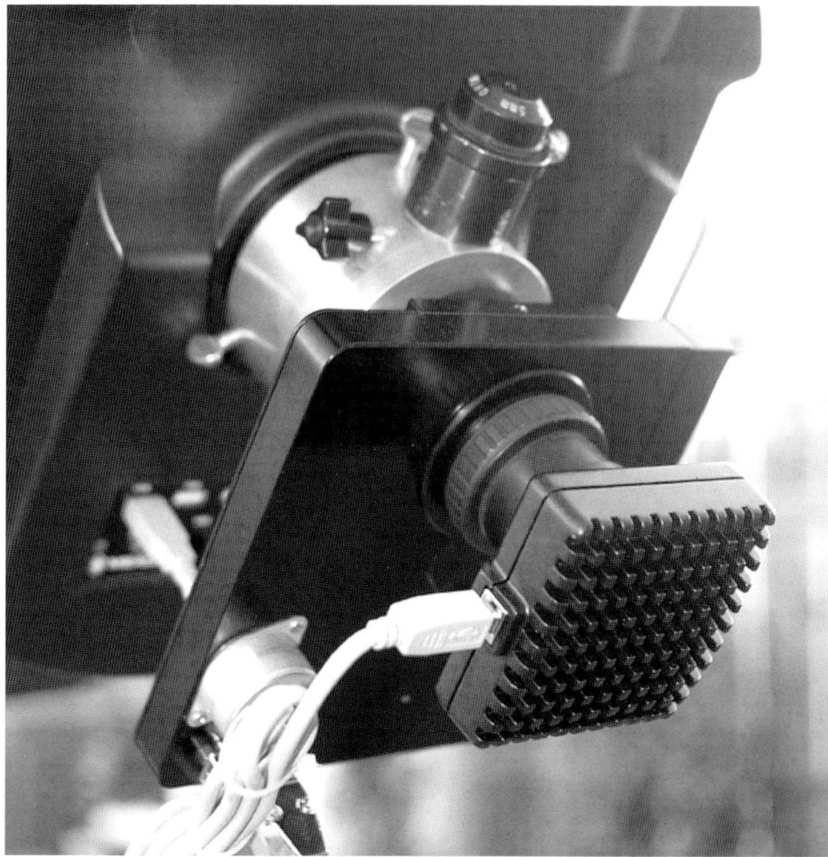

Fig. 7.6 *The standard Meade Deep Sky Imager is a "one-shot" color camera. It is shown here mounted behind an OPTEC IFW filter wheel on a Meade 16-inch SCT. Notice the "waffle-iron" casting pattern on the camera back designed to increase heat radiating area to help cool the camera. Photo by Jason Ware.*

follow a planetary image as it drifts across the field.

Using both a ToUcam and an LPI on the same computer will lead to some quirks in software. Running the ToUcam will make the computer forget the LPI and the next time the LPI is run, Windows will display the "New Hardware Found" message and search for drivers. While the installation disk is not needed, the reinstallation does cause a delay. Using a ToUcam after the LPI does not cause this problem. Another quirk is that a USB mouse cannot be used at the same time the LPI is used.

7.5 Meade Deep Sky Imager

For imaging deep-sky objects, Meade chose to create two separate cameras with greater capabilities than the LPI. The result was the Deep Sky Imager (DSI) and the Deep Sky Imager Pro (DSIP). Both utilize USB 2.0, and unlike modified long-exposure webcams, no additional serial or parallel cable connections are required to operate them. The DSI provides "one-shot" color imaging with a 510 x 492 pixel array Sony color CCD featuring pixels 9.6 microns high and 7.5 microns wide. The unit has 16-bit per channel color output, resulting in 48-bit color images. The DSIP utilizes a monochrome CCD with 2.3 times the sensitivity of the one-shot color version. A color filter slider containing parfocal 1¼-inch infrared, red, green, and blue filters enables color images to be made. The filter housing spaces the camera away from the telescope focuser, so parfocal eyepieces used for rough focus and target centering will need an eyepiece extension tube.

The DSI is capable of exposures between 1/10,000 second and one hour, although lengthy exposures will be hampered by noise. While the camera is not actively cooled, its cast aluminum body has a "waffle-iron" style back that acts as a heat sink to help reduce sensor temperature through convection cooling. The camera should be set up in the open for about 15 minutes prior to imaging to allow it to cool to ambient temperature. The 10-ounce DSI can also be used as a capable autoguider.

The Autostar camera-control and image-processing software allows successive images to be automatically stacked, averaged, and dark-frame subtracted for on-the-fly image processing. A live onscreen histogram allows tailoring the exposure for best results. A very useful software feature with the DSIP is the ability to de-rotate successive images. This allows image sequences taken with an altazimuth telescope to be stacked without displaying field rotation.

Another interesting feature is the software's ability to control a Meade Autostar-capable telescope while imaging a target. This option allows the camera to take a seamless mosaic of an area four times larger than a single view. The telescope is commanded to make small incremental movements between exposures, which are then stacked to slowly create a larger 1280 x 960 pixel image.

The usability of the DSI is somewhat hampered by its need for adequate computer resources. Windows XP must be upgraded with Service Pack 2 or the DSI software will not run due to driver recognition problems. USB 2.0 is essential, and a powerful computer is necessary. Meade's Autostar Suite software opens and executes slowly with a P-III 700 Mhz machine. A display mode of at least 1024 x 768 is also needed as important control buttons on the Autostar screen will be off the screen with 800 x 600 displays. Modern computers with 2 GHz processors and at least 512 Mb of RAM are capable of operating several DSIs at the same time, allowing one

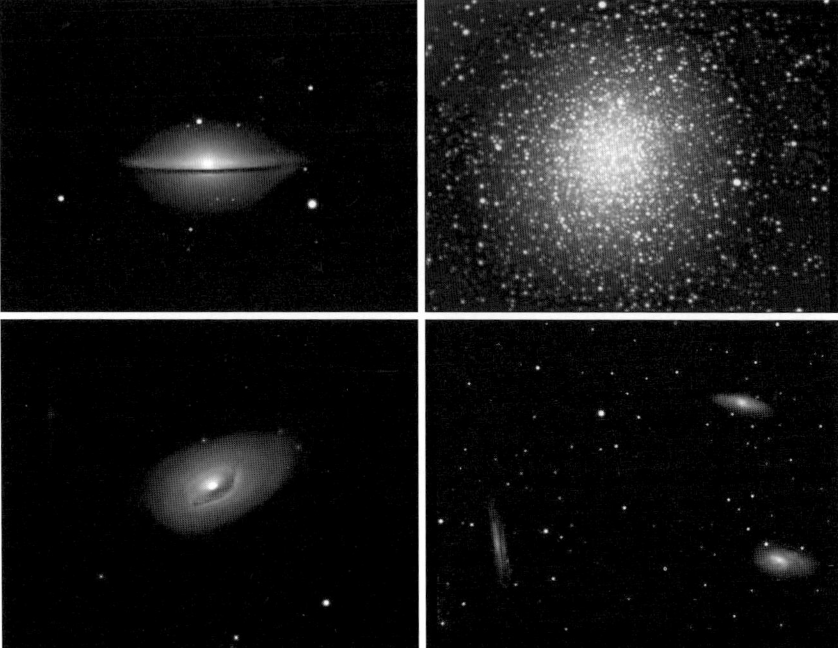

Fig. 7.7 *The three galaxy images above, M104 (upper left), M64 (lower left), and M65, M66, and NGC 3628 in Leo (lower right), were taken with a DSI camera on a 10-inch LX200GPS telescope. The Omega Centauri star cluster (upper right) was imaged with a DSI on a Celestron 80mm refractor. Exposure data were not recorded. Photos by Dave Street.*

to image while the other acts as an autoguider.

Image noise increases with elevated sensor temperature. Experienced users have recognized that the monochrome DSIP is sensitive to temperature changes of as little as 3°F. If the ambient temperature rises or falls by more than 3° during an imaging session, it changes the noise signature of the sensor, so additional dark frames should be taken. The "one-shot" color DSI does not seem to be as temperature sensitive.

Meade offers a clip-on battery-operated fan that will keep the camera about 5°C cooler than without it. This temperature drop will cut image thermal noise by almost half.

Matt Taylor, a Meade products beta tester, developed the Outback Cooler, an active cooling system for the DSI. The Outback Cooler is not a Meade product, but is actually manufactured by Steven Mogg, the well-known supplier of webcam telescope adapters. Taylor did not develop it specifically to reduce the temperature of the DSI sensor, although the device does this quite well. The Outback Cooler was originally developed

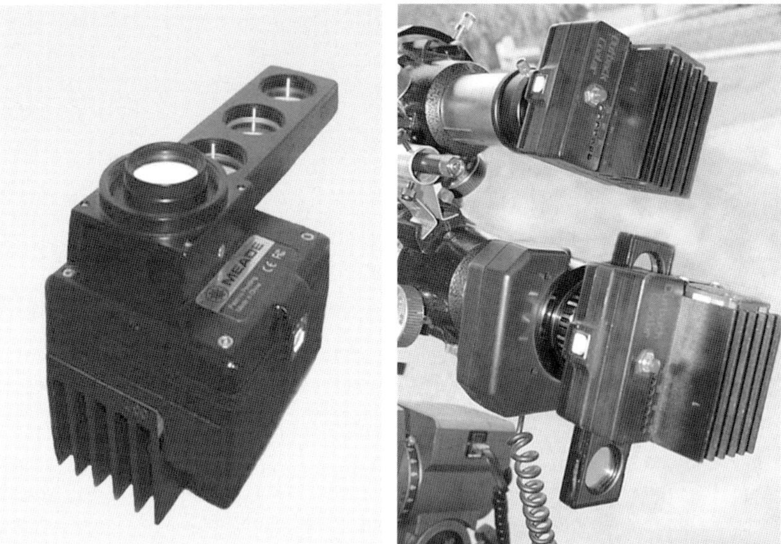

Fig. 7.8 *The Meade SDI Pro (left) has a more sensitive black-and-white CCD and uses color filters to create color images. The unit is shown with the Outback Cooler installed in place of the standard "waffle-iron" camera back. Both the Meade DSI used as an autoguider (right) and DSI Pro have Outback Coolers installed to increase the thermal stability of the cameras. Photos by Matt Taylor.*

Fig. 7.9 *These two 30-second dark frames taken with an Outback Cooler-equipped DSI camera have been equally stretched to emphasize image noise. The image on the left was taken at ambient temperature and displays large amounts of noise. That on the right was taken with the cooler turned on and displays significantly less noise. Photos by Matt Taylor.*

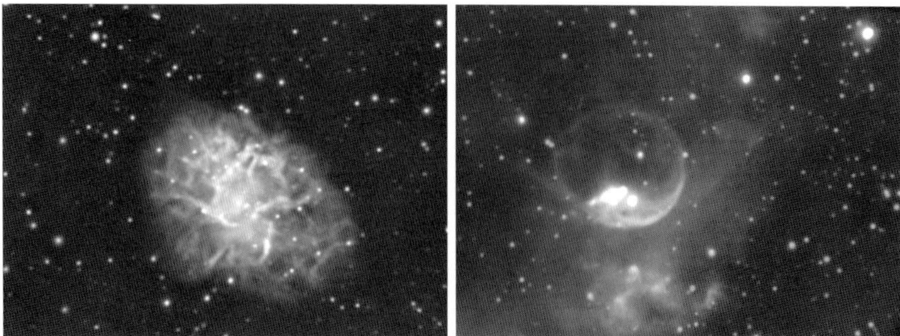

Fig. 7.10 *M1 and NGC 7635 were both imaged using a Meade DSI camera equipped with an Outback Cooler to lower the sensor temperature and shot through a 12-inch LX200GPS with a focal reducer, producing an f/3.52 optical system. M1 is a combination of 10 x 8 minute exposures taken with a DSI camera and 5 x 8 minute exposures taken with a DSI Pro camera. The Bubble Nebula, NGC 7635, is a combination of multiple color channels exposed with a DSI Pro camera using 100 x 1 minute exposures for luminance, 25 x 3 minutes in H-alpha, 25 x 3 minutes in O-III, and 25 x 3 minutes each in red, green, and blue for a total of 475 minutes exposure. Photos by Matt Taylor.*

to stabilize the sensor's temperature so as to minimize the need for repeated dark frames as the temperature changed.

The Outback Cooler replaces the original "waffle iron" back cover on either the DSI or DSIP and uses 12 volts DC to cool the CCD to 15°C below ambient temperature. The Cooler typically draws 2 amps of power and will stabilize the temperature to within 0.25°C. The device adds only 250 grams of weight to the DSI and will not affect telescope balance. A single button on the Outback Cooler cycles through preset temperature settings and controls the fan. LED status lights display the desired and achieved temperature settings.

7.6 Orion StarShoot Solar System and Deep Space Cameras

Orion Telescopes, one of the leading aftermarket suppliers of astronomical accessories, offers two versions of their new StarShoot astronomical camera. Both were designed in cooperation with SAC Imaging, a long-recognized name in electronic astronomical cameras. The StarShoot Solar System Color Imaging Camera (SSCIC) is a webcam-style model utilizing a ⅓-inch progressive-scan 8-bit CMOS sensor with a 640 x 480 pixel array. The StarShoot Deep Space CCD Color Imaging Camera (DSCIC) uses a cooled Sony ⅓-inch ExView HAD ICX259AK color 16-bit CCD, and while the camera is based on a "one shot" color webcam-style design, it performs more like a true astronomical CCD camera. The DSCIC provides a 752 x 582 array with 6.5 x 6.25 micron pixels. The resulting 4.9 x 3.6 millimeter imaging area provides a larger field of view than most webcam-style imagers, typically 8.4 x 6.25 arcminutes with a resolution of 0.65 arcminute per pixel on an 8-inch *f*/10 telescope. It is recommended that telescopes of less than 1000 millimeters focal length (with focal reducer if necessary) be used to more effectively utilize the camera's inherently narrow field of view. The new Orion cameras offer affordable high-performance imaging systems from a recognized astronomy dealer.

Both new Orion cameras connect to a computer using only a USB cable; no parallel cable is needed for long-exposure control. The DSCIC also has a separate 12-volt power line dedicated to the thermoelectric cooling system, but it is not mandatory for camera use, and indeed, the cooler may not be needed on cold nights. Both cameras will operate on either USB 1.1 or 2.0 systems, although the latter is recommended. Both cameras will be slower with USB 1.1 and the SSCIC may revert to 320 x 240 resolution. Unlike the CPU processor power-hungry Meade cameras, the StarShoot models will operate on any Pentium computer with Windows 98SE or above as long as USB 2.0 is available. The included MaxIm DL Essentials camera-control and image-processing software does benefit from increased computer RAM, however.

Both Orion cameras feature a removable infrared filter that mounts behind its removable 1¼-inch nosepiece and any 1¼-inch eyepiece filter can be installed.

Orion warns that the StarShoot cameras require 0.8-inches (20 millimeters) of additional inward focuser travel from the point at which standard eyepieces reach good focus. This is because the imaging sensors are recessed within the camera bodies. A parfocal ring is supplied with each model to allow establishing coarse focus with a 10mm eyepiece before installing the camera on the telescope.

The SSCIC allows exposures ranging from 1/1000- to ½-second with frame rates as high as 30 per second with the appropriate shutter speed.

The DSCIC offers an exposure range from 1/500-second to 9.3 hours. Such an extended exposure is impractical, but the thermoelectric cooling does allow longer multi-minute exposures than possible without it. It draws about 0.3 amps during operation while the 12-volt thermoelectric cooling system requires a separate power source and draws an average of ½ amp per hour. The Deep Space camera is heftier than the Solar System camera and at 14 ounces, it may require some telescope rebalancing. With appropriate control cables, the Deep Space camera can be used as an autoguider.

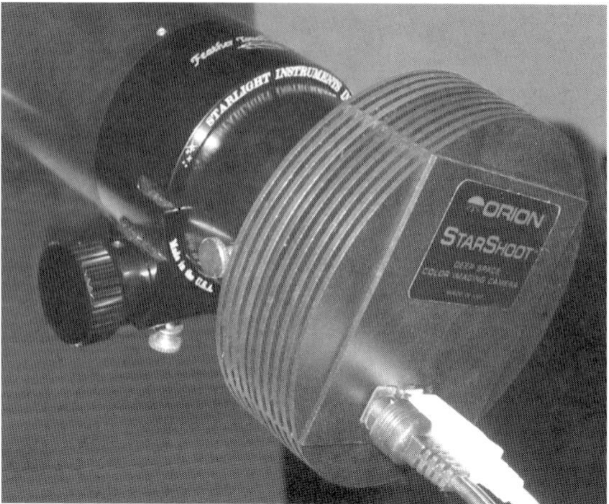

Fig. 7.11 *Orion Telescopes has entered the imaging field with their new StarShoot series of cameras. The Deep Space Color Imaging Camera shown here is an affordable "one-shot" internally cooled model that produces results competitive with other webcam-style long-exposure capable astrocameras. Photo by Craig Stark.*

7.7 SAC IV and SAC8.5 Cameras

Cameras made by SAC Imaging are not modified webcams, but are discussed here because they fall under the loose category of webcam-style cameras. The popular SAC IV and SAC8.5 cameras, for instance, are designed with proprietary electronics unique to the camera model.

The SAC IV solar system camera is unique in possessing a much larger imaging sensor than standard webcams. The ½-inch Micron MT9M001 CMOS color sensor has a 1280 x 1024 pixel imaging area utilizing 5.2-micron pixels and performs on-chip 10-bit A/D conversion. This gives the sensor a large 6.82 x 5.45-millimeter imaging area creating 1.3 megapixel images—or greater than four times the imaging surface available from standard 640 x 480 pixel webcams. The sensor's 0.3 lux light sensitivity coupled with exposure speeds as long as ½ second, or frame speeds up to 30 per second, allow it to be a good planetary camera.

The SAC IV comes bundled with AstroVideo image-capture and processing software. It requires USB 2.0, Windows 2000 or above, and minimally a 600 Mhz, and preferably a one gigahertz processor. Although the imaging area is over 400 percent more than a "normal" webcam, the SAC IV utilizes a low-power-consumption CMOS sensor and needs only 325 milliwatts of power when in use.

The SAC8.5 camera, which evolved from the SAC8 long-exposure

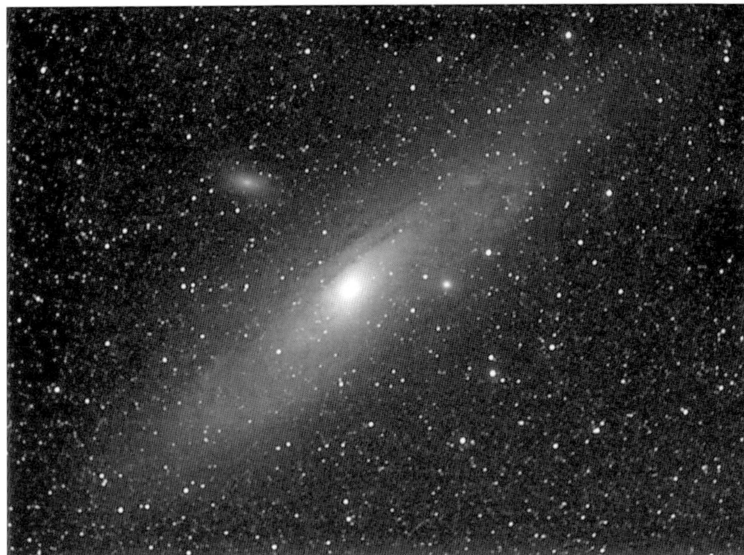

Fig. 7.12 *In his "first-light" effort with the Orion Deep Space Color Imaging Camera, Craig Stark coupled it to a 22mm f/4.5 aperture Leica camera lens and captured 20 three-minute exposures of M31 to create this image of our galactic neighbor. No dark or flat frames were used. Photo by Craig Stark.*

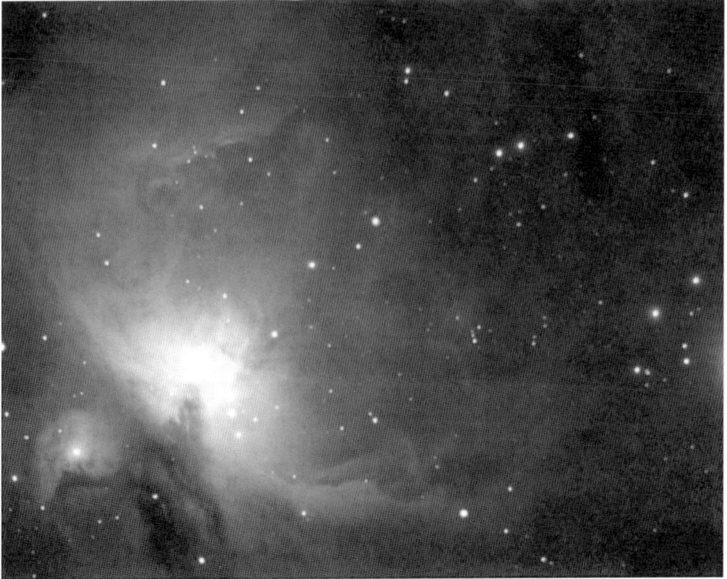

Fig. 7.13 *M42 was imaged with an Orion StarShoot Deep Space Color Imaging Camera mounted on a Williams Optics ZenithStar 66mm Petzval f/6 refractor. 40 unguided 30-second exposures were stacked to produce the final image. Photo by Craig Stark.*

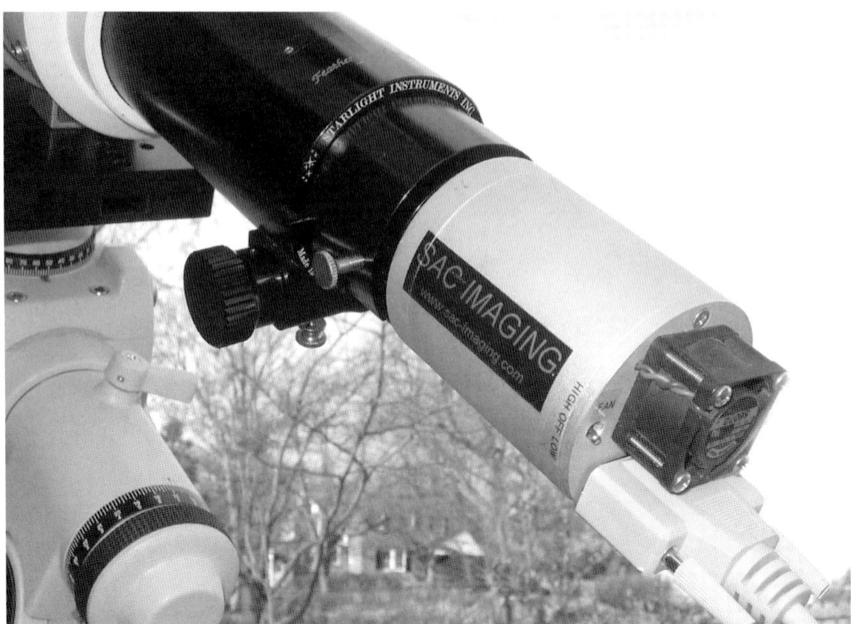

Fig. 7.14 *SAC Imaging produces a series of webcam-style cameras that are not actually modified webcams. These models use proprietary electronic circuitry and although some, such as the SAC8 shown here, use webcam-style CCD sensors, they are designed to be true astronomical cameras. Photo by Craig Stark.*

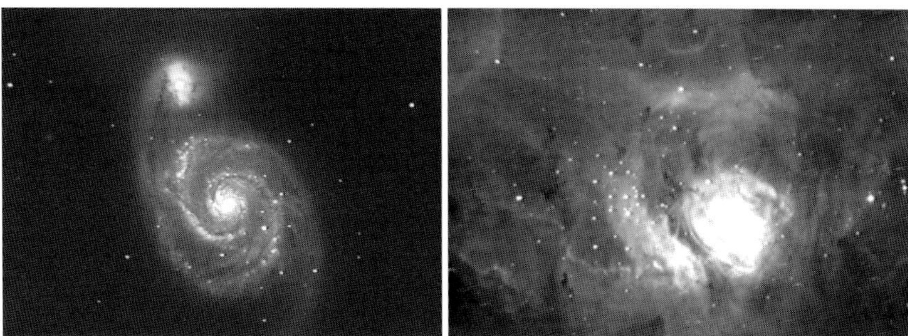

Fig. 7.15 *(Left) The Whirlpool Galaxy, M51, was imaged using a SAC-8 camera on a 10-inch LX200GPS at f/6.3 using a SAC focal reducer. A combination of 30 x 40 second and 10 x 40 second exposures was used to create this image. Photo by Tim Tasto. (Right) The Lagoon Nebula, M8, was imaged using a SAC8 camera and 0.6 focal reducer with an H-alpha filter on an Orion 80ED refractor. Six exposures, ranging from three to six minutes each were stacked to create this image. Photo by Dave Street.*

camera, now uses only a USB connection; a parallel cable is no longer needed for exposure control. A separate 12-volt connection is needed to power the sensor's Peltier cooler. The camera uses a Sony ⅓-inch ExView

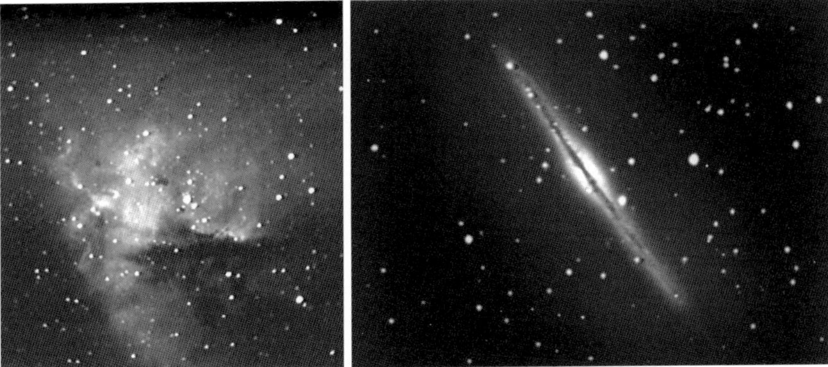

Fig. 7.16 *(Left) The Pac-Man Nebula, NGC 281, was imaged using a SAC8 camera and a SAC 0.6 focal reducer with an H-alpha filter on an Orion 80ED refractor. Three image stacks exposed at 10 x 8min, 5 x 10 minute, and 4 x 15 minute were combined to create the final image. Photo by Dave Street. (Right) The edge-on galaxy NGC 891 was imaged using a SAC8 camera on a 10-inch LX200GPS using a SAC 0.3 focal reducer. 50 separate 30-second exposures were stacked to create this image. Photo by Tim Tasto.*

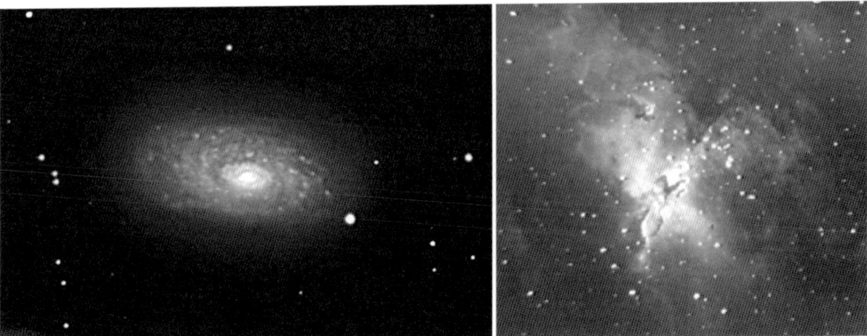

Fig. 7.17 *(Left) The intricate Sunflower Galaxy, M63, was imaged using a SAC8 camera at 320 x 240 resolution on a 10-inch LX200GPS telescope with a Meade f/3.3 focal reducer. 35 separate 30-second exposures were stacked to create the final image. Photo by Tim Tasto. (Right) The Eagle Nebula, M16, was imaged using a SAC8 camera with a 0.6 focal reducer and H-alpha filter on an Orion 80ED telescope. Six exposures, ranging from three to six minutes, were combined to create the final image. Photo by Dave Street.*

HAD CX254AL black-and-white 8-bit CCD sensor. The sensor has a 640 x 480 array, but possesses larger 9.6 x 7.5 micron pixels. The larger pixels translate into increased sensitivity, with the camera capable of detecting a 0.0003-lux illumination level. The Peltier-cooled sensor is capable of exposures ranging from 1/1000 second to unlimited time. Output can be JPG, BMP16, 32-bit FITS, or AVI video. Like the SAC solar system camera, the SAC8.5 requires USB 2.0 and Windows 2000 or higher, and the camera is also compatible with MAC OSX. A 20-foot USB cable is provided, one of the longest supplied with a commercial astrocamera.

Accessories for the SAC8.5 include a filter wheel, allowing the rapid change of RGB or CYM filters for color imaging, and a 0.5 focal reducer. The Steven Mogg supplied reducer can be converted to a 0.3 reducer with the addition of an extension tube.

Chapter 8 Image Processing

It only takes a few minutes to capture video of a planet, the Moon, or a deep-sky object, but it can take hours to process it into a satisfactory image. It is common for me to take nearly twice as long to process an image as it did to set up the scope and capture the raw video. This is the nature of digital and webcam astrophotography.

Our goal in any kind of digital astrophotography is to maximize the signal-to-noise ratio to make the most pleasing picture. In previous chapters we have seen how noise can contaminate digital images. The best way to deal with noise is to cool the sensor (usually with a fan on a webcam) and average multiple frames. In this chapter we will discuss the processes used to turn a webcam's video output into a finished picture. But first some advice: when you begin your webcam adventure do not get carried away with the heavy details. Learning how to get every available photon, and to squeeze the last bit of data out of a video sequence will come with experience. The joy of the webcam is that you can achieve good results quickly. Your first efforts may not be superb, but they will be a success that you can build on without getting bogged down with trying for perfection right from the start. Experience is the best teacher; master the basics and your interest level will lead you too more advanced techniques.

8.1 The Laws of Astronomical Image Processing

In their excellent book, *The Handbook of Astronomical Image Processing*, Richard Berry and James Burnell discuss the "laws of astronomical image processing." These laws are not the mathematics or methodology of the procedures of image processing, but an acknowledgment of its realities. Berry and Burnell's basic laws are simple in their own right, but they lead to profound possibilities as well as limitations when we try to process the images we digitally capture at the telescope. Although these laws were derived while performing CCD imaging, they are equally applicable in the field of webcam astrophotography. I think it is important that people entering digital imaging through this portal should also be exposed to them.

Before we recount some the Berry and Burnell laws, we need to understand the basic principles of image processing. These include:

- Image processing is not a miraculous process, but is instead the application of rules of mathematical numeric manipulation.
- Image processing involves discarding some forms of image information in order to enhance other forms of image information.
- Image calibration is needed to eliminate the noise signature of the imaging sensor from the final image.
- Brightness scaling is performed to eliminate unwanted sky background and excess star brightness in order to more effectively display galaxies and nebulae.
- Unsharp masking and deconvolution are applied to eliminate unwanted blurry portions of the image that hide desired detail.
- Image information is always lost during image processing and enhancement. The final result may be more pleasing to look at, but will have less raw data.

The first of the Berry and Burnell laws is that astronomy is photon-limited. In a nutshell, if there are insufficient photons arriving at the imaging sensor, the target will not be discriminated from the sky background or sensor noise. The nature of the astronomical target, atmospheric conditions, the size and optical configuration of the instrument used for imaging, and the response characteristics of the imaging sensor all come into play and at some point limit one's ability to acquire the desired image.

Their second law is that artifacts happen. Artifacts are imperfections in the image caused by outside factors such as dust on the sensor, vignetting in the optical system, optical imperfections and focusing errors, reflections within the optical system, electron noise in the sensor . . . the list goes on and on. Long-exposure webcam astrophotography is more prone to artifacts than lunar and planetary imaging and we strive to suppress or eliminate artifacts during processing by calibrating our images with dark frames. But the fact remains that any image may have many artifacts in spite of our best efforts to eliminate them. One should learn to recognize artifacts and not mistake them for reality.

Law number three is never trust one image. This law is not important if the object is to simply get pretty pictures. In fact, liberties and license are often taken in the processing of images destined *only* for visual enjoyment. However, if we are doing science or deliberately seeking something new, the accuracy of what is recorded on the image becomes very important. Digital images are prone to many kinds of artifacts that mimic astronomical objects. A cosmic ray strike, sensor noise, or even misapplied image processing can create realistic apparitions that do not appear on star charts.

There is always the possibility of something new appearing in the sky, but the watchwords are corroboration and verification. If it appears on two consecutive images, the probability of reality increases, but this is still not concrete proof of existence. If it appears in two images taken by separate instruments, the probability of reality increases significantly to the point of seeking verification from other sources.

The fourth law states that image processing always discards information, no matter if its goal is to do science by position or magnitude measurement, or to simply provide pictorial information about the image. The very nature of calibrating a digital image by subtracting a dark frame changes it, hopefully for the better. Further processing, or enhancement, throws away more image information in order to increase the visibility of other image information. Contrast stretching and brightness scaling alter image information, while deconvolution, unsharp masking, and image scale resampling throw away or distort other image information. If done properly, the resulting images appear more pleasing to the eye, but they no longer have the same amount of data as the original image.

The final Berry and Burnell law states that a well-processed image cannot be improved. Sky backgrounds set to black have thrown away image information about the dark portions of the image while stars that have been brightened to saturate pure white no longer posses true brightness information. Increasing the contrast of mid-range tones to improve visibility of wide-scale faint detail has similarly reduced the range of available data contained in the raw image. After processing, you can better see what there is, but it is in a more restricted range of shades and tones than was contained in the original. Further attempts to process the image eliminate more of the limited data, resulting in posterized colors, washed-out highlights, and garish over-enhancement of minute detail as pixels are stretched more toward black or white.

Keep Berry and Burnell's laws in mind as you approach your image processing. Seemingly miraculous transformations can be made, but you have to balance the results within the framework of what is in the raw data and what is mathematically possible to achieve with that data. Over-processing an image cannot substitute for the lack of data.

8.2 Don't Overdo It

The first look at a new image-processing program will be intimidating. My advice is not to tackle the program with the intent of immediately performing complex operations. The possibilities can be overwhelming. Which one to choose, what will it do?

For starters, learn the basics of video frame selection and alignment in RegiStax. Once you are comfortable with navigating the basic program operations, then approach the more complex operations one at a time, becoming familiar with each before progressing to the next. Look at the process as a building block operation.

As fast and powerful as today's desktop computers are, it still takes time for them to perform complex operations involved in image processing. An AVI video with several thousand frames takes a noticeable amount of time even for a fast processor to crunch through. Such operations can take several minutes on my 700 Mhz P-III desktop. While not a state-of-the-art multi-gigahertz computer, it is powerful enough to preclude any thought of replacing it soon. Therefore, time-consuming automated processing steps are usually performed while I am engaged elsewhere and not hostage to the lengthy computations.

After combining and aligning the selected frames in a video, there are basic operations that have to be performed to separate the resulting image from background noise, to correct colors and brightness, and to enhance faint detail. But it is easy to get carried away during this process. Oversharpening an image of the Moon can actually result in less detail in the final image because the sharpening process will drive pixel brightness too far toward black or white. Subtle crater walls bloat into donuts and artifacts begin to appear that overwhelm finer details on the lunar surface. With deep-sky objects, image-processing steps that are applied too vigorously can give the resulting image an unnatural color or brightness.

It must be remembered that celestial objects are naturally colorful, but their inherent faint appearance in a telescope renders them essentially colorless because of the limitations of human vision. The objective in our image processing is to render these objects so that they appear as if they were sufficiently bright to be seen in full color with human vision. So how do we know when an image is properly processed to show this? We don't. Most of our perception of what a celestial object is supposed to look like is guided by a lifetime of seeing photographs that have been enhanced in the darkroom to show colors, or that have been reproduced in a publication that introduces its own set of color variations during the printing process. The basic fact is that most celestial objects other than the Sun and Moon display relatively muted colors, so do not try to recreate an overly contrasty and excessively colorful scene just for the sake of duplicating what you have seen elsewhere. Let's face it, it took the photons from a distant nebula or galaxy thousands, even millions of years to reach us. After such a journey, they deserve all the respect we can give them!

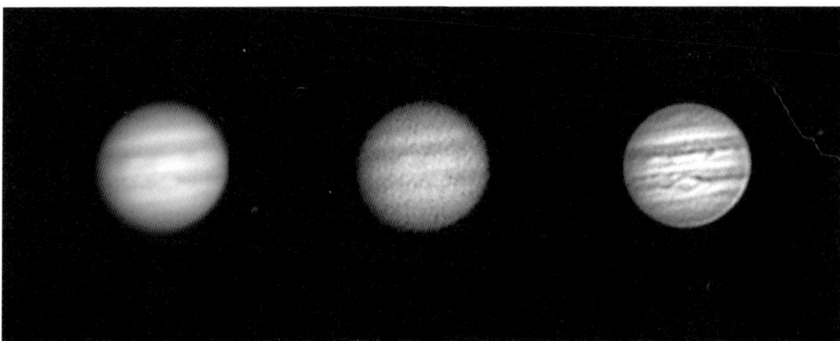

Fig. 8.1 *The importance of stacking many webcam planetary images is shown in the above three images. The left one is a single frame of a 600-frame AVI of Jupiter taken with a ToUcam 840 through a Celestron-8 telescope at f/10. It clearly shows the effects of poor seeing. At center is the same single frame after attempting to enhance it with RegiStax wavelet processing. The image lacks enough detail and data to permit further enhancement. The right image is a stack of the entire 600-frame AVI with the same wavelet settings that were applied to the center one. The stacked image has enough data and detail to allow enhancement of planetary detail. Photos by Robert Reeves.*

8.3 Color in Astronomical Images

One step in image processing that causes confusion is color correction. This often arises when astrophotographers have a wrong expectation about what color a celestial object should be, then try to make their image match that erroneous expectation. The colors of various objects are:

- Non-emission objects are naturally pale
- Hydrogen-alpha and H II emission objects are crimson
- Galaxy cores are yellowish or reddish
- Starburst regions in galaxies are blue
- Reflection nebulae are blue
- Regions of the Milky Way can be either blue or yellow.

Often a user will try to correct color by only adjusting the overall color balance. But there are several steps that must be done first regardless of what processing program is being used. First, you need to adjust the image density, and then remove any color bias through a process called "color normalization." A color bias is present if the black points do not match in all color channels. To remove color bias, set the black point in each color (the far left histogram adjustment) just short of the pixel hump in the histogram. If the background is neutral, then the color bias is gone, but if the target of the photograph is still off-color, then color balance is the problem. Color balance can be addressed by changing the white point in each color channel (the far right histogram adjustment). It is possible to introduce a color bias by applying improper adjustments to the color balance. The user must also be careful not to be overly aggressive in color balance adjustment because it can accentuate image noise.

Noise is a limiting factor in how much color correction can be applied to an image. Stacking a greater number of video frames to increase the signal will reduce noise, but ultimately the image will be limited by noise in the weakest color channel.

Perhaps the most widely used image-sharpening tool is unsharp masking. A form of this is used in the wavelet processing steps applied by RegiStax after video frames have been combined and aligned. The wavelet controls allow the operator to change the filter values applied to the image to achieve a greater or lesser sharpening effect.

8.4 Dark Frames and Flat Fields

Short-exposure webcam photography of bright lunar and planetary targets typically requires little image processing beyond using RegiStax for stacking and wavelet processing to enhance details. However, dim images of very faint planets at high *f*-ratios will benefit from additional noise reduction through dark-frame subtraction.

Long-exposure webcam astrophotography is considered to be basically a simpler form of CCD astrophotography and does require some additional image processing. But before processing can be applied, the image needs to be calibrated in order to remove noise and imperfections. To some extent, both forms of astrophotography, webcam and CCD, share the same basic forms of image calibration, although typical webcam work does not employ many of the advanced processing steps used with CCDs. At a minimum, all long-exposure webcam work should employ dark-frame subtraction to remove dark current and bias inherent in the imaging chip. Unlike with CCD work, individual bias frames are rarely used in webcam astrophotography. The image sensors used in webcams are inherently noisier than astronomical CCDs and the contribution of bias is minimal compared to dark current. For best results, the images should also be flat-fielded to remove artifacts created by the optical system. Purists will pursue additional processing steps, but these two basic operations will usually produce results a webcam imager will be proud of.

Let's look at some of the generic procedures used in the basic calibration steps mentioned above. Three things must be done to produce a calibrated, long-exposure deep-sky webcam astrophoto that can be further enhanced through image processing to stretch contrast, accentuate color,

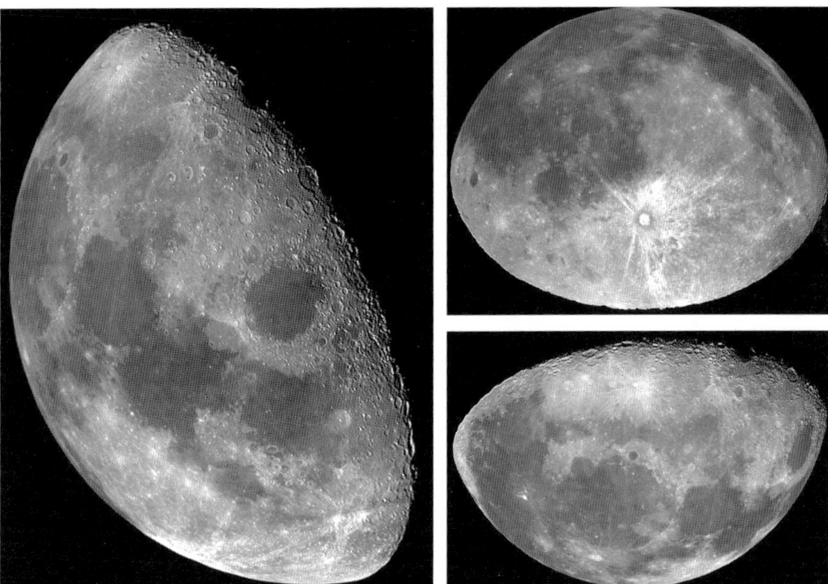

Fig. 8.2 *Photoshop has a 3-D plugin that allows images to be "pasted" onto a sphere, then the sphere rotated to any orientation. This allows astronaut-like views of the Moon that are not possible from our Earth-bound perspective. The plugin is on the Photoshop CD, but is not installed during the default installation procedure. In Photoshop CS, go to the "Resources and Extras" CD. Browse to: Goodies\Photoshop CS\Optional Plug-Ins\Filters and copy the "3D Transform.8BF" file into C:\Program Files\Adobe\Adobe Photoshop CS\Plug-Ins\Filter. Reopen Photoshop for the filter to become available. To use it, it is located under <Filter>, <Render>, <3D Transform>. Photos by Zac Pujic.*

or sharpen detail.

- **Acquisition of "light frames."** Light frames are the actual exposure of the astrophoto target. Unfortunately they contain more than just the image of the star cluster, galaxy, or nebula that we desire. Also present is noise in the form of the sensor's dark current and inherent bias as well as any image contamination from defects in the optical system, such as reflections inside the optical path or shadows created by dust on the imaging sensor.
- **Dark frame acquisition.** Before sensor noise can be subtracted from an astrophoto, it must be identified and isolated. This is done through the process of dark frame acquisition. Applying dark frames to an astrophoto during processing helps cancel the effects of bias and dark current in the image. Also, subtracting the dark frame from the image resets the numerical pixel values, especially the zero point, so that the values are where they should be. Dark frames are obtained by simply capping the telescope so that no light reaches the image sensor, and then taking a series

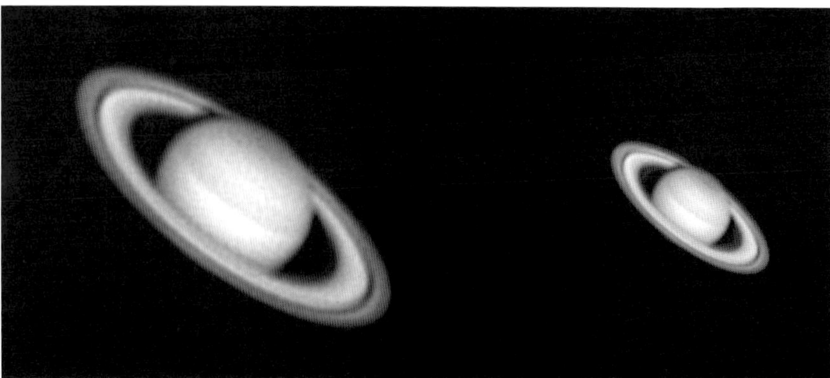

Fig. 8.3 *Upsampling, or enlarging the scale of a digital image, will yield no additional real detail and will often make the subject look blurry. Although the detail in the above images taken with a ToUcam 840 through a Celestron-8 telescope at f/10 is identical, the smaller-scale original looks sharper and better focused than the enlarged version. Photos by Robert Reeves.*

of exposures equal in length to the actual astrophoto exposures. Because dark current doubles with every 6°C increase, it is desirable that they be taken with the camera at the same temperature as when the astrophotos were taken. Late at night when the temperature is lower, take more dark frames any time you take a break for any reason. You need not be at the telescope—since guiding or other attention is not needed while taking a dark frame sequence, the camera can perform this duty alone.

No two dark frames will be exactly alike. Each will show the same hot pixels, but the random noise will be slightly different. For this reason it is best to take many dark frames. Averaging five dark frames will reduce noise by 55 percent; averaging 16 will reduce noise by 75 percent. Taking a greater number of dark frames becomes counter productive because the valuable observing time it uses up produces little added noise reduction.

The dark-frame procedure is the same for both short-exposure and long-exposure work. For thermal stability take the dark-frame video immediately before or after the deep-sky exposure. For best results, take at least four videos and average the resulting dark frames into a master dark frame containing only sensor dark current and bias. This dark frame will be subtracted from each deep-sky exposure in the video sequence to remove noise from the image.

- **Flat field acquisition.** While dark frames allow us to remove noise inherent in the imaging sensor, flat fielding allows us to remove noise created by other factors such as optical vignetting, reflections within the telescope or lens, dust-induced shadows on the sensor, and pixel-to-pixel inconsistencies in the sensitivity of the sensor. To acquire flat-field frames, use the same optical setup that you will use to take the light images. Duplicate the actual "light frame" exposure conditions; including using the same focus, because changing focus may affect vignetting. Set the exposure to allow about 50 to 75 percent sensor saturation and capture a video sequence of clear twilight sky near the zenith. If this is impractical, alternative methods are to aim the telescope or lens into an internally-illuminated "light box" or at an evenly-illuminated white screen. The important thing is to insure that the method used evenly and completely illuminates the camera field. If this is the case, then any variations in the density of the flat-field image will be due to the problems that we want to remove from the light images. It should be noted that if a "one-shot" color camera is being used, the flat frame must be processed with a median filter (radius 2, power 0.3) or else the flat-field image will remove all color from the light frame. To halve the noise within the flat-field image, take four and average them into a master flat field. In practice, the master flat field is divided (not subtracted) into the light frame in order to minimize image defects.

Image calibration is the process of applying a master dark frame and master flat field to each frame of the video being aligned and stacked. Exactly how to do these steps depends on the software being used, but for now we will concentrate on the theory of what happens when they are applied. Simply put, the dark frames are first subtracted from the light frames to remove sensor noise, and then the results are divided by the flat frames. The process is mathematically intensive, but fortunately most software packages have wizards or automated routines to handle these functions. The webcam user has access to the above calibration procedures—bias, dark frame, and flat-field subtraction, in the freeware program RegiStax.

8.5 Gray Scale and Color Space

When performing image processing, it is important to understand how colors are represented by the camera and handled by the computer and printer. Gray scale is a single color channel containing only shades of gray. More commonly, this is called black-and-white. Modified webcams that have had their sensor replaced with a more sensitive monochrome sensor produce images in 8-bit gray scale where values ranging from 0 to 255 present 256 steps of gray between pure white and pure black.

Color webcams use the RGB color space, which consists of the red, green, and blue additive primary colors. It is like human vision in that all three colors added together equal white. This system is also used by computer monitors to create color. Twenty-four-bit color is created by combining three separate 8-bit channels, each displaying 256 shades of a primary color. Together, all three channels are capable of displaying more than 16 million individual shades of color. Gray scale can also be created from RGB images when the individual RGB values for the same pixel in all three channels are adjusted to be the same, but this creates a file size three times larger than true gray scale without increasing the dynamic range of the image.

When any two of the primary RGB colors are combined, they create what is called a secondary color. For instance, green and blue make cyan, blue and red make magenta, and red and green make yellow. These colors make up the subtractive color space where the three colors, cyan, magenta, and yellow combine to make black. This color space is employed by inkjet printers that use a light and dark shade of cyan, a light and dark shade of magenta, yellow, and pure black to create color images using the CMYK color space. The printer software accepts the RGB output from a graphics program and converts it into the CMYK color space used by the printer. CMYK is a device-dependent color model and printed colors will vary depending on the printer, ink, and paper characteristics. For this reason, printer color profiles are supplied by the manufacturer so their inks will perform consistently when different types of paper or printing resolutions are used.

The range of colors that a color space can reproduce is called the color gamut. The RGB color space can display a wider range of color shades than is possible for the CMYK color space to produce and therefore RGB has a wider gamut. Although the CMYK gamut is smaller than RGB, there is rarely any problem printing color images with today's inkjet printers.

8.6 Basic Processing Operations

The processing operations performed on video frames to create a final image can be broken down into categories. The basic operation groups are: geometric transformations, point operations, linear operations, and image operations. We call upon various functions in these operation groups to combine video frames into a single image then manipulate the resulting digital file in order to eliminate unwanted data that are masking the desired image data, or change the range of digital data to render it more visible to the eye.

- Geometric transformations are functions that move the entire image while leaving its content and appearance intact. An example is translation, in which the image is shifted a specified distance and direction from its original location. Here, the image is literally moved up or down, side to side, flipped backwards, inverted, or a combination of these. Rotation is another geometric transformation, where the image is turned in either direction around a specified axis that can be located anywhere within it. Scaling is a transformation used to resize the image to a different overall size in pixels. Any combination of these geometric transformations can be applied. The most useful application of geometric transformations is the process of aligning a series of images so they can be stacked and combined into a single image.

- Point functions are those which find the desired range of pixel values needed to display image information and alter the values of those pixels to make the pictorial information they contain more visible. Point operations are therefore the most powerful computations that can be applied to an astronomical image. Point functions include working with the histogram to remap the pixel values, stretching brightness values, and setting the dark point.
- Linear operations generate pixel values based on a specified pixel's neighboring pixel values. This is useful in smoothing an image to reduce the effect of noise or in sharpening an image to give it a more focused appearance by increasing the contrast along a border. Smoothing filters, called low-pass filters, have their uses in astronomical imaging, but their side effect is an overall reduction in image detail, therefore they are rarely employed.
- Image operations include functions like dark-frame subtraction, aligning, and stacking images. Without the mathematical magic performed by image operations no form of digital astrophotography as we know it today would be possible.

8.7 Good Results with Basic Equipment

Basic image calibration involves subtracting a dark frame from each video frame to remove the effects of noise present in the sensor. Fortunately for simplicity's sake, short exposures of solar system objects usually need no calibration. The noise reduction properties of the image-stacking process that is used to average hundreds of video frames to create the final single image are all that is needed. Images with a modified webcam that are taken simply for fun, or in pursuit of "pretty pictures," only need the "amp-off" selection to limit amplifier glow, and basic dark-frame subtraction to control noise. Additional steps are, of course, performed during image processing to enhance desired features and suppress artifacts. Long exposures taken with a modified webcam for science such as astrometery, photometry, or spectroscopy need to be calibrated to a higher standard, but shooting science images is beyond the scope of this book. However, the principles of dedicated astronomical CCD calibration laid out in *The Handbook of Astronomical Image Processing* by Richard Berry and James Burnell also apply for scientific webcam astrophotography. Users who pursue that activity are encouraged to read this magnificently informative book.

Among the many processes used to produce webcam astrophotos, the most common is video-frame alignment and stacking. There are a number of freeware and shareware programs available for this but that most commonly used is RegiStax, written by Cor Berrevoets. RegiStax is freeware, but its free status should not mask the fact that it is an amazingly powerful program and without it, amateur solar system webcam astrophotography would not be in the advanced state that we find it today. RegiStax can combine and align hundreds of video frames to reduce noise and apply advanced wavelet image processing algorithms to achieve incredible detail not possible with single-frame digital or film images.

The novice must be aware that no amount of image processing with any software package will ever remedy poor focus, or blurry images caused by inadequate collimation. The crisper, better focused, and brighter the image is from the telescope, the better the resulting image will be after processing. Here, the old computer adage "garbage in — garbage out" still holds true.

8.8 Importance of Stacking Images

Stacking means mathematically combining two or more images to improve the signal-to-noise ratio. For noise suppression in deep-sky images, stacking is actually part of a two-phase approach. Dark-frame subtraction first reduces the noise in each video frame, while the stacking of all video frames further reduces noise.

The classic reason for image stacking is to increase the signal-to-noise ratio. The less noise there is in an image, the easier it is to apply contrast stretch during image processing to accentuate dim details. A secondary benefit is achieving increased image dynamic range.

Dynamic range, simply stated, is the greatest possible difference between the brightest and dimmest recorded pixel values. Values greater than the brightest possible value saturate the sensor to white and tonal values are lost. Values less than the lowest possible values simply do not record at all and appear as black. If pixel values are too low to produce any signal, the solution is to extend the exposure. However, longer exposure also increases values at the high end of the dynamic range and may drive pixel values to saturation, again resulting in loss of detail.

When imaging deep-sky objects, the dynamic range of the recorded

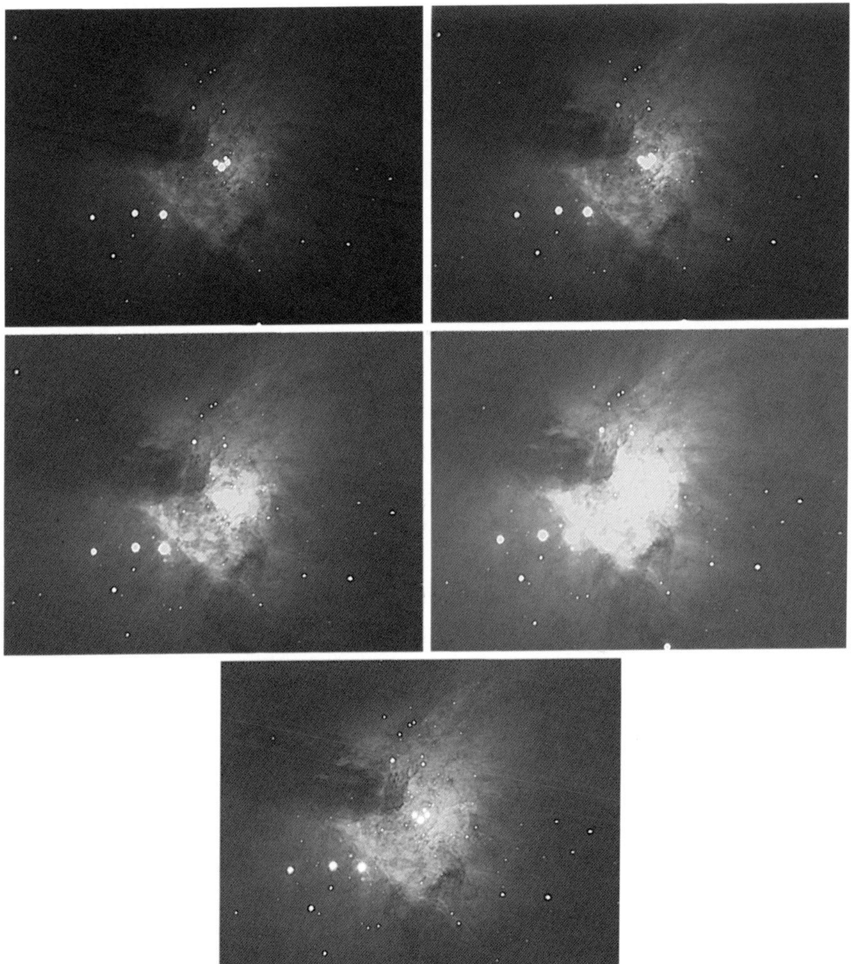

Fig. 8.4 *The bit-depth, or amount of image tonal information that an 8-bit webcam image can hold, is limited compared to 12- or 16-bit conventional digital cameras. One way to gain bit-depth to show greater dynamic range is to stack multiple exposures. The upper four images of M42 were taken with an Atik ATK-2HS camera through a Celestron-8 telescope operating at f/5. Exposure varied between ⅕ to 6 seconds. Combining the images in Photoshop (bottom image) allowed the faint outer detail of the Orion Nebula to be seen, while preserving the inner detail near the Trapezium that is washed out in the longer exposure. Photos by Robert Reeves.*

image is often less than the potential dynamic range of the sensor. No single exposure time is perfect for all deep-sky objects. An exposure brief enough to prevent stars from saturating will be too short to record dim nebulae, while exposures long enough to record the nebulae will saturate the stars. Another problem is that most webcams are 8-bit per color channel devices. That is, a monochrome camera will produce 256 shades of gray while color cameras produce 256 shades for each of the red, green, and blue color channels. When all three channels are combined, they reproduce more that 16 million colors but they are still limited to the 8-bit dynamic range. Additionally, a bright skyglow background will create an exposure bias that takes away some of the 8-bit data range and leaves even less available magnitude range for the camera. However, we learn to make the best of what circumstances have given us and the effects of skyglow can largely be removed during processing, although the resulting image will have less dynamic range than a similar shot taken under a dark natural sky. Conventional digital cameras are usually 12- to 16-bit devices. This means they have 16 to 256 times more digitized pixel values representing brightness levels than an 8-bit webcam. Thus exposure times with 12- to 16-bit cameras can be lengthened to record dimmer areas and still have sufficient dynamic range to accommodate bright stars before they saturate. Eight-bit webcam sensors do not have this extra range, but stacking many images can help achieve the needed higher dynamic range.

8.9 How Stacking Works

The number of possible digitized pixel values increases linearly with the number of images stacked. This helps in the following way: when taking a series of images that are exposed so as to not saturate the brightest parts of the image, there is the likelihood that the dimmest portions are grossly underexposed. But when images are stacked, the dimmest values accumulate into values high enough that they exceed the bottom of the dynamic range. At the same time, the overall dynamic range of the image increases so as to make room at the top of the range. In a sense, the brightest possible values increase enough to stay just ahead of the increased pixel values of the brightest portion of the image. Now the stacked images contain not only increased pixel values for both the dimmest and brightest portions of the image, but also a greater possible range of values so that the bright portions do not saturate.

The reason image stacking reduces noise is that when two images are combined, the signal from real objects like stars and galaxies doubles while the signal arising from random noise only increases by the square root of two. This holds true as even more images are combined. As Table 8.1 shows, the signal increases by the sum of all images while the noise increases only by the square root of the total number of stacked images.

As we can see, the process reaches a point of diminishing returns where an appreciable increase in the signal-to-noise ratio requires an

Table 8.1
Improvement in Signal-to-Noise Ratio with Increased Frames

Number of Stacked Frames	Signal-to-Noise Improvement
2	1.4 (square root of 2)
4	2 (square root of 4)
9	3 (square root of 9)
25	5 (square root of 25)
100	10 (square root of 100)
1000	32 (square root of 1000)

unmanageable number of video frames. In fact, it requires a 10x increase in the number of exposures to gain a 3.2x improvement in the signal-to-noise ratio. It is not uncommon for USB 1.1 webcams to accumulate 500 to 1000 images during a video sequence, resulting in a very noticeable improvement in image noise. Newer USB 2.0 cameras are capable of higher uncompressed image throughput making it possible to quickly accumulate many thousands of separate images. But the resulting AVI file sizes will hit the file size limit of many computer operating systems, which allow a maximum of just over 4000 frames.

In practice, the improvement in the signal-to-noise ratio between stacking 1000 images (32x) and 4000 images (63x) is not worth the extra bother considering that four times the number of images, file storage space, and processing time are needed to gain little more than twice the noise reduction.

Increasing exposure time will also follow the same formula to improve the signal-to-noise ratio with deep-sky images; that is, doubling the exposure will improve the signal-to-noise ratio by 1.4x (as long as the increased exposure does not saturate the image).

Stacking also adds the luminance levels of each frame together and essentially increases the overall exposure time to that of the sum of all exposures. However, the "fly in the ointment" with this is that noise is combined with each exposure and combining many short exposures may not be as effective as combining fewer longer exposures.

There are a number of stacking methods employed by popular image processing programs: addition, averaging, and median. These operations are performed one pixel at a time on all images in the stack before the process moves to the next pixel, and then repeats the operation on every image in the stack.

- **Addition** is just that, the values from each pixel in an image are directly added to the values from the same pixel in the next image. This works well for boosting the pixel values in very dim portions of the image but causes problems with values that are near the 8-bit maximum employed by webcams. Adding values from the bright portions of several images quickly exceeds a value of 255, so that portion of the stacked image is saturated and loses detail. To prevent this, the final image produced by the stack requires the application of a scaling factor. This is a constant or integer value divided into the sum of the stack to reduce the pixel value below the upper 8-bit limit of 255. With the scaling division, the output from the addition stacking method becomes little different from the output of the averaging method.

- **Averaging** is the most common method used for image stacking. The values for the same individual pixel on all images are added together and then divided by the total number of images being combined. When all the images to be combined are taken with the same exposure, the averaged values will be the same as the value on one original image, but the noise is reduced by a factor equal to the square root of the number of images stacked.
- **Median** combine looks at the values for the same pixel in all the images to be stacked, then adjusts them so they are midway between the highest and lowest values. The result ends up similar to the average combine but with a useful difference; individual images with bright or dark defects such as cosmic ray strikes, airplane or satellite streaks, or random black spots will be eliminated as the pixel values are adjusted to the median of all images. This technique is good for eliminating defects in static images of deep-sky or solar system objects, but would not be the proper choice for stacking a video sequence seeking transient targets such as meteors or flashing Earth satellites.

8.10 Monitor and Printer Output

The first place one usually sees a digital image is on a computer monitor. Ideally, image processing should be done on cathode ray tube (CRT) monitors, not LCD screens or laptop screens. The CRT monitors may seem like bulky "old technology," but they still generally have better color quality and produce a wider range of colors and contrast than most LCD monitors.

A monitor should be turned on for a half hour prior to doing any critical color evaluation with it. If its control panel allows choices, use the 5000K color temperature setting or one as close to that as possible. Most monitors ship from the factory with a much higher color temperature setting and tend to display cooler colors.

Monitor calibration is a link in the chain of image processing tasks that is often ignored. The image is processed until it looks good on the screen, but if the monitor is not calibrated, the image will not look the same when it is printed. Nor can we count on the image displaying well when viewed on a different monitor. Monitors tend to darken with age and subtle details will be the first to fade, making their calibration a critical part of image processing. Not all can be calibrated. A monitor must have a gain control for each color channel in order to allow calibration. Inexpensive models may not have such refined controls.

A useful device for calibration is the Colorvision Spyder, a device that attaches directly to a CRT or LCD monitor with small suction cups and analyzes its color output. The Spyder is available at many retailers in the $150 range. Anyone who is serious about achieving accurate color in monitor-displayed images, and hard copy from a printer, should look into using one of these devices.

A majority of prints produced by astrophotographers are printed on inexpensive home-use inkjet printers. The cost of these devices has dropped to amazingly low levels which seem disproportional to the outstanding photo-like prints they can produce. But as advanced as modern inkjets are, they have some limitations when trying to reproduce images the same way they are displayed on a computer monitor. The two devices produce color in two fundamentally different ways. Monitors produce all shades of color with the additive red, green, and blue (RGB) colors while printers produce all shades of color with the subtractive cyan, magenta, and yellow (CMY). To achieve deeper blacks, modern printers also use black (K) as a fourth color in a system known as CMYK. Because the two color systems are fundamentally different, there is a small range of colors that will display on a monitor but not print from an inkjet printer. This is technically known as the colors being "out of gamut." For the most part, most users either never encounter this phenomenon or its effects are so subtle they never notice.

Once a printer is producing pleasing images, one should try to use the same brands of paper and ink in the future to maintain the good results. A particular brand of printer is designed to use a certain type of ink and paper for optimal output. If a brand of paper or ink is used with a different surface porosity or chemical makeup, the ink applied by the printer will display differently. In the search for economy in printing I have tried several aftermarket brands of ink and paper and found the basic truth is that for best results, the paper and ink recommended by the printer manufacturer will consistently produce the best results.

Image-processing programs and printers are capable of producing prints with various printing dot resolutions. Today's printers advertise being able to print up to 2880 printing dots, or discrete points of different ink, per inch. But image-processing programs often output images to a printer at far fewer dots per inch (DPI), with 300 being considered a high-quality output. Setting a printer to 2880 DPI will rarely produce better results than 1440, and sometimes even 720 DPI. Using maximum printer resolution is essentially wasting ink to create a print that the eye cannot discriminate from one created with a lower DPI setting. The user is encouraged to experiment with different printer resolutions to find which produces acceptable results without wasting ink and increasing the printing time.

The expected viewing distance for a print also factors into the DPI setting for the image-processing program. Prints up to 11 x 14 inches should be printed at 300 DPI because they will be held in the viewer's hands and examined close up. Larger prints are usually intended to be viewed from greater distances and can use lower DPI settings. Prints 20 x 30 inches can be printed at 200 DPI while really large prints in the 30 x 40 inch range can be printed at outputs as low as 100 DPI.

Chapter 9
Polar Alignment and Guiding

Accurate polar alignment is absolutely essential for successful webcam or digital camera imaging of any kind. Needless to say, if you plan to image faint deep-sky objects, your telescope's mount will need to track faithfully for extended periods of time. But it is equally important for lunar and planetary imaging because the small field of view in a standard webcam is approximately equal to that of a 5mm eyepiece. The use of a Barlow further reduces the size of the field. Without good polar alignment, a planet can drift out of the field of view during the course of a several-minute AVI sequence. This can be alleviated by making manual guiding corrections during the exposure sequence, but good polar alignment eases the tedium of constantly chasing the elusive target.

Video frame stacking software, such as RegiStax, attempts to track a chosen image feature from one frame to the next. If the image is relatively steady, slight shifts caused by poor seeing will be tracked throughout the video sequence. But the field of view in a video shot through an untracked altazimuth mount or a very poorly-aligned equatorial mount will continuously drift in one direction, a target such as a planet will quickly exit the field within a few seconds, resulting in a blank video. An untracked video of the Moon will similarly result in an unusable image because the selected target feature used to stack and align successive video frames will slip out of the field of view.

9.1 "Tracked, but Not Guided"

This chapter will discuss accurate polar alignment of telescopes and how to guide long-exposure astrophotos, regardless of what type of camera system is used; film, conventional digital, or some sort of webcam. However, we will begin it with a description of a technique that seems to counter the need for precision alignment and guiding that is the norm for film astrophotography. This technique assumes the target is imaged using a series of short exposures that will be stacked in an image-processing program. It further assumes there will be some frames in this series that will be unus-

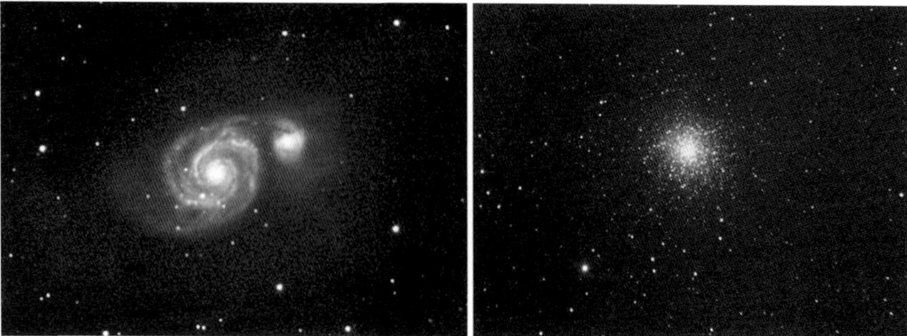

Fig. 9.1 *A series of multi-minute exposures of deep-sky objects requires accurate polar alignment to prevent star trailing. The Whirlpool Galaxy, M51, (left) was imaged through an Orion 120ST refractor using an Atik ATK-2HS camera with manual guiding. The M13 globular cluster (right) was imaged using the same camera on an Orion 80ED refractor. Two sequences of exposures, 10 x 30 seconds and 10 x 120 seconds, were combined to create this image. Photos by Jim Ferreira.*

able because of trailing and will be discarded from the stacking process.

Many of today's telescope drives are accurate enough that with good polar alignment they can track a deep-sky object for 30 seconds to a minute without the need for guiding. This has added a new phrase to the astrophotographic vocabulary: "tracked, but not guided". Accurate tracking allows the digital photographer to use software to stack a series of successive 30-second exposures to simulate a much longer exposure. Each successive exposure must be framed the same way to insure that there is minimal image drift between them so that a stacking program will be able to register each image in the stack. The beauty of the digital imaging process is that if any of the shorter exposures are degraded by tracking error, they can be eliminated from the image stacking. Snipping out a minute of poor tracking is not possible with a long-exposure film image, but it is a routine part of digital imaging.

The following describes how I polar aligned my Losmandy GM-8 mount and captured the image of the planetary nebula M57 shown in Figure 9.2 using an Atik IIs modified webcam on a Celestron-8 telescope. Portions of what I will describe run counter to discussions about polar alignment in my previous books on this subject (*Wide-Field Astrophotography*, and *Introduction to Digital Astrophotography*), but this fact only shows how webcams have fundamentally changed some aspects of deep-sky imaging.

Only a narrow portion of the sky is visible from the driveway of my home because of trees and roofs: from about 40 degrees above the northern horizon to about 30 degrees above the southern, and about 30 degrees east and west of the meridian. A large oak tree blocks Polaris, but it is imprac-

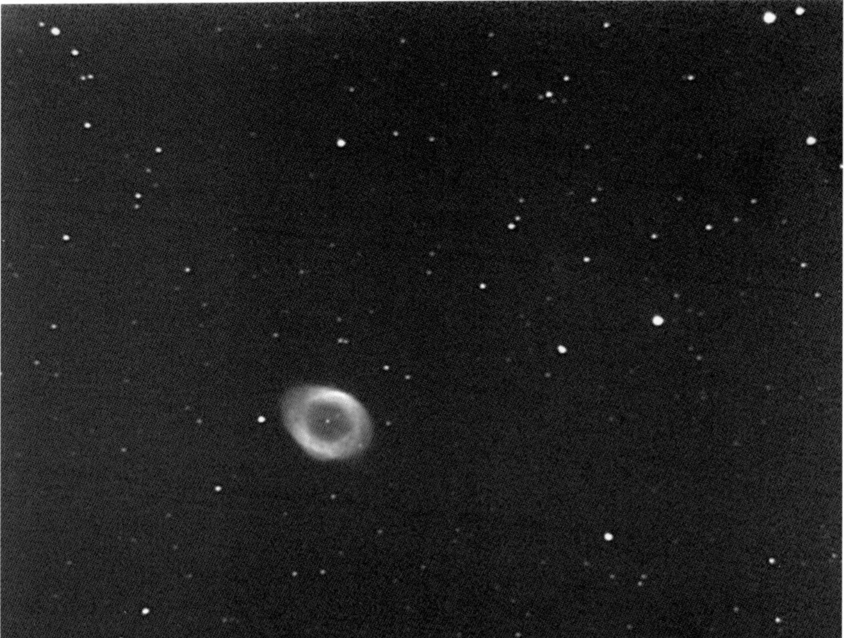

Fig. 9.2 *The Ring Nebula, M57, was imaged using the "tracked, but not guided" technique. From 200 separate seven-second unguided images taken at f/5 through a Celestron-8 telescope, 117 were tracked well enough to stack into this image. Photo by Robert Reeves.*

tical to say the least to roll several hundred pounds of telescope into the grass to sight Polaris through its pole-finder scope. Since my house is built facing due north (the irony of having a house that is better polar aligned than my own telescope has not escaped me) my solution is to just visually sight along the west wall of my house and align the polar axis of my mount so it is parallel to that wall. Those who lack a north-facing house for reference can use two widely separated paint spots on their driveway, or a distant landmark that accurately indicates true celestial north from the spot where their telescope is usually set up. The polar axis elevation was set to my latitude using an inclinometer and as long as I set up in the same area of the driveway, the elevation remains accurate. Thus the whole polar-alignment process takes less than a minute from the time I roll the telescope out of my garage. How accurate is this crude method? A target will remain on the small imaging surface of a webcam at prime focus in my 8-inch *f*/10 telescope for 15 minutes or more. Remember, this is equivalent to the field of view through a 400-power eyepiece. One should note however that while the above method does allow adequate tracking of an object, it does not prevent field rotation, which will be discussed later.

With alignment complete, it is time to acquire the target, focus, and set the camera operating parameters for the exposure sequence. In the case of the M57 image, a 0.5 focal reducer was used to lessen the effect of expected star drift because of the crude polar alignment. An IDAS light pollution filter was also used because the image was taken from the center of a large city and M57 was invisible through sky glow in the finder scope. After brief experimentation with exposure time, seven seconds was found sufficient to easily reveal the target while 10 seconds began to show oval stars. During this time it was determined that the target was slowly drifting from left to right across the field, thus M57 was offset to left of the field and an automated sequence of 200 seven-second exposures was programmed using K3CCDTools. Once the exposure sequence began, I snoozed for 20 minutes in a lounge chair while the camera worked away.

When played at normal speed, the resulting video sequence showed M57 drifting from left to right and zigzagging because of periodic error in the telescope drive gears. Because of this cyclical periodic error, sometimes the images tracked perfectly, at other times they either trailed left or right as the mount briefly sped up or slowed. The net result was that half of the exposures were well tracked while half showed some degree of egg-shaped star trailing.

Now the beauty of digital imaging comes into play. When using RegiStax to process the video sequence into a final still image, each frame can be examined and either selected for stacking or deleted from the images to be processed—thus only the well-tracked images are retained and stacked. The final result was that out of 200 exposures, 117 were tracked well enough to be stacked into the final image. That number is more than sufficient to average out image noise and build an adequate signal-to-noise ratio for satisfactory image processing.

The crude technique discussed above is sufficient to image objects bright enough to be seen through a telescope, but eventually, the astrophotographer will want to image dimmer, often invisible, targets that require longer exposure and tighter tracking tolerances. In these cases, the polar alignment principles and guiding techniques discussed in this chapter will need to be applied.

9.2 Principles of Polar Alignment

A permanently mounted telescope only needs to be polar aligned once. As long as the mount is not disturbed it should retain the alignment. Portable telescopes however need alignment every time they are set up. Most of the procedure can be accomplished in twilight before it is dark enough to begin the evening's celestial photography.

The concept of polar aligning a telescope involves making its right ascension axis exactly parallel with the Earth's rotational axis. Once this is accomplished, a drive motor can turn the telescope around the right ascension axis at a rate that matches the apparent movement of the stars. The easiest way to visualize this is to think of the mount's right ascension axis (the polar axis) as aiming at Polaris (for those living in the Northern Hemisphere). However, Polaris is actually about a degree away from the true pole. Indeed, star trail images will show Polaris tracing a tiny arc around that point. The right ascension axis must therefore be aimed at the *true* celestial pole using techniques that will be discussed shortly.

The drive rate at which a telescope tracks the stars is thought of, in general terms, as one complete rotation per day. However, the stars do not circle the celestial pole in exactly 24 hours—the actual time required is 23 hours, 56 minutes, and 4 seconds. This is the sidereal day, the time it takes the stars to rotate 360 degrees in the sky. The sidereal day is almost four minutes shorter than the standard 24-hour day of civil time because of Earth's motion around the Sun. As the Earth moves along its orbit, the Sun seems to move west-to-east among the stars. Thus every 23 hours, 56 minutes, and 4 seconds the stars have returned to their starting positions, but the Sun appears a little farther east among them. It takes an additional four minutes for the Sun to return to its original starting position (usually taken as local noon). Because of this daily four-minute displacement, the stars apparently shift position in the sky at the rate of about one degree per day, presenting a different sky to us over the changing seasons.

Because of the four-minute difference between a sidereal day and the 24-hour day of civil time, a good telescope clock drive will not rotate the instrument once every 24 hours but will move it slightly faster so as to exactly match the apparent movement of the stars. Some drives have additional settings to adjust their speed to accurately follow the Sun or Moon, each of which moves across the sky at a rate slightly different than the stars.

9.3 Telescope Mechanical Checks

There are a number of techniques than can be used to quickly align a telescope mount so it will perform adequately for astrophotography. In order for these techniques to work, however, the mechanics of the mount need to be in good order.

First, do not attempt to align a telescope set up on grass or soft dirt. Place a concrete stepping stone, a one-foot square piece of ¾-inch ply-

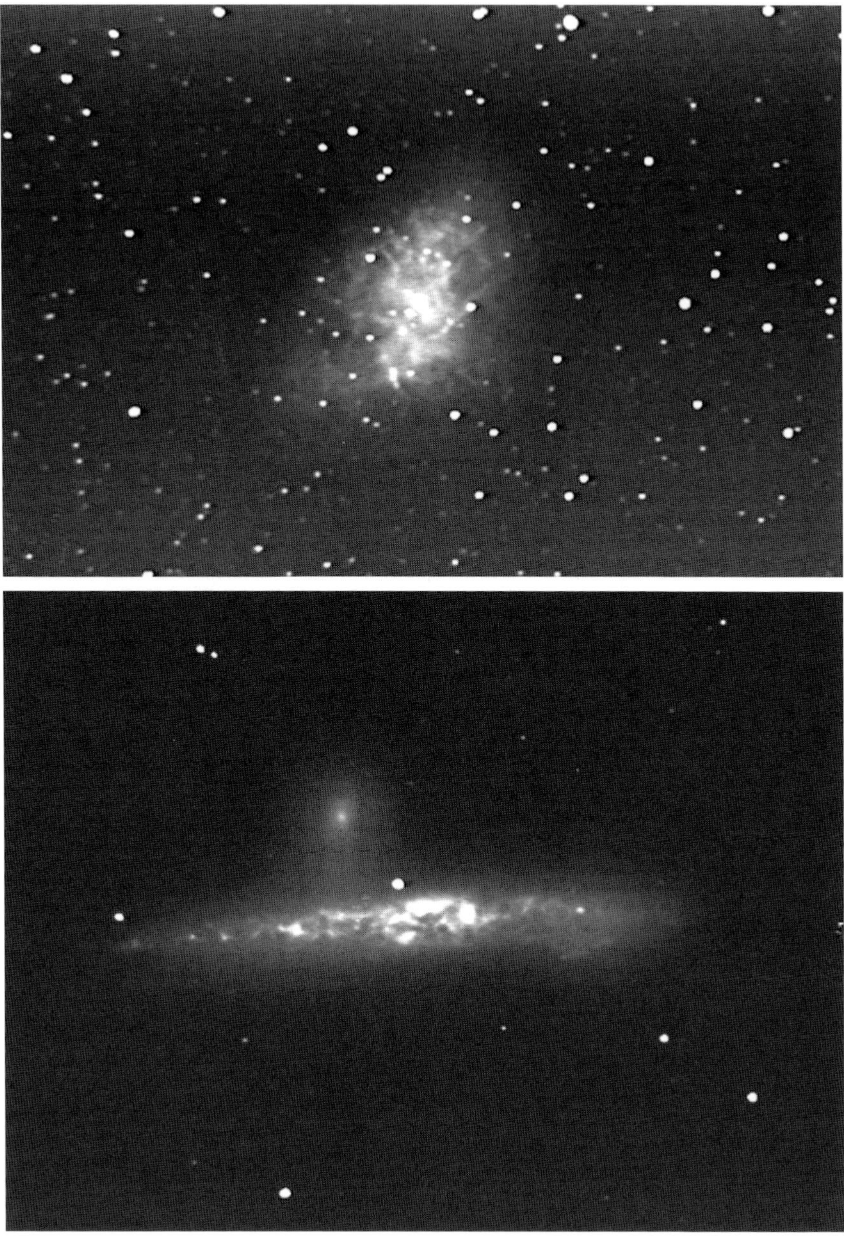

Fig. 9.3 *The Crab Nebula, M1 (top), and the edge-on galaxy NGC 4631 (bottom) were both imaged using a SAC8 camera and 0.3 focal reducer on a 10-inch LX200GPS telescope. Accurate polar alignment helped produce good images over lengthy multiple exposures. M1 is a combination of a 20 x 20-second and 20 x 10-second image stack while NGC 4631 is composed of a single 25 x 30-second image stack. Photos by Tim Tasto.*

wood, etc., under each tripod leg to prevent it from sinking into the soft ground.

Next, if a star diagonal is used, check that it shows the same star field as seen without it (many do not). We do this to verify that the diagonal is accurately positioned on the optical axis of the imaging setup. This is done by sighting a bright star through the telescope, then locking it so the drive tracks the star. Insert the diagonal and verify that the star has not moved. If the diagonal is at fault, it may have to be replaced, as few are adjustable. Not only does a misaligned star diagonal fail to show a field centered where the telescope is aimed, it is displaced from the optical axis and thus may cause aberrations in the image.

Check that the telescope's declination axis is precisely perpendicular to the right ascension axis. This done by using the setting circles as a reference to adjust the declination to 90 degrees. Next, lower the mount's latitude adjustment to zero degrees (horizontal) and sight an object along the horizon; this should be at least a quarter mile away for a fork mount and a mile away for a German equatorial mount. Rotate the tube around the right ascension axis and observe the target in the eyepiece; it should stay centered as the tube rotates. If the target makes an arc in the eyepiece field, then the declination setting circle may not read correctly or the declination axis may not be perpendicular to the right ascension axis. If the declination axis is not perpendicular to the right ascension axis, either the telescope tube and/or the declination axis will have to be shimmed. Hardware stores carry thin brass shim stock. In a pinch, an ordinary aluminum soda can can be cut to make shim stock. If thinner shims are needed, paper strips can be used.

Make sure the optics are properly collimated. If there is a means to lock the primary mirror on a Schmidt-Cassegrain telescope be sure to use it to insure that image shift due to "mirror flop" does not occur.

9.4 Basic Polar Alignment

As we have seen, quick alignment by roughly aiming the polar axis toward Polaris is adequate for visual observing, but it falls far short of the precision needed for long-exposure photography. This is especially true the closer the target is to the pole. When imaging close to that point, misalignment causes field rotation and/or image drift even if the mount perfectly tracks the motion of the sky. This is because the misaligned telescope and the sky are rotating on two divergent axes. Misalignment will also cause tracking problems away from the pole. For instance, if the polar axis of the telescope is just one degree below the true pole, a star on the celestial equator within two hours of the meridian will drift eight seconds of arc per minute. This amount of drift will call for frequent tracking corrections to keep the target within the narrow field of view of a webcam.

Prior to aligning any telescope mount, be sure to add all accessories that will be used such as guidescopes, cameras, and counterweights. Adding heavy accessories afterwards may cause flexure that will slightly shift the alignment, and/or change the rate.

Modern German equatorial mounts often have a small sighting telescope built into their polar axis that is equipped with a special reticle for determining the true pole (see Figure 9.4). Some pole-finder scopes, such as those offered by Losmandy and Astro-Physics, have a dual pattern engraved on the reticle to allow finding both the northern and southern celestial pole. In the case of the former, the reticle of the Losmandy pole finder is roughly aligned using Alkaid, the tail star in the Big Dipper, and the easternmost star in Cassiopeia. Other models achieve fine alignment by matching Delta Ursae Minoris and 51 Cephei with etch marks in their reticles. Once the reticle is aligned with the reference stars, the mount is shifted until Polaris is centered on its reference mark on the reticle.

Other pole finders have dual concentric circles engraved on their reticles that require that Polaris be placed between the circles, and offset at a certain clock angle that is dependent upon the time and date. Once the mount is properly aligned, Polaris should stay within the double circles when the mount is moved in right ascension. If it does not, either the polar axis is improperly aligned, or the reticle is misaligned. Be sure to use the appropriate Standard or Daylight Saving Time when setting up the pole finder.

Jason Dale's PoleFinder.exe program[1] reproduces an image of the view through any of the popular polar alignment scopes that utilize a Polaris circle. By entering the time and the geographic longitude, the program displays the proper location of Polaris relative to the Polaris circle in mounts that use such a device. The displayed graphic actually shows four versions of the Polaris circle; one for the current hour and three other versions for one, two, and three hours later (see Figure 9.5).

Polar scopes are generally accurate enough for long-exposure wide-field astrophotography but they lack the accuracy needed for high-resolution long-exposure work. For this type of imaging, the star drift method of alignment is necessary, as explained in Section 9.6.

Pole-finder telescopes must be analyzed for accuracy when they are new. Not every reticle is adjusted properly at the factory and rough ship-

[1] http://www.polarfinder.co.uk

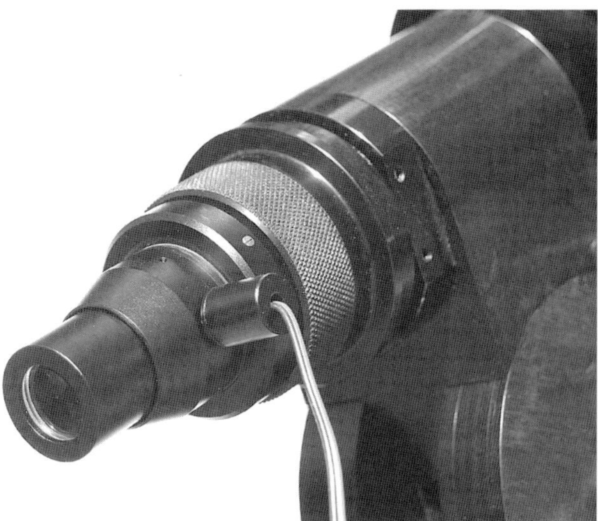

Fig. 9.4 *Modern German equatorial mounts often have a provision for a "pole finder" scope inside the mount's polar axis. A polar scope, like the one shown here on the author's Losmandy GM-8, greatly simplifies the polar alignment process for casual imaging of solar system objects. Photo by Robert Reeves.*

ping could cause misalignment. Use the drift alignment method for finding the true pole the first time this device is used to verify its accuracy.

If your German equatorial mount is not equipped with a pole finder, an alternative is to rest a finder scope directly on the polar axis using a machinist's v-block. The true pole can then be found using any of the popular star charts or star charting software that shows the dim stars that are closer to the pole than Polaris.

If the above methods are not applicable to your telescope mount, then another technique can be employed. First, roughly point the polar axis towards Polaris. This step can be done while it is still daylight by using an inclinometer to set its elevation to the local latitude (See Figure 9.6), and by using a good magnetic compass to adjust the mount's azimuth to within a few degrees of north. **One caveat: if you are considerably east or west of the true magnetic north line, you may be as much as 10 or 20 degrees offset from the Pole Star. Once the stars become visible, reposition the telescope's azimuth more closely to Polaris'.** Next, aim the telescope toward a bright star of known right ascension near the celestial equator. Set the right ascension setting circle to the current year's R.A. for that star. (Keep a list of bright stars and their coordinates for this purpose.) Now position the telescope using the setting circle to aim at 2 hours, 32 minutes right ascension and +89 degrees 16 minutes declination (or the

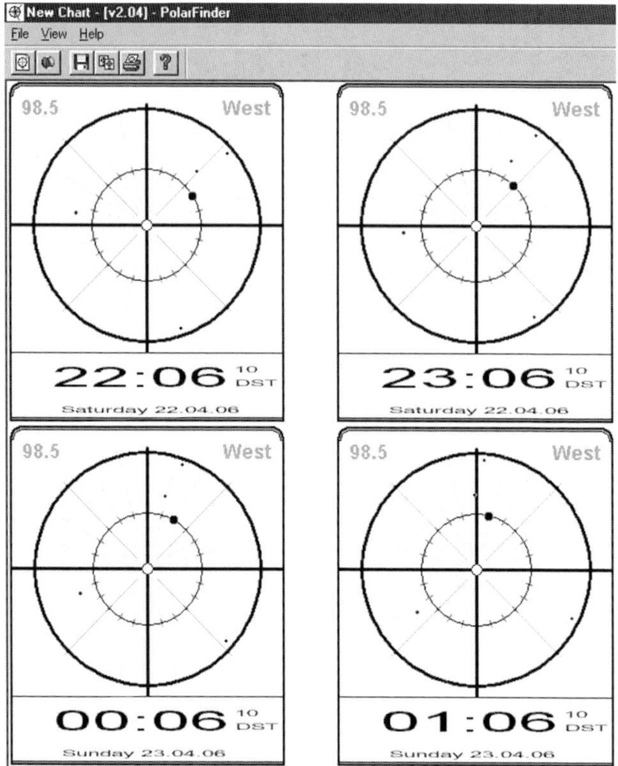

Fig. 9.5 *Telescope mounts with an alignment scope that displays a Polaris circle will benefit from Jason Dale's PoleFinder program. Entering your geographic location and the desired time into PoleFinder generates a chart showing the appropriate location of Polaris in a polar scope for the given time, and for the next three hours, at one-hour intervals. Screen shot by Robert Reeves.*

current year coordinates for Polaris). Leave the setting circles alone and move the entire mount in azimuth and elevation to sight Polaris in the telescope.

Repeat the process again by sighting on a bright star near the celestial equator and proceeding as before. If you can shift between the equatorial star and Polaris using setting circles alone, no further alignment is necessary. This method will also work for the Southern Hemisphere by using the star Sigma Octantis at 21 hours, 09 minutes right ascension and –88 degrees, 57 minutes declination (or the current year coordinates for Sigma Octantis). Beware, however, that this process sets the mount's polar axis on the refracted, not the true, celestial pole.

While performing alignment in twilight, beware of the star Kochab. This star in the Little Dipper's bowl is about 15 degrees from Polaris and is almost as bright as the Pole Star. Ordinarily there is no mistaking the lat-

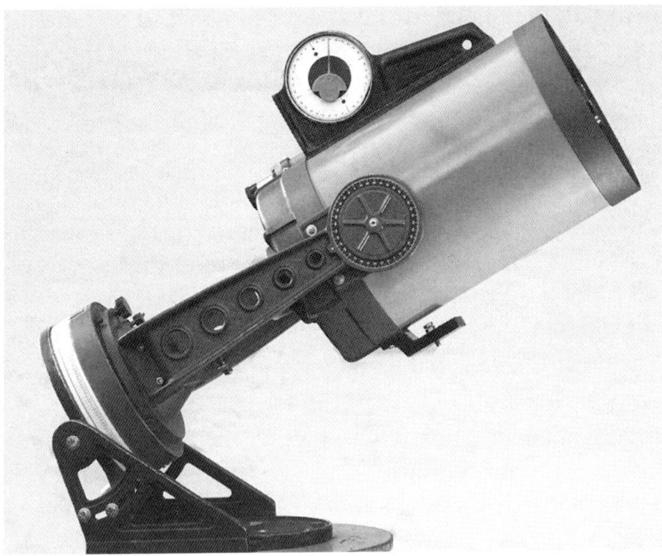

Fig. 9.6 *A rough preliminary polar alignment can be done in daylight by using an inclinometer to adjust the polar axis elevation, and a compass to point the polar axis toward north. Photo by Robert Reeves.*

ter, but when in unfamiliar territory one's sense of true north can be off a number of degrees. At certain times of the year, such as early evening in April or November, Kochab is at the same elevation as Polaris so beware of aligning on the wrong star, or valuable twilight setup time will be lost.

9.5 Other Factors in Polar Alignment

Star drift caused by misalignment can be corrected using the mount's slow motion controls to periodically update the telescope's tracking, but field rotation cannot be corrected with telescope controls. Field rotation is caused by the telescope's axis of rotation not being aligned with Earth's axis of rotation. The result of this misalignment will not be visually noticeable in the eyepiece, but a series of exposures of the same object over a period of time will reveal that all objects in the field of view have slowly rotated about its center due to the geometry of the telescope's misaligned axis of rotation.

Three things affect both declination drift and field rotation: the location of the true pole, the polar alignment point in the sky, and the celestial location of the target. Michael Covington, in *Astrophotography for the Amateur*, explains that if the polar alignment error is no more than a few degrees, both declination drift and field rotation are linearly proportional to the amount of alignment error and are geometrically interrelated. If the true pole, the alignment point, and the target form a right triangle, then the declination drift is greatest and field rotation is zero; but if the true pole, the alignment point, and the target lie in a straight line, then declination drift is zero and field rotation is greatest.

Covington's analysis shows that the declination of the target itself has much to do with which factor—declination drift or field rotation—comes into play. Declination drift is fairly independent of the target's own declination and quickly diminishes for those above 80 degrees. On the other hand, field rotation increases with higher declination; at 60 degrees, it is twice its value at the celestial equator, and it increases dramatically very close to the poles.

It is mathematically possible to predict the greatest amount of declination drift and field rotation for a given set of circumstances. In many cases the actual drift or rotation will be less than calculated. The maximum declination drift is

$$15.7 \text{ arcseconds} \times \text{exposure time (in minutes)} \times \text{alignment error (in degrees)}.$$

and the maximum field rotation is

$$0.01 \text{ degree} \times \text{exposure time (in minutes)} \times \text{alignment error (in degrees)}.$$

Generally field rotation will not be noticeable unless it exceeds 0.1 degree.

9.6 Star-Drift Polar Alignment

Permanent telescope installations, and long-exposure astrophotography, require higher alignment precision and thus the star drift method of aligning should be used. Wallis and Provine popularized the following procedure, with extra tips courtesy of Chuck Vaughn. Stars bright enough for the drift-alignment procedure can usually be seen about 15 minutes after sunset. If you begin then, alignment can usually be completed by the time it is dark enough to begin photography.

1. Level your tripod. This has no effect on a properly aligned polar axis, but leveling the mount makes it easier to perform the corrections in altitude and azimuth needed for proper alignment.
2. Roughly aim the polar axis toward Polaris.
3. Operating at about 200 power, place the crosshairs of an illuminated-reticle eyepiece on a star that is near the meridian and within five degrees of the celestial equator. Align the reticle so the star moves up and down the declination crosshair and left and right on the right ascension crosshair.

4. Ignoring for now any east-west right ascension drift by the star, look for any declination drift. The star is likely to noticeably drift in declination within 30 seconds.
5. If the star drifts northward, the polar axis is too far west of the celestial pole. To correct, rotate the mount in azimuth so that the star moves to the right. A star diagonal can confuse the perceived direction, so it is generally best not to use one while drift aligning.
6. If the star drifts southward, the polar axis is too far east of the pole. To correct, rotate the mount in azimuth so that the star moves to the left.
7. If the star drifts noticeably within five seconds, the polar alignment is off by at least 10 eyepiece field diameters. If it takes the star at least 30 seconds to drift, the setting is probably not more than one or two eyepiece fields off.
8. Once the north-south drift is minimized, place the crosshairs on a star near the celestial equator about 15 degrees above the eastern horizon. If the eastern horizon is obstructed, go to a star near the western horizon and reverse "above" and "below" in the next step.
9. If the star drifts northward, the polar axis is above the pole; adjust the elevation of the polar axis to move the star toward the bottom of the field. If the star drifts southward, the polar axis is below the pole; adjust the elevation of the polar axis to move the star toward the top of the field.
10. Once the corrections have been made for polar axis position, repeat the above procedure because each subsequent correction will slightly upset the previous adjustment. When properly aligned, and assuming the drive motor is turning at the proper speed, there should be no noticeable star drift for five minutes.

9.7 Improvements on Star-Drift Alignment

If you cannot find an adequate star on the celestial equator for drift alignment, stars at higher declinations will still work. A star at 10 degrees declination, for instance, will suffice because the cosine of 10 degrees is 0.985; thus a star at that declination will, in a given amount of time, still move 98.5 percent the angular distance of a star on the celestial equator. Indeed, you can easily drift-align on a star 40 degrees from the equator. The cosine of 40 is 0.766; thus such a star will still move 76.6 percent the angular distance in a given time that an equatorial star would. In practical terms, this means that if you let a star on the equator drift for three minutes, a star at 40 degrees declination should be allowed to drift for four minutes to achieve the same accuracy.

Thomas Krajci has worked out a quick technique for determining how much to move a polar axis's azimuth and elevation when making corrections indicated by the drift method. Tom's technique assumes the mount is already in rough alignment with the pole. You will need an eyepiece with a readable scale, such as Celestron's Micro Guide. Edmund Scientific also markets transparent scales that can be inserted into an inexpensive 12mm Kellner eyepiece.

With Krajci's method, you measure star drift at two locations—the meridian and near the horizon, just like the traditional drift-alignment method. Orient the eyepiece reticle and monitor the drift in declination. After five minutes of observation, you can quantify the amount of drift using the eyepiece scale and hence calculate how much to move the polar axis to correct the alignment. An important point to remember with this technique is that you measure the north-south drift on the meridian, but you must move the telescope to a star 90 degrees away (+ or - 6 hours from the meridian) when making the polar axis adjustment. Here, in detail, are the steps involved:

1. Roughly polar align to within a degree using the finder scope.
2. Find a star on the meridian, near the celestial equator, and measure its north-south drift in reticle divisions for five minutes.
3. Multiply this drift value by 46. This is how many reticle divisions you need to move the mount in azimuth (but don't move it yet).
4. Point the telescope to a star near the horizon along the celestial equator.
5. Now move the telescope in azimuth to shift the star by the number of reticle divisions calculated in step 3.
6. Remain on the star near the horizon and measure the north-south drift with the reticle for five minutes.
7. Multiply this value by 46. This is how much you need to move the mount's polar axis in elevation (but don't move it yet).
8. Move back to the star along the meridian near the equator.
9. Now move the mount in elevation to shift the star by the number of reticle divisions calculated in step 7.
10. Repeat the above steps if needed to further refine the polar alignment.

Krajci's method is independent of focal length used. The accuracy is limited by shifting Schmidt-Cassegrain optics. Mirror movement in an SCT must be eliminated before the 46-to-1 rule can be accurately implemented. A method for locking a Schmidt-Cassegrain mirror is explained in Section 5.9.

9.8 Using a Webcam to Assist in Polar Alignment

The alignment steps of the previous methods are performed visually, but a standard short-exposure webcam can assist with precision drift polar alignment. Jan Timmermanns provides the following steps using either a short- or long-exposure capable camera. His method as described here is for aligning a Newtonian telescope in the Northern Hemisphere. If a star diagonal is being used with an SCT, reverse all directions given.

1. Aim the polar axis toward Polaris as best you can.
2. Select a bright star near the meridian along the celestial equator and center it in the eyepiece.
3. Mount the camera and center the star on the computer screen.
4. Turn off the right ascension drive and observe the star motion. It should drift from right to left. If it drifts left to right, the camera is mounted upside down. Adjust the camera position so the star drift is a straight horizontal line. The crosshair option in K3CCDTools or GuideDog will assist this step.
5. Turn the drive back on and recenter the star.
6. If the star slowly drifts left or right, the mount is not misaligned in azimuth, but instead suffers from a poor tracking rate or periodic error. If the star drifts north (down on the screen), the mount is pointing too far west. If the star drifts south (up on the screen), the mount is pointing too far east. Only small adjustments are needed to tweak the east-west alignment of the mount. If the mount does not have a jackscrew to adjust its azimuth use a long pry bar for leverage to slowly and incrementally move the mount in azimuth until the star's up or down drift on the computer screen is eliminated.
7. Once the mount's azimuth is adjusted, aim at a bright star that is most convenient above either the east or west horizon. If, when viewing above the west, the star drifts north (toward the bottom of the screen), the polar axis is too low. If the star drifts south (toward the top of the screen), the polar axis is too high. When viewing a star above the east, these directions are reversed.
8. Once all alignment adjustments are made, the star should remain in the center "bull's eye" of the crosshairs displayed on the screen for several minutes (with exceptions for periodic error drift).

If a long-exposure capable camera is available, the accuracy of the alignment can also be checked and refined by taking 30-second exposures in the same sky directions as above, and then previewing them in K3CCDTools. Beware of slight drift in right ascension caused by periodic error, and watch for any shift between frames and correct accordingly using the same adjustments as above.

9.9 Guiding Corrects Alignment Errors

In astrophotography, guiding is the process of constantly correcting a telescope's tracking, both in right ascension and declination, to keep the target in exactly the same spot on the camera's focal plane. The errors being cancelled out by guiding may be caused by inaccuracies in the telescope's polar alignment, its mechanical drive components, electrical power fluctuations, atmospheric refraction, or atmospheric turbulence. Guiding enables the stars to be recorded as pinpoints on long exposures instead of streaks and trails.

There are two ways to guide a telescope, manually or electronically. With the former method we visually monitor a star with an illuminated-reticle eyepiece and manually correct the telescope's aim with a controller that slightly speeds up or slows down the telescope on the appropriate axis, to keep the star centered on the crosshairs. This method is relatively easy for exposures of several minutes, but the relentless need for close attention to the guide star can make lengthy exposures an uncomfortable challenge. Fortunately, individual exposures with a webcam are fairly short by nature, usually no more than several minutes at most. Long-exposure film astrophotography has embraced electronic autoguiders which automatically detect the need for corrections in right ascension and declination, and issue the proper commands to the mount's drive motors to compensate for the detected tracking error. However, dedicated autoguiders are expensive, ranging in some cases up to $2000, so their cost is an obstacle to the novice. However, relatively inexpensive long-exposure webcams make excellent autoguiders when used with any of the either free, or very affordable, autoguiding programs.

9.10 Problems With Accurate Tracking

No telescope drive is so accurate that it can track the stars for long periods without intervention. Even if the telescopes alignment and tracking were perfect, atmospheric scintillation and refraction would conspire to slightly shift the image. Fortunately, if properly polar aligned, most modern telescope drives are accurate enough to allow tracking for up to several minutes with no guiding corrections, as long as they are aimed near the zenith where atmospheric refraction is least. Mounts such as those by Astro-Physics, Losmandy, Mountain Instruments, Takahashi, and Software Bisque achieve an extraordinary tracking accuracy that was unheard of a generation ago. Such accurate mounts are important for the digital astrophotographer because of the nature of digital imaging. Current technology CCD sensors are limited by electronic noise to just several minutes' exposure, thus long-exposure digital images are really a stacked combination of shorter images. Each of these shorter exposures is brief enough that often they can be taken with little or no guiding corrections. Of course, good polar alignment is still needed to prevent field rotation, and to simplify image processing when the numerous short exposures are combined to form the final image.

If a telescope has a 120-volt drive motor, it is usually a synchronous one. Varying the frequency of the alternating current powering the motor controls the drive speed on such mounts. Newer electronically-controlled mounts use 12-volt pulse or stepper motors that allow more precise speed control and respond quickly to guiding commands.

Older mounts may not track as perfectly as those of the new generation, but a number of things can be done to improve their accuracy or make their known errors more predictable. The most common cause of small tracking errors is random manufacturing error in the drive gear. Each tooth in a telescope's drive gear is slightly different than the next. Consequently, it is not good to have the right ascension axis of a telescope perfectly balanced; a slight imbalance toward the east will make the drive have to "pull" the telescope from east to west and better compensate for irregularities in the gear teeth. Minimizing play in the mesh between the drive worm and gear is important to take up the backlash in the gear system. This will prevent the telescope from shifting if there is a gust of wind.

Periodic error is the cyclic speeding up or slowing down of the correct tracking rate resulting in alternating periods of fast or slow tracking. Periodic error usually occurs when the worm gear is not centered on its axis, but it will also occur if a perfectly machined worm and drive gear are not on the same plane. Periodic error repeats every cycle of the worm gear. This occurs every four minutes with a 360-tooth gear and every eight minutes with a 180-tooth gear. Thus for half of the worm's revolution, the drive will track well but the good tracking period is split by periods of faster and slower tracking. So about half of the 30-second exposures taken in a tracked-but-not-guided series will display slight trailing and have to be discarded.

Not all drives have the same periodic error. Mounts benefiting from modern machining techniques may display a cyclical periodic error of only a few arcseconds, while older mounts may have up to several arcminutes of periodic error. If periodic error is no more than three or four times the pixel resolution of the camera, it can be easily guided out.

Periodic error receives the lion's share of blame for tracking problems, but in reality only the good drive gears have true repeatable periodic

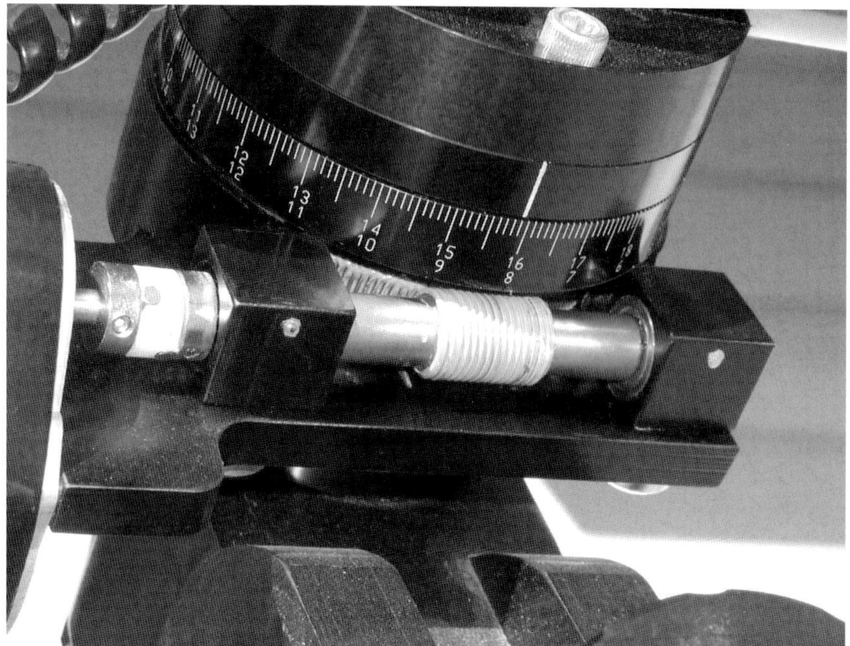

Fig. 9.7 *Gear lash in the drive worm gear, or improper worm gear alignment against the polar drive gear, will create tracking problems like those seen in Figure 9.8. Photo by Robert Reeves.*

error. The rest have random errors that are harder to predict. Newer mounts with electronic controls often incorporate Periodic Error Correction (PEC) that can be programmed to anticipate true periodic error and issue tracking commands to significantly reduce its amplitude. Although PEC can smooth out 90 percent of true periodic error, it can do nothing for random non-repeating errors.

To test your mount for tracking error—periodic, random, or both—use a long-exposure webcam to take a series of about 60 unguided 15-second exposures with the polar axis deliberately aimed several degrees east of the pole. This exposure sequence will span several cycles of the drive's worm gear. The deliberate polar misalignment will cause the stars to drift southward as they cycle back and forth from periodic error. If the resulting video clip shows the stars zigzagging in a smooth sine wave-like pattern, then the telescope drive is exhibiting true periodic error. However, if the stars have random zigzags or bumps, then the worm and drive gears need aligning, the telescope is grossly out of balance, or the gear set is poorly machined. Make sure tracking errors are actually drive errors and not flexure in the focuser or camera coupling. If the camera coupling and focuser shift just a few thousandths of an inch with long focal length systems there

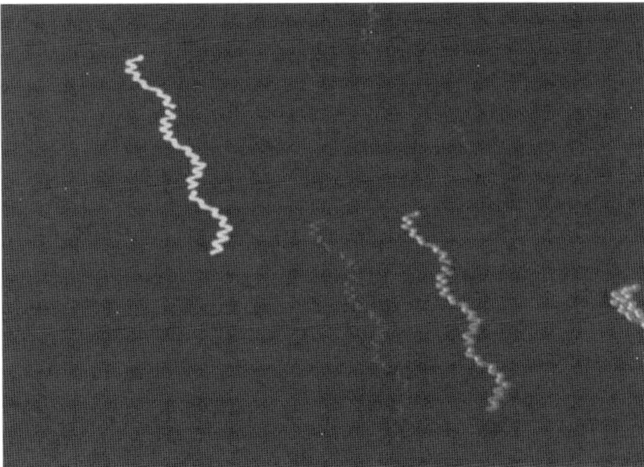

Fig. 9.8 *A good way to test the tracking performance of a telescope drive is to offset the polar axis by several degrees to deliberately create trailed star images while tracking, and then take a long exposure through the telescope. This combination of short exposures through an Atik ATK-2HS totals 24 minutes. The image shows the drive's sine wave-like periodic error that repeats at eight-minute intervals, but also reveals a problem with the drive worm that creates a tracking "jiggle" every time a worm tooth engages the drive gear. Photo by Robert Reeves.*

will be image trailing.

Nearly all telescope mounts now have a pushbutton controller for fine adjustments in tracking. Older telescopes must have a separate drive corrector to vary the frequency of the right ascension drive's motor voltage, and thus its tracking speed. Declination adjustments with older mounts can be made either manually or with a remote switch controlling a small motor on the declination knob. With four-button controllers that drive both right ascension and declination adjustments, orient the controller so the buttons align with the direction of their respective movement so there is no confusion over which button to push. If a direction is reversed on the controller, and the controller has no direction-reverse switch for that axis, turn it upside down and push the buttons from underneath.

If your mount has the capability, be sure to use the lunar tracking rate when imaging the Moon. As briefly mentioned in Chapter 6, the Moon moves through it's own angular diameter eastward every hour relative to the stars. Thus a lunar target will slowly slip out of the field of view if a standard sidereal tracking rate is used. The Moon's orbit around the Earth is an inclined ellipse and thus at times the Moon may be traveling either slightly faster or slower than the lunar tracking rate. Also, depending upon its position in its inclined orbit around the Earth, it can also drift slightly north or south.

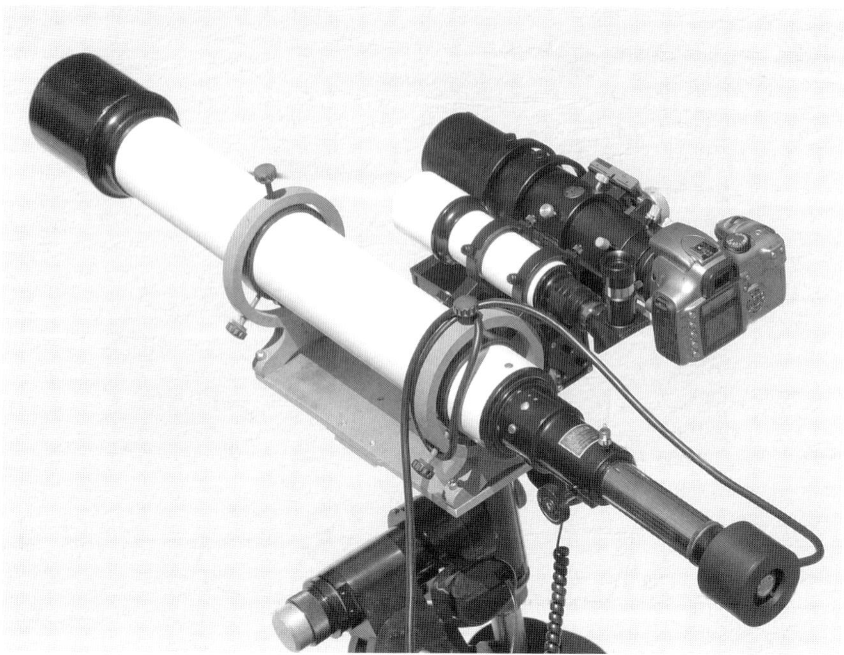

Fig. 9.9 *A long-exposure capable webcam can be used as an autoguider by utilizing a software interface that translates star movement on the CCD into telescope control commands. The author uses an 80mm refractor with an Atik ATK-2HS camera and GuideDog software to autoguide his Tele Vue 60is and modified Canon 300D. The long exposure-capable Atik ATK-2HS has never failed to find guide stars in a telescopic field. Photo by Robert Reeves.*

9.11 Causes of Guiding Errors

A majority of guiding corrections arise because of variations in the telescope's right ascension drive, or clock drive. Moderately well-aligned telescopes will need only occasional corrections in declination, usually always in the same direction, but the right ascension drive is prone to tracking errors caused by a number of factors. Lets examine the cause of these errors:

- **Imperfect polar alignment.** In order for an equatorial mount to properly follow the stars, its right ascension axis must be perfectly aligned with the Earth's axis of rotation. As we have seen in this chapter, there are several procedures that can be followed to accomplish this.
- **Periodic error.** The heart of a motor-driven equatorial mount is the worm and gear (or spur and gear depending on the design) assembly that actually turns the mount in order to follow the rotation of the stars. Tiny manufacturing errors in the worm and gear's machined mating surfaces cause recurring errors as the worm or spur rotates. These errors, called periodic error, cause the mount to slightly speed up, then slow down over

a period lasting several minutes.

- **Incorrect drive rate.** Drive rate is usually not a problem with today's electronically-controlled drive correctors, but the drive rate must match the rotation of the sky in order for the mount to track correctly.
- **Flexure.** The mounting between the telescope and a piggyback-mounted camera or separate guidescope must be absolutely rigid. Any mechanical play or bending between the two due to gravity as the telescope moves will cause the optical axes of the two instruments to diverge. This will cause stars to trail in the camera because it will no longer be aimed at the same spot in the sky as the guidescope.
- **Atmospheric refraction.** Refraction, or the bending of starlight, is greater the closer the star is to the horizon. If a star is descending toward the horizon, it will seem to slow down as refraction makes its apparent position seem higher. Conversely, if a star is rising from the horizon, it will seem to speed up as it approaches higher elevations. Shoot objects that only are above 30 degrees elevation, if possible, to minimize the effects of refraction.
- **Field rotation.** The wider the field of view of an image, or the lower the magnification, the more forgiving of guiding errors it will be. However if a telescope or tracking platform is not well polar aligned, field rotation will occur.

9.12 Choice of Guidescope or Off-Axis Guider

Modified webcams are theoretically capable of unlimited exposure length, but the practical limits imposed by thermal noise restrict the useful exposure to no more than several minutes under the best conditions. Unguided exposures of this length will likely suffer from trailing so a means must be used to guide them. If the camera is used to photograph through the telescope, it will not be possible to guide with the latter because the camera is in place of the eyepiece. There are two ways to guide when the camera is shooting through the scope—a separate piggyback-mounted guidescope, or a device called an off-axis guider.

A separate guidescope is usually a refractor that is smaller and lighter than the main telescope and rigidly attached to its mount. Since a camera is at the focus of the main instrument, a guidescope's sole purpose is track a guide-star and thus "steer" the main scope to accurately follow the stars. The primary problems encountered with using a separate guidescope are added weight on the telescope's mount, the extra cost of the second telescope, and the possibility of flexure between the two instruments. To work properly, a guidescope mounting must be light enough to not overtax the capacity of the main mount, yet strong enough to prevent the guidescope

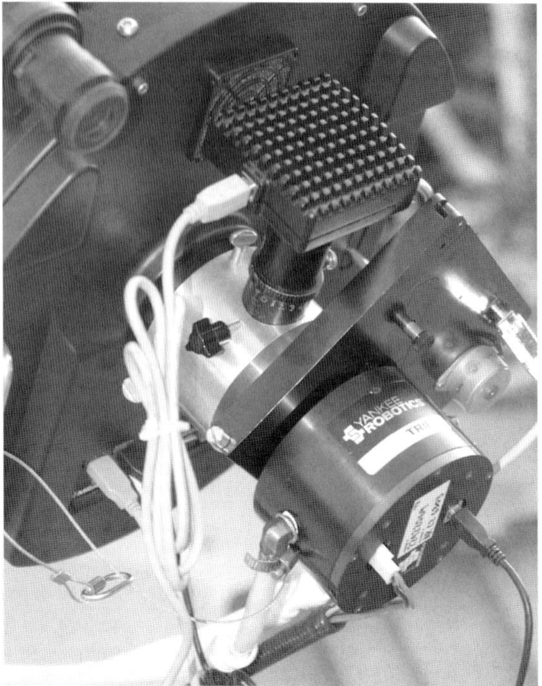

Fig. 9.10 *Experienced deep-sky imager Jason Ware reports that the Meade DSI camera not only works as a good autoguider, it actually outperforms the classic ST-4 autoguider that is regarded as "benchmark" for performance. Here, Ware uses the DSI with Maxim DL software to autoguide his Yankee Robotics CCD camera attached to a Meade 16-inch SCT. Notice the steel safety lanyard that secures the camera to the telescope, preventing disaster in case the assembly falls. Photo by Jason Ware.*

from shifting, drooping, or flexing in any way during the exposure. Any relative movement between the main and guidescope will cause oval, trailed, or hook-shaped stars on the image, even if the guiding has been performed perfectly.

Refractors are usually the choice for guidescopes because of their solidly mounted optical components. Reflecting telescopes are notorious for having optics that can shift slightly as the telescope moves during an exposure. The user should be especially wary of Schmidt-Cassegrain telescopes unless a means has been added to solidly lock the primary mirror as shown in Figure 5.9. Many of today's Maksutov-style instruments also to achieve focus, so the same caution should be applied to them if used as a guidescope.

Most practical guidescopes have a focal length about half that of the primary instrument and achieve the needed focal length for visual guiding by using a Barlow lens. Also, the primary telescope can use a focal reducer to decrease its focal length, with an apparent gain in photographic speed, and a further increase in the ratio of guidescope to primary instrument focal length.

The combination of eyepiece and Barlow lens used in a guidescope should be chosen so as to produce three or four times the magnification of the image in the camera's field of view. The greater the guiding magnification, the quicker it is to spot and correct any tracking error before it becomes large enough to affect the image in the camera. If an electronic autoguider is used, the focal length of the guidescope can be shorter than the focal length of the camera/telescope system taking the image. Autoguiders are, of course, tireless, constantly vigilant, and respond to slight deviations in tracking much faster than a human looking through a crosshair eyepiece.

If a telescope is too small to carry a piggyback guidescope, or if the optical configuration of the main telescope presents the possibility of shifting optical components, then an off-axis guider (OAG) can be used. The OAG method is cheaper than using a separate telescope and eliminates the differential flexure problem by having the guiding device view through the same telescope as the camera. An OAG is placed at the focus of a telescope and uses a small prism or mirror to "pick off" a portion of the image and divert it 90 degrees to a guiding eyepiece or autoguider. The pick-off prism lies just outside the field of view imaged by the attached camera, so there is no prism shadow or occultation in the camera's view. An OAG therefore both guides and takes the image using the same optical system. The advantages of an OAG are that it is lighter and cheaper than a separate guidescope, and it does not have separate optical components that can shift and flex relative to the main telescope. If the latter's own optics shift slightly, the view through the OAG will show this flexure, allowing corrections during guiding. An OAG can thus be used successfully with any kind of telescope, even a Schmidt-Cassegrain.

A major advantage of using an OAG is that a focal reducer can be installed between the OAG and the camera. This provides a wider, brighter field of view for the camera with a lower apparent focal ratio, while the guiding eyepiece or autoguider still sees the telescope's full focal length making it easier to minimize guiding errors.

OAGs eliminate differential flexure, but present their own unique challenges. Since the pick-off prism is at the very edge of the telescope's view it is gathering star images that suffer from coma, an aberration common to the extreme edge of a telescopic image. Compounding the problem is the fact that OAGs are used most often with SCTs, which have spherical primary mirrors that are very susceptible to coma. This, coupled with the small area of the pick-off prism, presents us with dim, fan-shaped stars with which to guide.

The image is framed and focused normally when using an OAG. The guiding eyepiece or autoguider is then slid in or out of its holder to focus it separately from the camera. At this point, there is usually no star near the guiding eyepiece reticle. If the target is small and there is maneuvering room in the camera's field of view, the telescope can be moved to a guide star if one is in the eyepiece field of view. If there is little room to shift the aim of the scope, another solution is the use of a projection reticle. This device, inserted between the telescope and any eyepiece, projects a virtual reticle onto the eyepiece field of view. It has several advantages, such as enabling the projected reticle to be moved to any star in the field of view, and usability with eyepieces of any power or eye relief—a boon for those who wear eyeglasses.

If there is no star visible at all in the field of view of an OAG, then it is time to hunt for one by rotating the entire guider assembly around the optical axis of the telescope. If a suitable guide star is only available when the OAG is rotated into a difficult-to-view position, then a star diagonal may be needed to conveniently access the eyepiece. This is not a problem of course with an autoguider, as it will work in any position.

Separate guidescopes have the advantage of presenting a sharp bright star to guide on. This reduces fatigue with visual guiding, or presents a crisper image for an autoguider to lock onto. Off-axis guiders on the other hand, by their design, only have access to the edge of the telescope's light cone. The image in this area suffers from field curvature, coma, and possibly other aberrations, rendering dim, distorted, and blurry star images. Autoguiders, like human eyes, prefer to view sharp round stars. So OAGs can thus present challenges to the user, but overall they are worthwhile. Indeed, sometimes they are the only guiding method possible with an imaging setup, due to weight limitations.

9.13 Webcam Autoguiding

If your telescope mount has a six-wire RJ-12 modular jack to accept a hand controller cable, then it will most likely accept inputs from an autoguider. On the popular Losmandy mounts, for instance, the port is labeled HC/CCD (hand controller/ccd) and looks like an overgrown telephone cable jack. Other mounts such as the Meade LX series and most electronic Celestron mounts also have ports to accept input from autoguiders. If the mount has a serial interface, such as that used for a computer planetarium program to control a GOTO mount, it can also be used for autoguiding control.

For nearly two decades, the "industry standard" autoguiders have been either the ST-4 and SBIG STV units made by Santa Barbara Instruments Group (SBIG). However, the more affordable ST-4 is no longer manufactured, and the $2000 SBIG STV is priced higher than the casual user may be comfortable spending. But the good news is that inexpensive webcams can perform functions similar to dedicated autoguiders costing ten times more.

A webcam alone is not capable of guiding a telescope, but using freeware software and an inexpensive parallel computer interface, a working webcam autoguider system can be created for about the price of a good eyepiece. It may be more convenient to use a turnkey autoguider like the STV, but a lot of nice accessories can be purchased with the nearly $2000 saved by using the webcam autoguiding system. Moreover, once one is assembled, it is easy to use in its own right.

There are many different webcams capable of performing autoguider duties, but the most popular are CCD-based models like the Philips ToUcam, SPC900NC, and the Logitech QuickCam Pro 4000. The CCD offers two advantages over a CMOS detector: higher sensitivity; and the camera can later be "SC" modified for long exposure, thus yielding greater sensitivity. Even unmodified, these webcams are capable, using an average telescope, of guiding on stars down to 6th magnitude. I find that if a star is visible in the 6x30 finder on my classic old Celestron-8, the webcam will be able to image it and the GuideDog autoguiding software will lock onto it.

When using a webcam for autoguiding, do not use the infrared filter normally used with webcam imaging. Stars radiate a lot of their energy in the infrared portion of the spectrum. When autoguiding, we are not concerned with focus degradation or colorcast caused by infrared, but instead we desire the infrared in order to see fainter stars.

A limitation of approximately 6th magnitude with an unmodified camera will restrict the selection of guide stars but will be adequate for wide-field piggyback astrophotography with a film or digital camera, particularly if the camera mount allows slewing the camera slightly if the field does not contain a bright guide star. If fainter guide stars are needed for telescopic astrophotography, an SC-1 modified camera using multi-second exposure times will enable guiding on stars down to at least 10th magnitude. The ToUcam is easiest to modify for the "SC" long-exposure option. The Meade Deep-sky Imager camera, as well as various models of the Atik and SAC webcam-based cameras, will also work as long-exposure autoguiders capable of imaging very faint guide stars.

An autoguiding software program calculates the centroid of the guide star image, that is, it mathematically locates the exact center of a multi-pixel star image. This allows the software to determine the star's location on the imaging sensor to a sub-pixel accuracy level. For this reason, the best accuracy is achieved with guidescope focal lengths short enough to produce "fat" stars with reasonably short exposures. It may seem counter-intuitive, but long focal lengths and short exposures that produce guide star images that only cover a single pixel will actually result in worse autoguiding because the guiding will jump from pixel to pixel as the centroid algorithm searches for a non-existent multi-pixel center. Some people have successfully autoguided wide-field astrophotos with a 200mm camera lens as a guidescope using software with centroid calculation.

A number of software packages offer autoguiding control with a webcam. The most popular are the freeware program GuideDog, created by Steve Barkes, K3CCDTools, and Maxim DL, an advanced imaging program. Although K3CCDTools is an excellent camera control program and offers autoguider functions, I prefer to use GuideDog to control the telescope and cannot comment on K3CCDTools' autoguider functions. The GuideDog software is free, very easy to use, and supports many models including long-exposure SC-modified cameras, Meade DSI, SAC, and Atik cameras. It can be configured to work with either serial or ST-4-style "relay" output.

The ST-4-style autoguider "standard" requires four guiding signals, one for each direction in RA and DEC, as well as a ground. In practice, the six-wire RJ-12 connection works well for this. Most telescope hand controllers interface with the mount using this type of connection with the sixth wire either left unused, or to power the controller's map light. To interface the RJ-12 modular cable with a computer parallel port, a Shoestring Astronomy GPINT-PT (Guide Port INTerface) is required. This is a pass-through connector that installs on the computer parallel port and permits the usual parallel peripherals to still be attached, but also has a modular port to allow the six-wire RJ-12 cable to be attached. If a long-exposure-modified camera is used, the parallel cable used for exposure control will plug directly into the open pass-through port on the GPINT-PT adapter. K3CCDTools version 2.2.4 or later also support the use of the GPINT-PT. The six-wire cable from the GPINT-PT plugs into the same RJ-12 jack where the hand controller plugs into the mount. An RJ-12 splitter adapter allows both the hand controller and the autoguider input cable to attach to the telescope so that both functions are possible. Splitters are available from Shoestring Astronomy. The advantage of having both controllers active is that objects can be centered using the normal telescope hand controller, and then once the guide star is locked on, the autoguider

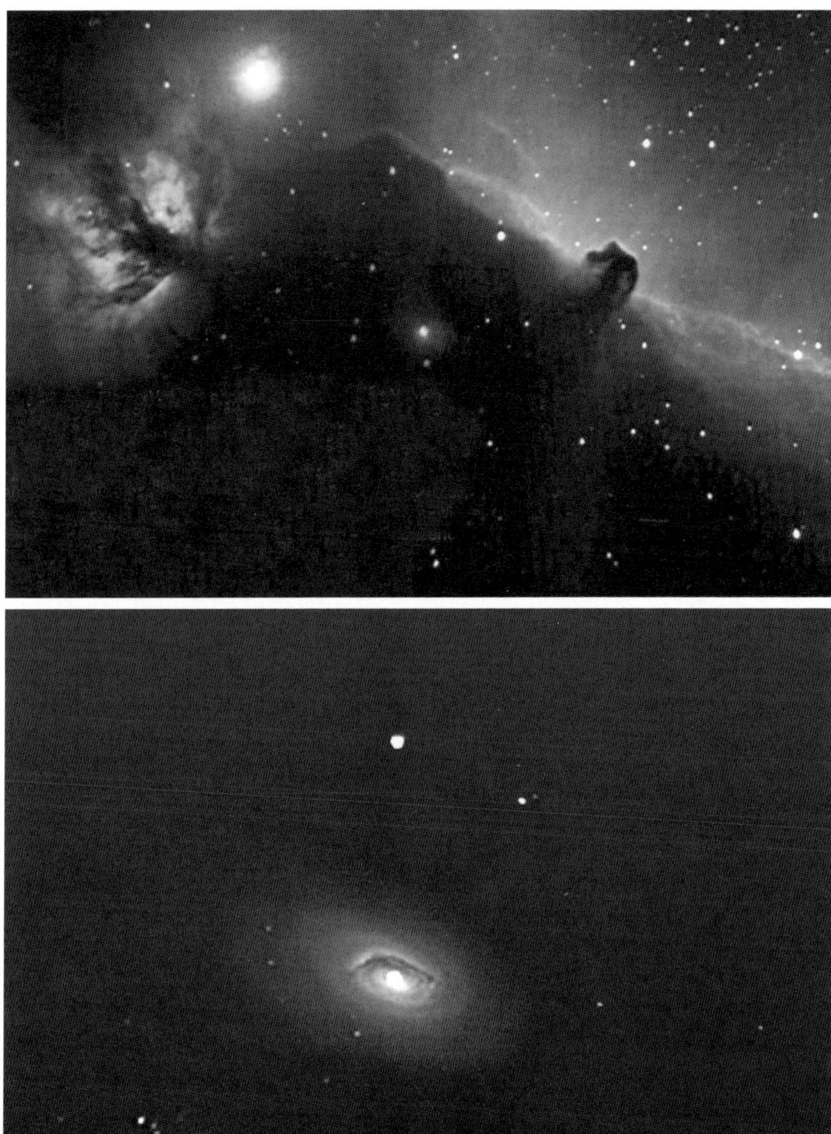

Fig. 9.11 *With today's inexpensive autoguiding equipment, long exposures that used to tax the astrophotographer's endurance are now routine. Good polar alignment is a must for such lengthy exposures. The Flame and Horsehead Nebula (top) were taken with eight separate 15-minute exposures using a SAC8 camera and Orion 80ED refractor. The Blackeye Galaxy, M64, (below) was exposed for 50 x 30 seconds with a SAC8 camera and 0.3 focal reducer through a 10-inch LX200GPS telescope. Photos by Dave Street and Tim Tasto.*

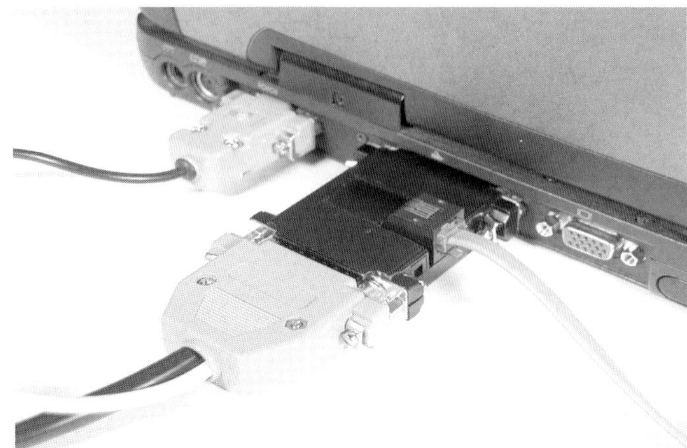

Fig. 9.12 *Shoestring Astronomy markets the GPINT-PT (Guide Port INTerface) parallel interface that connects an RJ-12 telescope-control cable connector to a computer's parallel port for autoguiding, and also allows a modified-webcam exposure control parallel cable to connect to the same port. Shoestring also offers USB adapters to allow telescope control and webcam exposure control cables to interface with newer USB-only computers. Photo by Robert Reeves.*

can be activated to control the telescope drive.

The Shoestring parallel port adapter is known to work with the following mounts: Losmandy GM-8 and G-11 (with and without the Gemini box), Meade LX-200, Meade LX-90, Celestron NexStar i-series, Celestron Celestar 8, Celestron Advanced Series GoTo, JMI NGT, and the Tom Osypowski Equatorial Platform. The EQ-style dual-axis hand controller can be easily modified to accept the Shoestring adapter. The EQ controller is sold as the CG series by Celestron; as EQ-1, EQ-2, EQ-3 and Sky View series by Orion, and as EQ3-2 and EQ-5 from Sky-Watcher. Adapter kits to modify the EQ controllers to accept the six-wire RJ-12 cable interfacing with the parallel adapter are available from Shoestring Astronomy for about $7.

The diminishing use of parallel ports on new computers has led to the need for USB-compliant interfaces. A USB-to-parallel adapter will not work with the GPINT-PT autoguider interface. However, if the mount being used is capable of receiving guiding inputs from a serial interface, a USB-to-serial adapter will work. If your computer is USB only and has no parallel port, Shoestring Astronomy has a number of new USB interface devices for digital astrophotography. The GPUSB interface will allow autoguiding with a standard webcam. The LXUSB interface will enable a long-exposure modified webcam to operate on a USB-only computer. The DSUSB interface allows shutter control on a SLR digital camera being guided by a webcam autoguider.

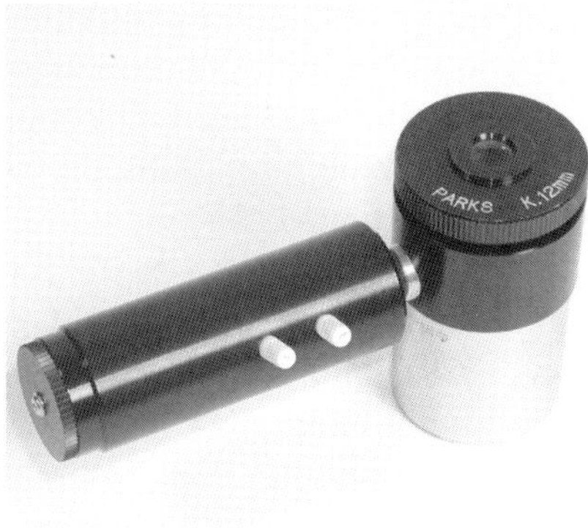

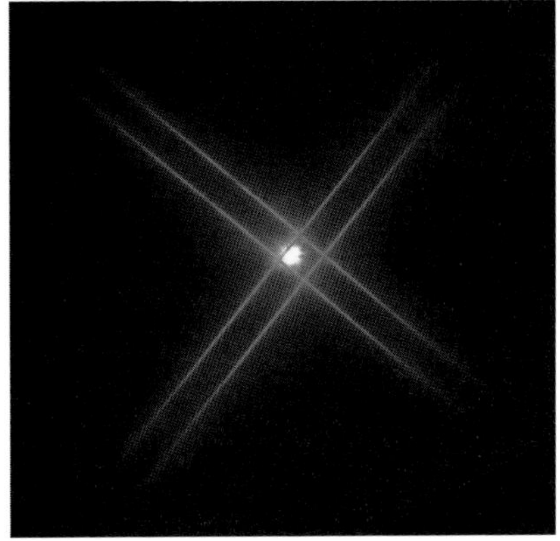

Fig. 9.13 *Manual guiding is accomplished with an illuminated reticle eyepiece. A red LED illuminates the crosshairs, enabling the guide star to be accurately tracked. Older guiding eyepieces can be retrofitted to wireless operation by adding the PulsGuide battery package. Photos by Robert Reeves.*

Autoguiders add complexity to the imaging system and require another computer to run the guiding camera, especially if it uses a USB 1.1 connection. But take it from someone who has manually guided countless astrophotos—you will never look back after switching to an autoguiding system.

9.14 Illuminated Reticle Eyepieces

The art of manual guiding using an illuminated reticle eyepiece is declining because of the wide availability of both dedicated autoguiders and the use of webcams and software for autoguiding. The increasing popularity of the "tracked, but not guided" technique also decreases reliance on manual guiding. However, thousands of illuminated reticle eyepieces are in existence as carryovers from the film era, and they are still applicable to guiding long-exposure webcam astrophotographs. Indeed, their use for webcam imaging is far less grueling than with film because each exposure is much shorter. For these reasons we will review these devices for use with webcams.

There are two types of illuminated reticle eyepieces: those with the reticle permanently built into the eyepiece, and those with a device called a projection reticle. The latter, as mentioned earlier in this chapter, is a device that installs between telescope and eyepiece and projects a virtual reticle onto the focal plane of the latter. This allows any ocular to become a guiding eyepiece, along with the advantage of selecting various magnifications and eye relief as needed. Most projection reticles are adjustable; that is you can move its location of the reticle within the eyepiece's field of view. This is a major advantage in manually guiding prime-focus photography because the camera's aim and subject framing, often restricted by the size and celestial orientation of the target, sometimes cannot be moved without losing part of the target.

Guiding eyepieces with built-in reticles come in a variety of styles; single crosshair, dual crosshair (resembling a tic-tac-toe pattern), or a grid pattern for estimating angular separation of objects in the field. Adding to the versatility available, they can have self-contained batteries, external power supplies connected by a wire, and blinking and rheostat options that vary the brightness of the reticle to allow viewing extremely dim guide stars that would otherwise be obscured. The majority of them have a standard 1.25-inch diameter barrel. The 12.5mm focal length dual-crosshair model is most popular and is widely available. Meade also makes a 9mm Super Plössl that has the advantage of wireless operation, higher magnification, and an adjustable reticle that can be moved via small jackscrews. Celestron offers the 12.5mm Micro Guide eyepiece that has a reticle with various angle, size, and measurement scales as well as a guiding bull's eye and crosshair. These scales may be useful for double-star measurements and other aspects of visual astronomy, but many find the view through the Micro Guide too "busy" for optimal guiding.

An important check when using a guiding eyepiece is to insure that the reticle and guide star are both at focus in the same plane. This can be checked by shifting the eye from side to side; the guide star should not move relative to the reticle. If it does change, this will create phantom movements of the guide star when the eye shifts slightly during the course of the exposure. To correct this, most guiding eyepieces have a diopter adjustment so the focus of the illuminated reticle can be set for your particular vision.

I personally recommend the dual-crosshair variety of eyepiece because the "box" formed at the intersection of the pair of crosshairs provides a reference not available with single crosshairs. This can be useful in estimating the acceptable amount of guide star drift before the image begins to trail. For the most part in piggyback astrophotography, if the guide star is confined to the crosshair box, images will be very well tracked. Align the reticle so the right ascension and declination movements of the telescope will be parallel to their respective crosshairs.

Another feature I recommend is the ability to dim the reticle's illumination. Nearly all illuminated guiding eyepieces have this capability, and it really comes in handy when a bright star is simply nowhere near the target field. I personally like to use one with a blinking option for dim stars. It is quite fatiguing to follow very faint guide stars for any period of time, especially late at night. Setting the reticle to blink off and on in a time cycle comfortable for you makes the job a lot easier. It allows the eye to briefly rest from the glow of the reticle and stay more sensitive to dim stars. The guiding process becomes a series of updates, as first the star is visible by itself, then the reticle briefly overlays it for reference, the blinks off again. This system is admittedly not for everyone and some photographers find it annoying. Another technique for very faint stars is to turn the crosshairs 45 degrees relative to the right ascension and declination movement of the star. This way the star will not creep behind a crosshair and be lost.

Most illuminated guiding eyepieces use a red LED. After staring at bright red crosshairs for a time, the eye can become blind to them, making needed corrections hard to notice. Some eyepieces allow switching to green light to rest the eye. If you suffer eye fatigue from lengthy use of red LED illumination, the red light source in most eyepieces can be changed for a green LED obtained from Radio Shack. Just insure the polarity is correct when installing the new one.

If the guiding eyepiece you have does not have a pulsing option, it may be retrofitted with a pulsing illuminator with a battery package offered by Rigel Systems which also converts the eyepiece to wireless operation. The retrofit package has 8-millimeter threads and screws into the side of the eyepiece in place of the existing corded illumination. Rigel Systems offers 10- and 14-millimeter adapters in case your eyepiece has a larger threaded opening.

Guiding eyepieces vary considerably in price with basic models running about $50 while higher quality ones with more options can cost as high as $200. Excellent guiding eyepieces are becoming available on the used equipment market, however, as more and more skyshooters upgrade to an autoguider.

Be aware that light from an illuminated reticle may backscatter through the OAG to the image sensor. Before your first astrophoto session, run a test to make sure light from the LED does not hit the CCD. Cap the telescope and do two long exposures, one with the eyepiece illumination off, then on. The CCD should record almost exactly similar electronic noise in both exposures.

9.15 Manual Guiding Techniques

The guiding tolerance for piggyback astrophotography using a camera lens on a webcam is far more lax than that needed for prime-focus work. But

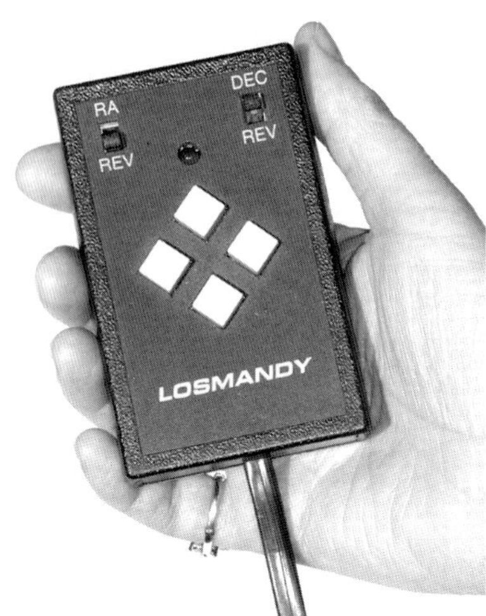

Fig. 9.14 *The easiest way to control a telescope with "old-fashioned" hand guiding is via electric slow-motion controls operated with a push-button hand controller. E-W and N-S buttons correct drift in the right ascension and declination axes. The small switches at the top of this hand controller allow reversal of the direction of motion of each axis, if needed, to match the view through the guiding eyepiece. Photo by Robert Reeves.*

even one night of nudging a guide star while doing wide-field work will instill great respect for those who manually guide high-power astrophotos requiring ten times the precision of piggyback imaging. A factor working in favor of the digital astrophotographer is that exposures are limited by electronic noise, which forces one to break up a normally lengthy exposure into a series of shorter exposures. Each is limited to several minutes, at best, and they will later be combined with image processing. This necessity of splitting the overall exposure into segments even allows one to take a brief break if needed. Any substandard exposures acquired during the break can be eliminated from stacking when the images are processed.

There are many different ways to manually guide. After some practice, your own style will begin to evolve. The main thing is to get comfortable. If the overall stack of exposures is going to be lengthy, make sure you are seated so the eyepiece will be accessible throughout the shot. It can rise or fall quite a bit as the telescope tracks an object across the sky.

After you are comfortable at the eyepiece, practice guiding for a minute before starting the exposure. See which way the controls move the telescope in response to the guide star's movement. Align the pushbuttons on the hand controller to correspond with the directions the star moved along the crosshairs.

With good polar alignment, the guide star will slowly and gently rock back and forth in right ascension because of the drive's periodic error. Unless an extremely accurate alignment has been achieved, a star will usually drift in one direction to require a correction every few minutes. This drift is usually constant and can be predicted after the first couple of cycles.

Some people like to slightly defocus the guide star to make it more visible against the crosshairs. This is not a good idea because it will introduce parallax effects that can shift the guide star's true position relative to the crosshair. This presents little problem with low-magnification piggyback work, but it can lead to erratic guiding with high-power photography.

Too much backlash between the drive worm and gear can be troublesome and create a "dead zone" where the telescope will not track properly and guiding corrections will not be effective. Almost all drives have some backlash. In right ascension it is usually taken up by the drive motor, which is continuously pulling the telescope toward the west. In declination, backlash can cause delay in corrections in the reverse direction, as the reversed motor has to take up the slack. Many electronically controlled mounts have an "aggressiveness" setting that speeds up direction changes in declination.

If declination play persists after you adjust the gear, there are several things that can be done to eliminate a correction dead zone. The excess slack can be taken up by slightly unbalancing the telescope in the north-south direction. If this does not work, deliberately move the polar axis slightly east or west of the true pole. This will cause the guide star to drift in only one direction, thus removing the backlash.

My suggestion is that if you are a novice at guided astrophotography and are planning to use an autoguider, you should learn to manually guide first in order to have an understanding of what the autoguider will be doing once you start to use one. Even if you don't pursue manual guiding for long, the illuminated reticle eyepiece used will be a handy accessory to help achieve good polar alignment.

9.16 Final Check List for Polar Alignment and Manual Guiding

1. If possible, use a refractor for polar alignment and guiding to prevent mirror shift from changing the true aim of the instrument.
2. Make sure the finder and telescope optical paths are parallel. Use a straight-through eyepiece on both when polar aligning, both for simplicity and because star diagonals can induce field aim-point errors.
3. Do not add or remove accessories after performing a critical polar align-

ment. Changing the weight carried by a mount, particularly a fork mount, will alter the "droop," and thus the aim of the polar axis.

4. When using a Schmidt-Cassegrain telescope for guiding, be sure to reach final focus while turning the focus knob counterclockwise. This leaves the focus adjusting screw pushed against the mirror and prevents it from settling backwards due to play in the focus mechanism.
5. Make sure the declination setting circle is aligned. On fork-mounted SCTs they can slip, but can be adjusted by loosening the bolt at the center of the declination circle.
6. The drift alignment method is time consuming, but it can accurately align the telescope's mechanical axis, regardless of its optical alignment.
7. The fatigue of continuously by monitoring a guide star can cause your eye to "see" non-existent crosshairs. Frequently switch from one eye to the other to prevent eyestrain.
8. If your mount has the option to do so, use the King drive rate at elevations below 30 degrees in the east and west. This special drive rate alters the tracking speed of a mount to partially compensate for the refractive effects of the atmosphere at lower elevations.

Chapter 10
Computers for Webcam Imaging

Webcam astrophotography can be performed quite well by anyone who uses a computer. One does not need to know everything about computers in order to take great celestial images, but they are important tools for digitally imaging the heavens. And the more you know about a tool, the more versatile it becomes as you master its capabilities and limitations. Thus, the webcam astrophotographer is like a race car driver who can gain greater performance from it through intimate knowledge of how it is constructed and how it works. This chapter, then, is directed toward the computer user who wants to "look under the hood" to learn a little more about how his or her computer and its peripherals can lead to better, and more enjoyable, astrophotography.

10.1 Desirable Computer Features

Unlike digital cameras that are standalone imaging devices, it is mandatory to connect a webcam to a computer in order for it to function. Not only does the computer provide electrical power to a webcam through its USB cable connection, it provides the operating program that controls the camera and the hard drive that acts as the camera's recording medium.

If your imaging is being done on your home's driveway or patio, it is possible to place a spare desktop computer and monitor on a cart and roll them to the telescope. But otherwise it is impractical to transport a desktop model to the telescope, especially out in the field, so most webcam photography is performed with a laptop computer. Since typical nocturnal conditions can be rough on a laptop, most imagers opt to use an older "disposable" model, so there is no great loss if something catastrophic happens. Pentium II laptops that are several years old still have plenty of horsepower to run USB 1.1 cameras and image processing programs. The newer USB 2.0 compatible cameras, such as the Philips SPC900NC, require a Pentium III 500 Mhz computer.

The key variables in laptop suitability are the amount of RAM and hard drive capacity. Most have at least 128 Mb of RAM which is sufficient to operate the camera and run the image processing programs. But, as your expertise in astrophotography increases, so does your reliance on computing power as powerful programs are employed. So the bottom line is to get enough RAM to cover future applications. That means it is best to bite the bullet and get 512 Mb of RAM to begin with.

The fundamental advantage of using a webcam to image solar system objects is its ability to accumulate images. It is not unusual for a two-minute video sequence of the Moon to take up several hundred Mb of hard drive space. This usually calls for a hard drive upgrade on older laptops that may have as little as four gigabytes of on-board storage. Fortunately, the smallest laptop hard drives available today have 20 Gb of capacity and cost less than a good eyepiece. Twenty Gb should be considered a minimum hard drive capacity, as a single evening of lunar or planetary imaging will fill it completely, and videos would have to be processed or transferred to CD or DVD to make room for further imaging the next night. An 80 to 100 Gb hard drive would be preferable because it has sufficient storage space to hold several evenings' worth of imaging. If your laptop lacks a CD-RW drive with which to archive your digital images, it may be more cost-effective to set up a wireless network with an existing desktop PC that does have this capability, instead of buying a whole new laptop or a USB external CD or DVD burner.

For users who do their astrophotography with long-exposure capable SC-modified cameras or store-bought long-exposure cameras, older laptops are preferred because of the current need for serial and parallel interfaces to operate them. Newer laptops rely on USB connections and often do not have parallel and serial ports. Cameras now on the drawing board will rely only on a USB connection, but unfortunately today's long-exposure capable webcams use parallel connections for shutter control. Users of USB-only laptops have to rely on USB-to-parallel converters.

Laptop display screens can be dazzling at night and ruin your dark adaption. Bright computer screens also make the owner unpopular at star parties. Some astronomical programs feature a "night vision" mode where the display is rendered a deep red, but the intensity of the screen brightness may still be objectionable. Some laptops have an option to dim the display, but this is usually not enough under dark conditions. A full-screen filter made from red-tinted plexiglass, rubylith foil, or deep red cellophane will go a long way toward making a laptop more vision-friendly in the night.

Almost all of the United States can see subfreezing nights, at least part of the year. Laptops performing astronomy duty need to be protected from dew and low temperatures. Power saving options that turn off hard drives

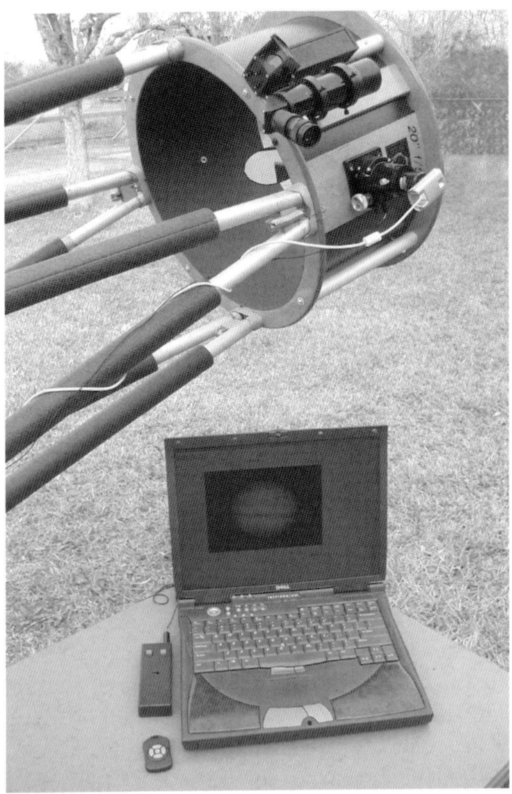

Fig. 10.1 *A laptop computer is no longer a luxury item for astrophotographers. In the era of digital imaging, a computer is essential for camera control, image acquisition and storage, telescope control, autoguiding, and image processing. Photo by Glenn Schaeffer.*

after a certain period should be disabled in order to keep the moving parts inside the laptop warm. Should moisture condense and freeze on the hard drive it will fail. An insulated carrying case or a heating pad designed to keep the laptop warm will maintain operating temperature on cold nights.

New laptops are becoming incredibly powerful with Pentium 4 models available with 3 GHz processors and impressive graphics. They are often considered desktop replacements, and can be very pricey. However, last year's high-end model is this year's "value" model, but it is still extremely capable of performing any function desired by the digital astrophotographer. Low-voltage consumption mobile processors like the Pentium M allow long battery life, averaging five hours.

A laptop should be considered as part of the imaging system, but not the final destination for images intended for enhancement in an image-processing program. For the finishing touches in any processing, do not use the laptop's display screen alone. The liquid crystal display used with even the best laptops still lacks the basic colors and dynamic range that a standard desktop CRT-type computer monitor offers. Initial image stacking may be done on a laptop, but the final corrections in color and shades should be done using a CRT display in order to see its full range. If the model you use has a powerful graphics driver, it is possible to plug in an external CRT monitor and use the laptop for the processing while viewing the results on the CRT.

If you are buying a new or used laptop for your webcam imaging, consider the following points:

1. Get as much hard drive capacity as affordable. It is amazing how quickly videos and images accumulate on it. Remember, each download of a two-minute AVI adds another one-quarter gigabyte of data to your hard drive.
2. Get 512 Mb of RAM if possible. The added cost of RAM upgrades at the time of purchase is minimal, but provides a significant increase in performance with image-processing programs.
3. Purchase the highest screen resolution model you can afford. The minimum desired resolution should be XGA, or 1024 x 768. Any less will make working with today's computer programs difficult. If you use your laptop for other graphics-intensive work as well as astrophoto work, consider higher resolution SXGA+ (1400 x 1050) or UXGA (1600 x 1200); or if you are going for the bragging rights of having a wide-screen laptop, resolution choices extend to WXGA (1400 x 900) and WVXGA (1920 x 1200), but with a premium price.
4. Get a combination CD/DVD-RW optical drive with which to archive your images. Remember, a disc full of digital images is the same as a plastic negative or slide file page in film photography. It is safekeeping for your images.
5. For a better price, consider slower processor speeds. Older 800 MHz to 1.33 GHz processors are still fast compared to what was available several years ago, and they are much cheaper than today's 3 GHz models. Consider what these processors do. Is it worth the extra expense to perform an operation in half a blink of an eye instead of just a blink?
6. Consider the keyboard configuration. A standard desktop keyboard has a 19mm key pitch with 3mm travel, and the space bar is centered under the "b" key. Some notebooks maintain this standard pitch while others downsize the keyboard for ultra-portable machines. Most users navigate an 18.5mm pitch with little problem, but keyboards with an 18mm or 17mm pitch are just too small for comfort.
7. Most laptops have at least two USB ports, but accessories like mice, camera control and interface cables, accessory keyboards and external drives, all compete for USB interfaces. Get at least four USB ports if possible to

avoid the need for external USB hubs that add to cable clutter at the telescope. A faster USB 2.0 interface is preferable because even if your present accessories are USB 1.1, future items will be USB 2.0.

8. Consider buying a spare battery to allow independence from power sources in the field. Also, today's lithium-ion batteries are high capacity for their small size, but have a limited life, being good for about 300 to 500 charge cycles. That is why two-year old laptops do not run as long as they did when they were new. A spare battery will be needed sooner or later.
9. Refurbished laptops can be a bargain. Check Compaq, Dell, or Gateway's web sites for current offerings on fully warranted used laptops at significant savings.
10. Be sure to accessorize your laptop with a mouse and DC-to-AC power inverter.
11. If the laptop is to be extensively used in the field, consider purchasing an extended warranty. For the extra $50 to $100 you have peace of mind that the nighttime elements will not permanently kill off your computer.

10.1.1 Monitors

Many astrophotographers use a high-end laptop as their only computer. Powerful models may have good graphics cards in them, but compared to standard cathode ray tube (CRT) displays, their flat-panel display screens are limited in dynamic range. This is most evident when trying to process a planetary video on a laptop using RegiStax, then viewing the result later on a desktop computer using a good CRT display. The image appears noticeably superior in the CRT.

The main problem with flat-screen displays is their inability to show the widest possible dynamic range of grayscale, particularly shadow detail. This is a vital factor in displaying astronomical photos which, by nature, have large areas of dim detail. Shadow detail falls into the display area known as "black level." This term is seemingly paradoxical because there is only one level of black, zero luminance, but it applies in principle to the dark area of the display's grayscale. Flat-screen displays tend to compress the grayscale, slightly clipping the high and low ends and creating a higher contrast image with reduced dynamic range.

With a flat-panel display, each pixel is never completely "off." On a CRT display, the screen phosphor will be black if there is no electron beam projected onto it to excite the emission of photons. But both types of display always have some level of signal applied to each pixel to prevent image burn-in, and black levels on a flat panel are inherently brighter than the phosphor on a CRT screen. The result is that a good CRT such as a

Prinston DisplayMate can achieve contrast ratios of 440:1 while a plasma display will achieve only 60:1. Careful measurement of the darkest grayscale levels on each display reveals that the CRT registers 0.2 nits. A "nit" is an industry-standard measurement of the luminance of a computer display. A similar measurement of the flat-panel display registers 3.6 nits, therefore it shows "black" 18 times brighter than the CRT. The net result is that on a flat-panel display showing a 16-step grayscale ranging from pure white to completely black, the two blackest steps blend together, limiting the dynamic range of any image shown.

Although with deliberate effort the human eye can perceive wide contrast ranges, in the practical terms of everyday viewing, it is limited to a contrast ratio of about 100:1. Under such circumstances flat-panel computer monitors look quite good. They are flicker-free, have bright colors and today's units have wide viewing angles making them very attractive in the store display. However, they still lack the full dynamic range of a CRT monitor.

10.1.2 Video and Image Storage

It is ironic that webcam astrophotographers work to achieve what are essentially some of the smallest images in the field of digital photography. Even when stored as a TIF or BMP file, each aligned, combined, and wavelet-processed image produced from a webcam AVI video is less than one Mb in size. However, while webcam users produce 640 x 480 pixel images for each completed exposure, they also face the need for the largest image storage capacity of all types digital astrophotography. A two-minute lunar AVI taken with a webcam featuring the normal 640 x 480 pixel resolution will produce an image file nearly 250 Mb in size. Although it is unlikely that you will store and save every webcam AVI you take, the best videos that show promise with better image processing will need a high-capacity storage medium. Today, the CD-R, CD-RW, and DVD-R are the mass storage media of choice for astrophotographers.

10.1.3 Compact Disc

The Compact Disc, or "CD," became a standard form of computer data storage media in the late 1980s. By 1993 the recordable CD had been developed, allowing the mass storage of up 650 Mb of data. At the time, this capacity exceeded that of most computer hard drives. The CD drive is now so ubiquitous that it is hard to remember when computers employed 3.5-inch floppy drives instead. However, early in their existence, CD and CD-RW drives were almost superseded by DVD technology. CDs gained an additional decade of popularity because inter-industry format wars significantly delayed the development of the DVD. During this time, the CD became established as the de facto removable data storage medium, and by 1997 the emergence of the reusable CD-RW posed a threat to various superfloppy drives, such as the Iomega Zip Drive.

The basic read-only CD consists of a 120-millimeter diameter disc of aluminum-coated plastic, upon which data are physically impressed as a series of pits. Data are read off the disc by a laser that detects the variations in reflectivity between the pits and the normal reflective surface of the CD. The information is stored on CDs as binary numbers, just as it is on a computer hard drive, with the pits being "1" and the blank areas "0." Data are written on the CD in a continuous track starting at the center and spiraling out to the edge. The track makes 22,128 revolutions around the disc with about 600 tracks per millimeter of radius.

Philips introduced the CD-R recordable medium in 1993. Initial CD-R discs were coated with a cyanine-based dye that changed reflective characteristics when exposed to a high-power laser. Data are stored on a CD-R when exposure to a laser darkens the dye coating it. Dark spots burned by the laser mimic the pits in a stamped CD and represent "1" while the normally-reflective area represents "0." The color of a CD-R disc is related to the specific dye used in the recording layer. The silver or gold color of the top on the CD disc comes from a silver or gold reflective layer behind the recording dye. Taiyo Yuden produced the green cyanine dye that appeared on early CD-R media. Mitsui Toatsu later produced the gold dye while Verbatim invented the blue dye. Today's silver discs use phthalocyanine dye invented by Ricoh in 1998 and are less sensitive to the degrading effects of ultraviolet, fluorescent, and sunlight.

The re-writable CD, known as the CD-RW, was introduced in 1997 and instead of organic dye uses a metallic phase-change material typically containing an alloy of silver, indium, antimony, and tellurium. Today, CD drives are fully compatible with either CD-R or CD-RW media and are capable of driving their lasers at three different power levels; write power, erase power, and read power. During the CD-RW recording process, the laser operates at its highest power, or write power. During the writing process, the laser selectively heats areas of the dye to above the alloy's melting temperature of 500 to 700°C and allows the atoms to move rapidly while the alloy is in a liquid state. If cooled quickly, the liquid state is frozen in place achieving the amorphous non-reflective state. The different reflective properties of these two states allow the recorded data to be read by a DVD player. In order to reuse the CD-RW disc, the laser operates at its medium, or erase, power. To erase the data in the amorphous dark spots, the phase-change alloy is heated by the laser to only 200° C, which is below the alloy's melting point but above the crystallization temperature, thus reverting the material back to a reflective state. The low, or read, power merely illuminates the CD-R or CD-RW disc so the data can be read.

CD drives read data at rates identified as a certain times "x," with x representing 150 Kb per second. Today's drives can read data at speeds as high as 52x. This high speed, coupled with deep discounts that lower the price of the CD recordable drives to less than a multi-pack of printer ink cartridges, has kept the CD as a viable storage medium even though the higher capacity DVD is now an affordable option as well.

10.1.4 Digital Versatile Disc

After a lifetime of about 10 years, the CD storage medium got a general face-lift, evolving in 1997 into what we know today as the DVD, or "digital versatile disc." During the CD's period of popularity the affordable capacity of computer hard drives increased by one hundred-fold, relegating the 650 Mb-capacity CD to near obsolescence, since it could hold only a fraction of the data produced by current machines. Clearly a new format was required, but the inability of major electronics manufacturers to agree on format specifications eventually led Microsoft, Intel, and Apple to issue an ultimatum; agree on a common format or do not expect any support from the computer industry. The result was that the DVD finally became established as a computer peripheral a decade after its initial development.

The DVD size format remains the same as the CD, a 120-millimeter-diameter, 1.2-millimeter-thick plastic disc. Indeed, a DVD can be visually mistaken for a CD, and fortunately nearly all DVD drives are backward compatible with CDs. The initial DVD-R, the recordable version of the DVD medium, had a capacity of 3.95 Gb, which is about 6 times that of a standard CD. This was increased in 2000 to 4.7 Gb for a single layer, and 9.4 Gb for a double layer, DVD-R. The initial discs could only be recorded once, while later DVD-RAM, DVD-RW, and DVD+RW versions can all be rewritten thousands of times. The widespread acceptance of the recordable DVD as a computer storage medium is evidenced by the fact that through the late 1990's, five times as many computer-based DVD drives were sold as were home DVD video players.

As with the older recordable CD, recording on a DVD is accomplished by using a high-power laser to change the reflectivity of the dye coated onto the DVD surface. On a DVD, this dye is coated over an injection-molded polycarbonate substrate that forms the body of the disc. As with the previous CD format, a pre-grooved spiral track is manufactured into the disc surface. This groove acts as both a guide for the recording laser and provides information for the DVD reader to determine the track's location on the disc so the reader can calculate the proper drive speed for that position along the spiral data track. While it is technically possible to manufacture double-sided, double-layer DVD recordable media with a capacity of 18.4 Gb, such devices are not yet on the market. Like a CD, the DVD can also store any type of data file—audio, video, images, data, or multimedia programs—and its greater capacity makes it attractive to the webcam astrophotographer.

Like familiar CD drives, the performance of a DVD drive is measured by its read speed in terms of "x," with 1x being 11.08 megabits per second. Thus the standard 1x data read speed of a DVD is about nine times faster than the 1x read speed of a CD drive. Like the CD drive, the DVD also uses a variable drive speed RPM to maintain a constant linear velocity of the data passing under the laser data read head. The data track is a continuous spiral beginning in the middle of the DVD and ending on the outer rim. Thus at 1x a DVD varies from 1632 RPM when reading data on the inside to 632 RPM at the outside along a spiral data track only 0.74 microns wide.

In order to greatly increase the capacity over a CD, a DVD recorder must imprint smaller data marks on its recording surface. To do this, the laser in the recorder uses a shorter wavelength and a longer focal length lens to focus a tighter beam. While a CD recorder uses a 780-nanometer infrared laser to record marks no larger than 0.834 microns, the DVD uses a red 635-nanometer laser to record marks as small as 0.4 microns.

Compatibility issues plagued early DVDs as DVD-ROM drives were incapable of reading re-writable DVD discs. And, the shorter wavelength 635-nanometer laser in DVD drives often would not read data written on CDs because the CD medium is optimized to reflect light from the 780-nanometer laser used with older CD technology. However, these issues had mostly been resolved by mid-1999 with the release of "third-generation" DVD drives.

The emergence of HDTV with its thirst for high data rates has created the need for still higher DVD data capacity and data transfer rate. The so-called "Blu-ray Disc" is an evolving technology to answer the need for higher DVD capacities. This driving technology will benefit the astrophotographer by allowing the convenient retention of good videos that in the past were deleted for lack of adequate long-term storage.

The Blu-ray Disc format retains the same dimensions of the standard DVD but decreases the thickness of the outer protective layer to only 0.1 millimeter. The thinner protective layer results in less aberration of the laser beam resulting in a data track width of only 0.32 microns, less than half that of regular DVDs, and also allows the minimum laser-imprinted mark size to shrink from 0.4 microns to 0.14 microns. The increase in data storage efficiency with the Blu-ray Disc format allows 25 Gb of data to be stored on a single layer disc and a staggering 50 Gb of data on a double layer disc with a data transfer rate of 36 Mb per second.

10.1.5 CD and DVD Storage Life

The details of how a recordable DVD works are technically interesting and its eventual replacement of the CD is grist for discussion among "techies". But the true question that astrophotographers should address is—are we placing too much faith in these two media as long-term storage for our hard-won celestial images? Advertising claims for the CD-R medium state "100-years archival life" or "one million read cycles." But repeated tests by various organizations have shown these claims are true only under very controlled circumstances. "Real world" conditions significantly shorten the advertised lifetime of a recordable CD, sometimes to as little as a few years.

The problem with recordable CDs and DVDs is the dye fades with age. Although newer versions of re-writable discs have more robust and fade-resistant dyes, older CDs are at risk of being unreadable in the near future. Tests of 30 different brands of CDs stored in a dark cabinet revealed that after two years 10 percent already showed aging problems and some were already unreadable. Once the dark spots burned into the dye fade, the read laser can no longer distinguish between the "1" and "0" and the disc appears blank. The dyes in these recordable media are sensitive to light. Exposure to sunlight, ultraviolet light, and humidity, as well as mechanical damage to the surface layer of the disc, will render a CD or DVD into a useless piece of plastic. Professional archivists and historians do not regard the CD-R as a permanent storage solution, relying instead on magnetic tape to store critical data. They view a CD-R as reliable for three to five years at best. Can users protect themselves from data loss by buying only top "brand-name" media? Not necessarily, because brand-name discs are often manufactured by the same subcontractors that produce "no-name" discs for discount retailers. On the other hand, it is equally possible to get a high-quality disc from a discount outlet. In other words, there is no absolute guarantee of disc quality through brand-name association.

But the long-term data storage problem is not as bleak as the above discussion would indicate. Today's re-writable CD and DVD media use metallic materials that change the phase of the light reflected off them. These materials are typically much more fade-resistant than previous organic dyes. The key to long data life is in the storage environment. Cool temperature (but not refrigeration), low humidity, and limited exposure to sunlight and ultraviolet radiation go a long way toward keeping data safe. It is also not a good idea to store labeled CDs or DVDs in a stack. It is possible for the ink printed on labels to react with the disc sandwiched next to it and affect the ability of the read laser to discriminate the spots representing data. Store important discs individually.

Once a disc has faded and the data imprinted on it are lost, you can't retrieve it. To prevent data loss over time, it is good to adopt the strategy employed by professional archivists: periodic recopying onto fresher, more robust media.

10.1.6 USB Interface

Today's webcams are USB devices, that is, they connect to a computer via a Universal Serial Bus cable. Because webcams are USB devices, some knowledge of the capabilities and limitations of USBs will be helpful. The fact that webcams are USB devices greatly simplifies their installation and use. The USB connection standard is designed to replace the RS-232 serial and parallel connections that often required the installation of cards in dedicated computer slots. Mercifully, USB is a true "plug and play" system and does not require the user to configure IRQ settings, DMA channels, and I/O addresses as was done with DOS and Windows 95. USB reserves a portion of the computer's bus bandwidth solely for use by peripherals and allows much higher data transfer rates than serial or parallel connections. A convenient feature is that USB devices can be "hot" installed or removed; that is, they can be attached to or disconnected from the computer without having to first shut it down, connect or disconnect the device, and then reboot the computer.

Another feature of the USB connection standard is its ability to chain more than one device to an attachment port. Multiple devices such as keyboard, mouse, scanner, printer, and camera can all share a common attachment port. Theoretically, as many as 127 devices can share the same port, though this would be impractical because the USB bus on a desktop computer will only supply 0.5 amps of power through each USB port. More flexibility in powering multiple peripherals can be gained with the use of a powered USB hub. These devices allow connection of seven devices to the same computer port while the hub itself supplies 0.5 amps to each connected device.

The electrical power distributed over a laptop USB port is relatively low, typically being only about 100 milliamps per port. Thus, one has to be careful about how many accessories are chained off one laptop port. A problem encountered with the low current available from a laptop USB port is that if a 15-foot or longer USB extension is used for remote camera

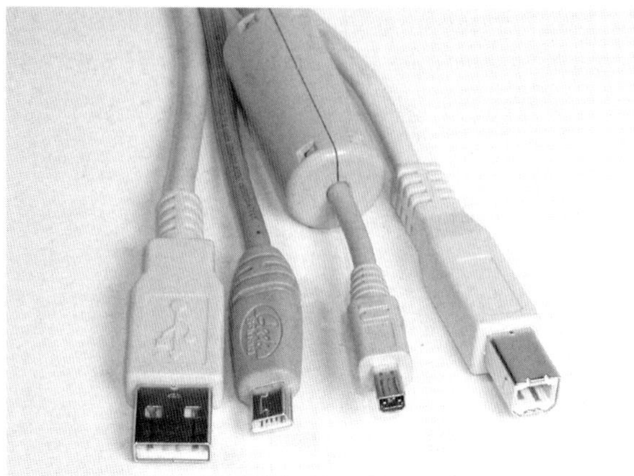

Fig. 10.2 *USB cables are the same for both USB 1.1 and USB 2.0 and all share the same style of connector for the computer end of the cable (left). The accessory side of the cable is available in a variety of styles, including the center two cables for the author's digital cameras and the larger connector at the right for a scanner/printer. Photo by Robert Reeves.*

operation, the device may fail to power up, even with a repeater-style extension cable.

There are two versions of USBs in use, the original USB 1.1 and the newer USB 2.0. The original allows data transfer over two channels, 1.5 Mbps and 12 Mbps, and is often called USB, Basic USB, or Full Speed USB. Devices such as keyboards and computer mice use the slower 1.5 Mbps channel in order to save bandwidth for devices that inherently need higher data transfer rates. The newer USB 2.0 is called High Speed USB and supports three levels of performance; 1.5 Mbps, 12 Mbps, and a 480 Mbps transfer rate that is ideal for scanners and cameras. It is easy to confuse the two standards, Full Speed and High Speed USB, when they are called by their "proper" names but there is no mistaking the fact that USB 2.0 is 40 times faster than the older USB 1.1.

Windows XP did not support USB 2.0 at the time of its 2001 release, but USB 2.0 support was added in 2002 via Windows Update and Windows XP Service Pack 1. Thus, USB 2.0 is now fully supported by Windows ME, 2000, and XP. Microsoft does not support USB at all in the original Windows 95 or USB 2.0 in the Windows 98 operating system. Windows 95B allows the use of USB 1.1 with the installation of the USB Supplement. In order to use USB 2.0 with Windows 98, an add-in PCI card and third-party software drivers are required. To tell if your computer supports the faster USB 2.0, open the Windows Device Manager and expand the Universal Serial Bus section. If 2.0 support is present, there should be an "Enhanced" USB controller listed.

Table 10.1
Typical Input and Output Speeds for Various Computer Ports

Type	Date Implemented	Speed (Mb/sec)
Serial port	1960s	0.2
Parallel port	1981	1.1
USB 1.1	1995	12
FireWire	1995	400
USB 2.0	2000	480
FireWire 800	2001	850

USB 1.1 and 2.0 both use the same cable and connectors to attach devices to a computer. Most peripherals are now USB 2.0 and are backward compatible with USB 1.1. However, if US 2.0 devices are plugged into a USB 1.1 system, their speed will default back to the slower 12 Mbps speed of the older system. Also, if a USB 1.1 device is plugged into a USB 2.0 system, all devices on the same hub will default to the slower USB 1.1 speed.

Here are some of the advantages of USB over older serial or parallel interfaces:

- One standard rectangular "Type A" connector to attach to the computer, although there are several different types of "Type B" connectors used at the peripheral end of the cable.
- "Hot" installation and removal with the computer running.
- USB devices can be chained together to allow more than one device to use a single USB port.
- USB-equipped computers automatically install and configure the drivers for the device being connected, although at initial installation you may be prompted for the Windows installation disk.
- Most small USB devices are powered by the USB cable itself, so separate power cords are not required.
- USB 1.1 transfers data at least 10 times faster than traditional serial ports and USB 2.0 is 40 times faster than USB 1.1.

Here are some of the characteristics of USBs that you should be aware of:

- Limited power from a laptop port. A laptop will only supply about 100 milliamps of power to run a USB device. This is sufficient to power a webcam, but may not be enough to operate other peripherals chained to the circuit.
- The maximum cable length for a USB device is 5 meters, or a little over

16 feet. Any longer and the ability to power the device or receive data will be compromised.

- A repeater cable extension may help with longer cable distance.
- If multiple USB items are connected to a single port, use of a powered USB hub is preferable to insure all devices receive sufficient power.
- USB 2.0 is backward compliant with older 12-Mbit USB 1.1 components, but the connection will revert to the slower USB 1.1 speed. For example, USB 1.1 hubs will work on USB 2.0 compliant computers, but the data transfer will be at USB 1.1 speeds. Alternately, a USB 2.0 hub will work on a USB 1.1 compliant computer, but again the data transfer rate will be at USB 1.1 speed.
- Up to two separately powered USB hubs can be cascaded together with little problem. If two USB 1.1 hubs are connected to a USB 2.0 port, both hubs will operate at full 12 Mbps speed.

A cautionary note about USB connections: the user should insure that the USB connection port at the computer is indeed unpowered when the computer is shut down. USB ports on some bargain-priced computers are known to remain powered up even if the machine is switched off. Any USB connection that remains "live" will continue to power up an attached webcam, so devices such as cooling fans on SC-modified webcams will continue to run. Power to USB connections on "bargain" computers is only off when the computer's power cord is unplugged or power is shut off at a surge protector powering the computer.

10.1.7 Computer Mouse

Although laptop computers are designed to operate without an external mouse, using a real mouse for controlling the computer is a great convenience. Using the cursor touch pad on a laptop in freezing weather is not fun. Not only do fingers suffer, but also the touch pad can become erratic as the temperature drops. A wireless optical mouse is preferred. This not only eliminates additional wires laying around in the dark, but an optical mouse will work on irregular surfaces, even the top of you pants leg, when open space is limited or dew is starting to coat your observing table.

10.2 File Formats

The image medium in webcam astrophotography is entirely electronic. There is no film negative or transparency containing the image that you can hold in your hand as with conventional photography. Instead, the images are stored in digital form as binary data on an electronic storage device like a computer hard drive, a USB jump drive, a digital camera memory card, or a CD or DVD. Because the image storage is completely digital, there must be an agreed-upon, widely-accepted standard for how the data are stored in order for users to access it and reconstruct a visible image. The order in which digital data are laid out in a file so others can access it is called an image file format. Thousands of file formats have evolved over the past decades, but only a handful have survived to become internationally accepted format standards.

The electronic image file formats used to define and store graphical information play a critical part in webcam astrophotography. The de facto image-processing program for this activity is RegiStax. This program accepts the AVI video files produced by a webcam or other video-producing camera and, after stacking and aligning the video frames contained in the AVI file, outputs a single still image in either of five formats; BMP, JPG, FIT, TIF, or PNG. Because we can encounter any of these formats in our webcam astrophotography adventures, we will briefly examine all of them.

10.2.1 AVI Format

The single most important file format in webcam astrophotography is AVI, an acronym for Audio Video Interleave. This is the video format utilized by various basic and modified webcams. Any video sequence photographed with a webcam will be written to an AVI file for subsequent processing into a stacked and aligned single image.

Introduced in 1992 by Microsoft as part of the Video for Windows technology, AVI is not a true format in its own right, but is instead a "wrapper" package that contains audio and visual data and, if needed, the proper algorithms for decompressing the audio-visual files. An AVI file is divided into two mandatory sections called "chunks." The first chunk is the file header that contains information such as width, height, and number of video frames. The second chunk holds the audio-video data and, if needed, a software module called a codec (compressor-decompressor) to translate compressed video data into viewable images.

Video files created by cameras using the AVI format can be quite large. Each full-resolution 640 x 480 video frame is 900 kilobytes in size and it does not take long for a lot of them to accumulate. For instance, a standard USB 2.0 webcam such as a Philips SPC900NC operating at 30 frames per second and 24-bit color will produce 26.37 Mb of data every second. A 30-second video will therefore produce a file size of 790 Mb.

One of the unfortunate facts of computer life is there is no set standard for the codec used to compress a large AVI into small enough size for convenient transmission over the internet. The software developer is responsible for including the translation codec within the files and any number of schemes can be used for file compression such as Motion-JPEG, Real Video, MPEG, MPEG4, DivX, among many others. Motion-JPEG will compress each video frame separately, but the more efficient MPEG algorithm achieves better video compression by working with the differences between consecutive frames. This takes advantage of similarities from frame to frame that do not change, such as stationary backgrounds. Windows comes with built-in support for many different codecs, but many video clips exist that utilize less-accepted standards and if the codec included in the video is incompatible with the video player on the computer, the dreaded Windows error message stating that the video selected could not be played will display.

Fortunately for the webcam user, compression schemes are not used in the acquisition of astronomical videos in order to retain maximum image detail. Therefore when we record an AVI file of a celestial object, we are recording the full video frame with no form of compression. While this leads to some huge video files on our hard drive, programs like RegiStax have no problem reading each frame and allow us to work with all the detail the camera is capable of recording.

10.2.2 BMP Format

BMP is the native raster bitmap image file format used by the Windows operating system. The format was developed in 1986 by Microsoft and, because of the sheer dominance of Windows, BMP remains the favorite for storing 1-, 4-, 8-, and 24-bit image data, which in turn displays monochrome, 16 colors, 256 colors, or 16,777,216 colors. The simplicity of the format makes it a good match for the 8-bit data output of webcams, and the BMP format is the normal stacked image output from programs like RegiStax.

BMP images are stored as binary data preceded by an ASCII header that identifies the file type, file size and dimensions, bits per pixel, and compression. A color table may or not be present depending on the image bit depth. Bitmap file headers always start with the ASCII character "BM" which identifies the file as bitmap. After the header, the image data are stored line by line in byte order. There are similarities in the basic simplicity of BMP and the TIF file format explored later, but there are also curious differences. The BMP format places the first line of the data order at the bottom instead of the top of the file and 24-bit color data are stored in BGR order instead of the familiar RGB.

10.2.3 FITS Format

The Flexible Image Transport System, known as the FITS format, was developed in the 1970s as an astronomical data archive and interchange format to allow data to be electronically transferred from one institution to another and be universally readable on a variety of equipment. Contracted to FIT when used as a file extension, the format had its roots in technological advances at Kitt Peak National Observatory and the National Radio Astronomy Observatory. Digital images from these observatories needed some universally readable form of distribution to other institutions. In its original form, FITS was designed to comply with standard format applied to 9-track half-inch magnetic tape that called for records to be stored in successive blocks of 2880 bytes. Scientific data were stored in a binary array preceded by an ASCII text header containing a description identifying it as an image or other form of data. In 1982, the International Astronomical Union endorsed the format and in 1988 formed the FITS Working Group that has expanded it to include other types of electronic data. Although the format is evolving, the overriding principle applied to its development has been backward compatibility so as to always be readable, no matter how old it is. This has led to the rule of "once FITS, always FITS" to preserve astronomical data and promote its electronic readability in the future. Because of this enforced compatibility, it has become the de facto standard for all astronomical data interchange.

For the most part, in using webcams to capture pleasing images of celestial objects, we use the default BMP format employed by RegiStax and other programs like iMerge mosaic. However, some aspects of this branch of celestial imaging deal with the gathering of scientific data through photometry, astrometery, and spectroscopy. To this end, RegiStax allows the export of images in the Basic FITS image format. FITS is designed to be flexible enough to handle virtually any kind of image file array as along as it is properly defined in the FITS header. Amateur astronomers gathering data with a webcam deal mainly with 8-bit unsigned integers, but the format is robust enough to handle more complex data like 16- and 32-bit signed twos-complement integers and 16- and 32-bit IEEE floating-point numbers.

Since FITS is a "specialty" scientific format, it does not include "FITS viewers" or software packages within the FITS files designed to decode the data into an image. Nor do most standard pictorial image processing programs like Photoshop and Paintshop Pro support the FITS standard. Fortunately, astronomical image processing programs like AIP4Win and Images Plus do support FITS, and an excellent FITS plugin is available for Photoshop. The FITS Liberator V1.01 for Photoshop is available as freeware.[1] A majority of FITS images are 16-bit and thus will only display on Photoshop CS, but 8-bit FITS images will display on Photoshop 7.0 and earlier.

10.2.4 JPG Format

JPEG stands for Joint Photographic Experts Group and the acronym JPEG (pronounced "jay-peg") is shortened to JPG when used as a file extension. Astronomical cameras do not normally use this format for image storage. However, because JPG efficiently compresses image file size, this format is very popular for the email transmission and web display of fully processed images that originated in another format. It is not unusual for an image that began as a 200 Mb AVI video sequence and then was processed into a 900 Kb BMP still image, to be converted into a 30 Kb JPG for display on a website.

JPG is not a true file format in its own right, but is instead a series of algorithms that is designed to compress either full-color or gray-scale scenes with tonal gradation across the scene. JPG is traditionally an 8-bit file format, using three color channels to create true-color 24-bit images. It utilizes what is known as "lossy" compression, that is, a certain amount of image information is lost during the compression process. One of its useful properties is that the user can select the degree of compression, and thus the amount of image data that is discarded in the process. There is no magic solution to provide both high image quality and small file sizes. To achieve one side of the equation, the other must be sacrificed. The smaller the resulting image file, the lower the quality of the saved image. Because of its "lossy" nature, a JPG should never be used as an archive format.

During the compression process the JPG algorithm splits the image into blocks of pixels, then looks for matching blocks. If the quality setting is high, and thus the compression level is low, the algorithm will seek an exact match but at the expense of a large image file. But if slight differences, most of which are not noticeable, can be accepted, then the algorithm can usually reduce a file to about a fifth its original size. Because the "lossy" compression schemes perform destructive compression, JPG always creates an image that is different from the original and the differences become greater as the level of compression is increased.

Each time a JPG image is altered and saved there is a chance of degradation because of the lossy compression routine. However, simply viewing and closing or copying an existing JPG image without performing any modification will not further degrade the image. If a JPG original is going to be extensively modified in an image-processing program, it is best to

[1] www.spacetelescope.org/projects/fits_liberator.download.html

convert it to a lossless format such as TIF or PNG, or a native format like Photoshop PSD, while doing the manipulation. Only when the image is completely processed should it be saved again as a JPG.

Large featureless areas like the black sky background in an astronomical photo compress very well with JPG. However, it has a hard time with images containing a great deal of intricate detail. Two images with different content but having the same pixel size will thus compress into two different file sizes. Sharp straight edges, such as a row of black pixels against a row of white ones, tend to be blurry in JPG images.

10.2.5 TIF Format

The Tag Image File Format, or TIFF, is a universally accepted cross-platform raster graphics format used by the graphic arts and publishing industries. Its name is shortened to TIF when used as a file extension and is pronounced "tif" just like it is spelled. Because of its wide acceptance, TIF is a good file format for archiving images because it will likely be around a long while. The TIF format is called "lossless" because, unlike JPG, it does not discard image data in order to compress the file size. Instead, it incorporates a lossless LZW compression algorithm. LZW is named for Israeli researchers Abraham Lempel and Jacob Zif, who pioneered computer file compression technology in 1977, and Terry Welch improved on the technique. The compression scheme works by recognizing repeated strings of identical data and replacing multiple instances of a data type with a single datum. LZW is quite effective with solid colors, and featureless areas compress much better than detailed areas. The process is less effective with 24-bit continuous-tone photographs. The process was patented but the latter expired in 2003.

TIF was designed by printer, scanner, and monitor developers and is universally supported by a wide variety of software and hardware. Various color spaces such as RGB, CYNK, and grayscale are supported. But the format's flexibility also creates problems because, though it is easy to program a TIF file writer, it is difficult to program a truly universal TIF file reader. This is because there are many ways images can be stored: there can be one or multiple images, or they can be complete images or broken down into strips or blocks.

A word of caution about archiving original images in the TIF format: never apply unsharp masking to original images. If an image is later resized, the unsharp masking halos are lost and the image becomes fuzzier. Resharpening the image a second time leads to loss of image data. Processed display images can be sharpened, but the archived original should not be sharpened.

10.2.6 PNG Format

A newer file format called PNG, for Portable Network Graphics, is becoming popular, but it is not yet incorporated into digital camera software. It is discussed here because it is a file output option with the popular RegiStax video-processing software. Pronounced 'Ping", this format was originally created in 1995 when the quasi-public domain status of the LZW compression scheme came to an end. PNG uses a form of royalty-free compression called the ZIP method which is similar to the "deflate" method in PKZIP. It also uses a special processing filter that greatly improves compression efficiency across the gradient data contained in 24-bit color images. This filter causes PNG to be a little slower than other image formats that use compression. The resulting 24-bit color file sizes are about 25 percent smaller than those compressed with TIF LZW.

A major advantage of the PNG format is that it contains embedded automatic gamma correction that works across all platforms. This means that images in this format will display equally well on Macintosh and Windows PCs and workstations, all of which have different gamma values. Images will thus display on all systems without appearing too dark, light or contrasty. Another PNG advantage is an interlacing routine that allows a preview image to display after only 1/64th of it has been loaded.

10.3 Troubleshooting

It is important to install the camera drivers first before plugging the camera into the computer for the first time. If the latter is plugged in first, Windows will recognize a new USB device and attempt to install its own drivers, possibly resulting in a blue screen. If this happens, unplug the camera and uninstall the webcam software. Then reload the software and follow the prompts presented by the installation program. Most webcam software comes with a utility that checks the camera and computer interface for proper operation, after installation.

In the early days of webcam astronomy, a phenomenon known as the "green screen" would lock up the webcam and end the imaging session. The camera would stop responding and only a green, or sometimes pink, screen would be visible in the preview window. This was caused by a crash of early versions of the webcam drivers. The phenomenon is rare now, but can still happen. The only real solution is to reboot the entire computer and begin again. Sometimes closing the camera control program, unplugging the camera, and then plugging it back in and restarting the control program works; but if a USB hub is being used to operate a mouse or other devices from a single USB port on the computer, then the USB hub may go down as well, forcing a complete reboot of the computer. With the advent of USB 2.0 connections, green screen problems have mostly disappeared.

Sometimes a Philips ToUcam will not respond when it is plugged in and the control software initiated. The first thing to do is shutdown and reboot. If the camera will still not activate, uninstall and reinstall the webcam drivers, then install an alternate control program such as K3CCD Tools and activate the camera from within that program. Sometimes it is necessary to select "WDM" mode under the device menu, then select the ToUcam webcam in the dropdown list in the device menu.

Another circumstance that may lock up the computer is using the manual stop imaging option to end a video capture while using ToUcam 840K software on an older, slower laptop. On my own laptop, the program becomes unresponsive after about half a dozen video captures while using this option. The program works well with fixed-length timed video sequences but eventually fails when making manually stopped sequences. When the program freezes, the only solution is to reboot the computer. To avoid such problems, install the superior K3CCDTools program and use it to operate the camera.

At this point it may be useful to say that rebooting is a viable solution to a lot of imaging woes. It is not unusual to have several programs running at once on a computer and there is a tremendous amount of data flow, computational activity, and hard-drive access happening during a video capture sequence. Program conflicts and errors can occur and cause strange things to happen. If the problem is a rarely occurring event, rather than waste good observing time trying to unravel a conflict I simply reboot and carry on. Of course, if a serious program conflict occurs often, then some troubleshooting is required to determine which program causes the problem and to limit its use while capturing video.

The order in which several webcam-related programs are opened may also affect the ability of the camera to properly function. For instance, if programs such as the Philips VRecord camera control, or alternative camera control programs like K3CCDTools or WcCtrl are opened first, the QCFocus focusing program will not recognize the camera because the camera driver is already being used by one of the other programs. QCFocus must be opened first, the camera properly focused, then QCFocus closed before K3CCDTools is started.

A majority of long-exposure modified webcams are connected through both parallel and USB ports. Image data still flow through the USB port, but the long-exposure controls operate through the parallel port. In some cases, a camera will not function if the parallel port is set to autoselect and the LTP-1 port is incorrectly configured. In this case, the user must open the computer's BIOS settings and manually configure the LPT-1 port. Users will need to refer to their computer's documentation for BIOS access procedures. If the computer tends to lock up when using an SC-modified Vesta or ToUcam, try changing the computer BIOS settings for the printer to ECP instead of the Standard or EPP settings.

A feature of many long-exposure modified webcams is what is known as the "amp-off modification." This turns off the image signal amplifier during long exposures to prevent amplifier glow from fogging the image. However, the amplifier circuit must be turned back on just before the exposure ends or the camera's operating program will not read the image correctly. K3CCDTools uses a default "amp-on" lead time of only 10 milliseconds. This may not be long enough for some cameras. Increase the setting progressively until the camera works properly. Some models require as much as 350 milliseconds of amp-on lead time.

When imaging Jupiter or Saturn, the user may occasionally encounter what is called the "onion ring effect" where the outer edge of the planet appears as a series of tightly packed rings. This is caused by underexposure. If the image exposure is close to the point of saturation and the condition persists, increase the gain. When using K3CCDTools, set the gain at about 50 percent for Jupiter and about 75 percent for Saturn and lower the exposure.

EMI, or electromagnetic interference, is usually of little concern to the webcam astrophotographer, but it occasionally affects a particular imaging system. EMI can appear as non-random image noise such as repeating patterns of horizontal lines or banding. Electromagnetic radiation, unless it is very powerful, is unlikely to interfere with the digital data from the camera. EMI will most likely affect the camera's analog signal processing with a sine wave from a power supply or interference from a nearby shortwave radio transmission.

If large amounts of EMI are present, it will be difficult to configure your imaging system to avoid it. A basic recommendation is to keep all connecting cables as short as possible. Long cables act like antennas and can pick up EMI just like a long wire can pick up radio signals. A radio transmission can show up as pattern based on the transmitter's modulation. Another basic rule is to never run power cables parallel with and near to the camera signal cable. If they must be close, arrange them so they cross each at the greatest possible angle. Inexpensive power inverters are known to be very noisy and popular anti-dew heaters use a controller that creates a square-wave power that can cause interference. A noisy telescope drive motor may create a 50- or 60-Hz regular pattern. If EMI is present in the

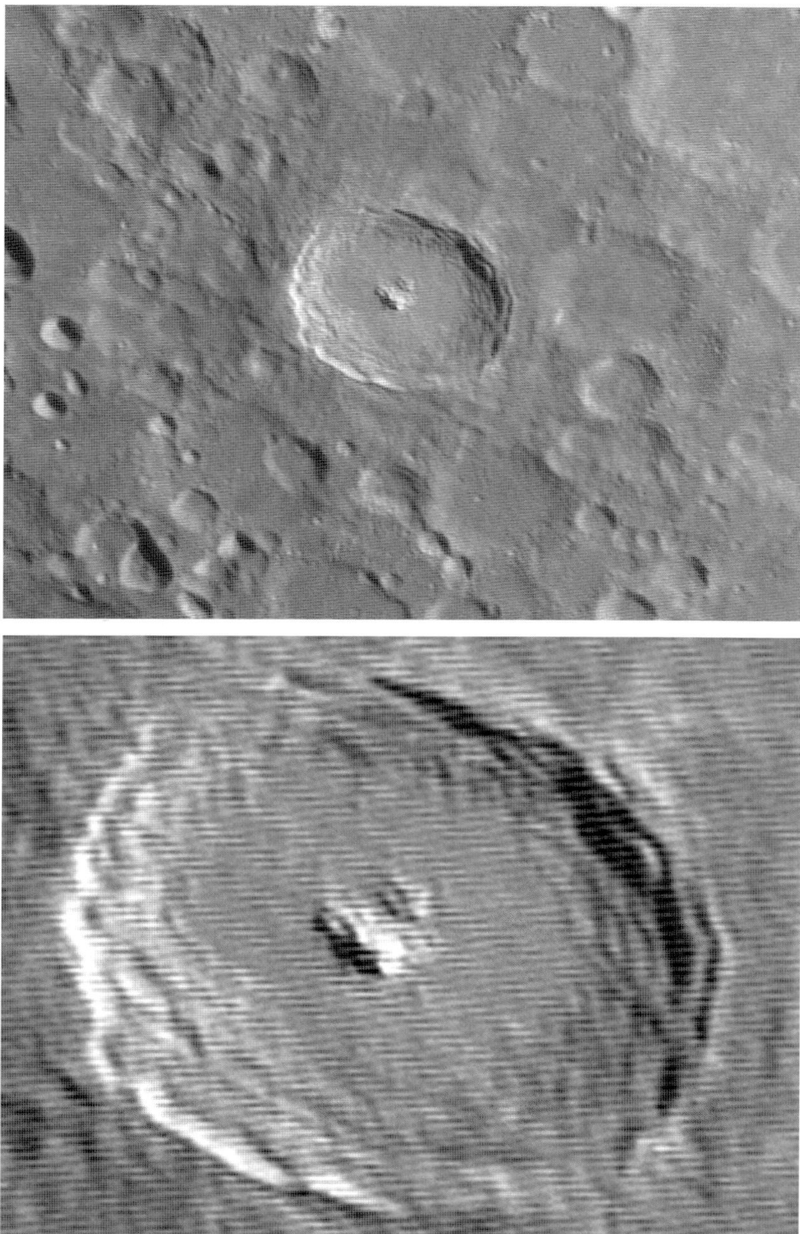

Fig. 10.3 *Electromagnetic interference (EMI) can manifest itself as a banding that repeats across the face of an image. Camera and electrical cables bundled in close proximity can induce interference. The enlargement of the Tycho image shows an example of a repetitive EMI pattern. Photos by Robert Reeves.*

image, try shutting down nearby electrical devices to try and isolate which is causing the interference.

If you are imaging with a wireless-capable laptop within range of a wireless Internet connection or wi-fi hot spot, beware of email programs that automatically check for mail at set intervals, or anti-virus programs that "phone-home" for virus updates. These programs may start up during an imaging session and lock up the video capture program. This is known to affect K3CCDTools. Close the Internet connection as a normal preparation step for imaging.

Chapter 11
Environmental Factors in Celestial Imaging

11.1 Seeing

In the age of space-borne telescopes providing us with unprecedented high-resolution images from their lofty perch above the Earth's turbulent atmosphere, astrophotographers are constantly reminded of how much the air above us limits our pursuit. It's a fact that we can't live without this blanket of atmosphere, but at the same time astrophotographers often can't live with it either! The great French astrophotographer Jean Dragesco reminds us of the quote by a famous optician, "The worst part of a telescope is . . . the atmosphere."

Early in any amateur astronomer's career, the realization comes that all clear nights are not equal. Telescopic images may be clear and steady one night, while the next they dance, shimmer, and cannot be sharply focused. The quivering of star images due to atmospheric turbulence is called scintillation. The relative amount of scintillation is commonly known as "seeing." The better the seeing, the more stable the star image is. Numerical scales exist to grade seeing, with 10 being perfect movement-free star images while 1 represents a star rendered into a total blur by turbulence. The atmosphere affects shorter wavelengths to a greater extent. Also, the closer an object is to the horizon, the greater the amount of air and turbulence the object is sighted through. Thus, the seeing deteriorates increasingly with the viewing angle from the zenith and especially in the last ten degrees or so. This is the astronomical equivalent of a distant terrestrial view being distorted by "heat waves" seen down a long highway on a hot day.

The atmosphere creates trouble for the high-resolution skyshooter because the refractive index of air changes with temperature. As air masses of differing temperatures mix, their respective refractive indices constantly change, which distorts the incoming wavefront from the celestial object. This results in a wavefront arriving at the telescope or camera as an irregular surface, not the smooth state in which it was emitted. The temperature change needed to create such refraction is amazingly small. A temperature change of only one degree Centigrade in a layer of air only 15 centimeters (six inches) thick is enough to cause stars to oscillate 0.2- to 0.3-arcsecond in an 8-inch telescope, and will render the star as a complete blur in larger instruments.

Seeing is the single most important factor in achieving high-resolution imaging. Veteran webcam user Damian Peach has summed up the three most critical factors required for high-resolution imaging: "good seeing, good seeing, and good seeing." Unless the seeing is at its best, you will not achieve the resolution that your imaging system is capable of. It is a fact that most urban locations have bad seeing. This cannot be helped because cities and towns are covered with roadways, parking lots, rooftops, and industrial facilities that radiate heat into the atmosphere, creating turbulence. Also, observing downwind from large buildings—even your own home—will place you in "disturbed air" that swirls around the structure, creating eddies and air currents that distort fine lunar and planetary detail. Pleasing images can be achieved under marginal seeing conditions, but the best optical performance requires the best seeing.

Webcam users who image the Moon and planets at their telescope's prime focus, or even at higher magnifications using Barlows or eyepiece projection, are at the mercy of poor seeing more than practitioners of other types of astrophotography. This is because of the webcam's inherent small field of view. Seeing is rarely an issue with skyshooters who only use camera lenses. Film and digital camera users who shoot through the telescope have a much larger field of view due to the camera's larger imaging surface. While seeing affects their images as much as it would a webcam, the wider field of view can, to some extent, mask the effects of poor seeing because the image does not have to be enlarged as much for viewing. As we saw in Chapter 3, however, telescopic webcam users are operating with very narrow fields of view—equal to that of a high-power 5mm eyepiece used on an 8-inch *f*/10 SCT.

Because telescopic webcam users are at the mercy of seeing, they need to understand it—what degrades it, how to predict when it will be best, and what to do to improve it. Seeing is influenced by many environmental factors, some we can control, others we cannot. The quest for good astronomical seeing has evolved into a near science, and experienced observers learn to be amateur meteorologists to understand the factors that affect it. The importance of this to our imaging efforts cannot be overstated because no amount of post-exposure processing can "undo" poor seeing.

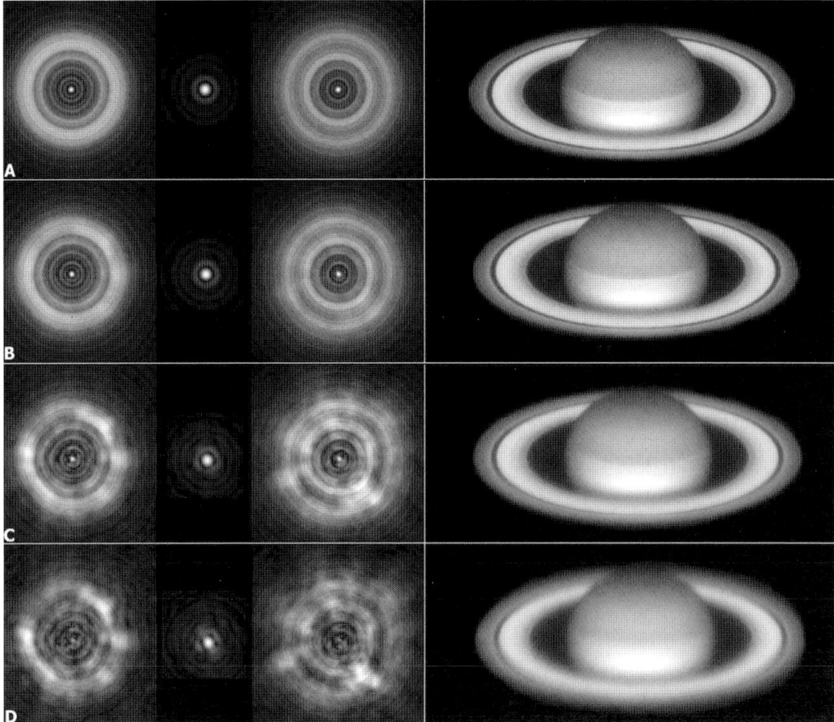

Fig. 11.1 *These simulated views of a bright star and Saturn were created with the Aberrator 2 program written by Cor Berrevoets and demonstrate the detrimental effects of atmospheric turbulence. The simulation assumes a perfectly collimated and focused 10-inch, ⅛ wave, reflecting telescope with a 30 percent obstruction. The intra- and extra-focal star images have been defocused by five waves. Rows A, B, C, and D display, respectively, no atmospheric turbulence, 0.05 waves, 0.15 waves, and 0.25 waves of turbulence. It should be noted that 0.15 waves is still considered fair on the Pickering seeing scale and demonstrates that it does not take much turbulence to degrade high-magnification lunar and planetary images. Photo by Damian Peach.*

11.2 Macro-, Meso-, and Microscale Weather Influences

Atmospheric turbulence can occur anywhere within the telescope's line of sight, which has traditionally been divided into two areas: the boundary layer, or region of the lower atmosphere extending in some cases as high as 3,000 feet where the ground and topography exert influence on the flow of air, and areas of high-altitude turbulence caused by the jet stream or frontal passages.

In meteorology weather conditions are considered on three scales. The largest is the macroscale that monitors weather conditions over an area greater than 300 miles away; the mesoscale that examines weather within 10 to 300 miles; and the microscale that looks at conditions within 10 miles. The influence of climate within each of these three zones affects the astronomer's view through the atmosphere.

11.2.1 Macroscale Influence

On the macroscale, a number of circumstances can quickly degrade local seeing. The phenomenon with the harshest effect is the jet stream, which is a band of strong wind, sometimes exceeding 300 miles per hour, at an altitude of 30,000 to 40,000 feet. The jet stream is a shifting river of air that changes course from day to day. Wind velocities drop off quickly with increasing distance from its core, but the eddies it stirs up mix various temperature layers in the air and create instances of extreme turbulence. There is nothing we can do to limit the effects of the jet stream except wait for it to go away.

Weather fronts are also macroscale events that have a detrimental effect on seeing. When two air masses with differing temperature and moisture content collide, the colder mass undercuts the warmer, forcing it aloft. Because of convection, turbulence near an advancing weather front can be great.

Thermal instability caused by surface heating and cold temperatures aloft creates vertical mixing of the atmosphere over macroscale areas. Air initially undergoes vertical movement from heating, and then falls after cooling, creating turbulence for the astronomer. The opposite of thermal instability is a thermal inversion where the air near the surface is cool and the air aloft remains warm. This usually leads to a stable atmosphere free from significant turbulence.

High-level turbulence can be viewed directly in the telescope by placing the edge of the Moon or a bright star in the field and defocusing the instrument the following way. When the focus is racked sufficiently away from the telescope's primary optic, we can directly observe the movement of the atmosphere across the image of the celestial object.

11.2.2 Mesoscale Influence

Low-level winds are one of the greatest influences on seeing within 300 miles of an observing site. Basically, the greater the wind speed, the worse the seeing. Using a weather chart showing pressure contours, one can gauge the expected severity of turbulence by noting how close together the pressure contours are in the local area. The closer they are, the higher the pressure gradient, which in turn creates higher wind velocities and more atmospheric turbulence.

The terrain roughness in the local observing area will also have an influence on air movement from winds. Smooth plains will disturb the passage of air less than rough or mountainous ground. The less the surface disturbs the passage of air, the smoother the boundary layer of air near the ground will be, thus promoting better seeing.

11.2.3 Microscale Influence

The seeing quality at any given locale can be heavily influenced by local terrain. Features such as buildings in a city or hilly terrain in the country will either disturb local airflow or create disturbances by local heating or cooling of the surrounding air. Observing from the roof of a building in a city will experience degraded seeing from the effects of local heating from the building itself (by radiation of either heat absorbed from the Sun, or heat produced within the building). Furnace or air conditioning vents on the roof of a nearby building will also degrade local seeing. It is possible to see the rising column of hot air from furnace vents or chimneys of neighboring houses. Local seeing will also be disturbed by convective airflows at the base of a hilly area where nocturnally-cooled air flows downhill.

The ground types in the immediate observing area can also adversely affect seeing. Paved parking lots may afford wide views of the sky in a city, but daytime heat stored in the asphalt and concrete is reradiated at night and will create turbulence similar to "heat waves" seen shimmering above a hot road. Large areas of exposed rock in a rural setting will also reradiate the day's heat buildup at night, creating local turbulence. Daytime seeing is especially problematic because constant solar heating stirs the atmosphere, which makes solar photography difficult. Generally, wide areas of flat grassy terrain that absorb little daytime heat and quickly dissipate the remainder at night are observing areas with the best seeing.

Even the area within several feet of the telescope can have a significant effect on seeing. Imaging over a hot roof can present an impossible situation where high resolution is never realized. Air movement near the telescope also plays a large role in the degree of resolution achieved. In the summer I usually have a fan blowing on me while doing piggyback astrophotography. The fan is for both comfort on a warm night and to make myself a "moving target" for any hungry mosquitoes. The image resolution achieved with wide-field astrophotography is so low that the moving air from the fan has no affect on either the seeing or the stability of the telescope. However, using a fan in this fashion during high-resolution lunar and planetary imaging would produce enough local air turbulence to noticeably degrade images.

Instrument turbulence, or "tube currents," are caused by daytime heating of the telescope and optical components. This affects the figure of a lens or mirror, and creates heat waves inside the tube of larger telescopes that can distort light rays in the optical path. Covering an instrument with a tarp by day may actually make the problem worse because the tarp absorbs heat. Many astrophotographers instead cover their telescopes with aluminized Mylar "survival blankets" which reflect much of the incoming solar radiation. If a telescope is protected from the Sun during the day and is in thermal equilibrium with outside air temperature, instrument turbulence will not be a problem. Traditionally, older telescopes were painted white to help control heat absorption. More recently, manufactures have produced instruments that are painted black, because in fact the latter have been found to cool off quicker after dark, as long as they were not exposed to the Sun during the day. However, a black telescope is difficult to keep thermally stable during the day when observing the Sun. If the temperature inside the instrument becomes three to four degrees Centigrade higher than outside temperature, the images produced will be inferior. Closed tube assemblies like Schmidt-Cassegrain designs do not allow circulation of air within the telescope and optics take a long time to cool down. Telescopes with internal fans to speed cooling will reach thermal equilibrium quicker. And although SCTs are prone to dewing on their corrector plates, imagers should resist the temptation to clear dew with a hot blast from a hairdryer since the heating may distort the corrector plate and create "disturbed air" in front of the telescope. A long plastic dew cap is best to maintain thermal equilibrium.

Local turbulence can be controlled in several ways. One technique is to raise the telescope above the ground. A 10-foot elevation will significantly decrease the amount of "ground clutter" or turbulence created from warm local objects. However, mounting an instrument on a high stable platform may be impractical in many cases. Another useful technique is to set up in areas of uniform local terrain, such as a wide-open grassy field with no changes in local topography to create unequal local heating. Astronomers often build their own observatory to keep the telescope ready for use at all times. But the observatory itself can also create heating problems and turbulence across the open slit of a dome, or the heavy walls of a roll-off roof observatory, can radiate heat that will affect seeing. Covering the building with white titanium dioxide paint can help minimize the amount of heat saturated in the observatory's structure. Domed structures will always have more problems with local turbulence than roll-off roof ones because of heat flowing out the slit. Roll-off structures that are raised from the ground work best for reducing local turbulence created by heat. Some well-known astronomers live in locations that are prone to better seeing than others. Don Parker, the high-resolution planetary imager from Coral Gables, Florida, benefits from a steady sea wind and stable ocean

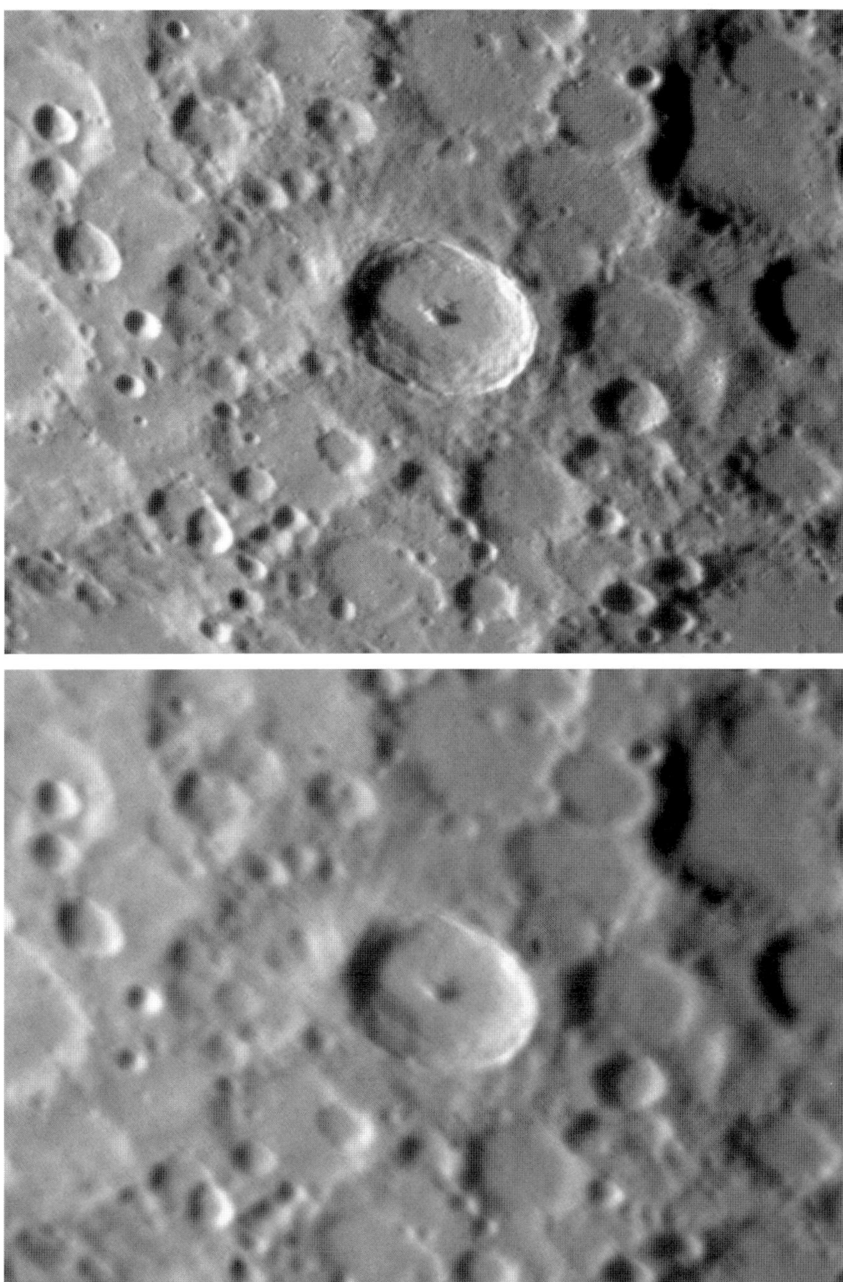

Fig. 11.2 *These two views of the crater Tycho were imaged several minutes apart and demonstrate that local seeing is a volatile commodity. Both were taken with an 8-inch SCT at f/20 with an Atik ATK-2HS camera. 600 video frames were stacked to create each image. During the session, seeing conditions deteriorated within a matter of minutes due to the arrival of a rapidly moving weather front. Photos by Robert Reeves.*

temperatures, which provide him with stable seeing nearly all year.

11.3 Good Results in Spite of Poor Seeing

Surprisingly good webcam results can be achieved under moderately turbulent seeing. An image of the Moon which appears to continuously shimmer and dance will still turn out well as long as one factor is present: the seeing must not be so poor that it actually defocuses or blurs the image to totally obscure detail. As long as the image is simply gently rippling, the editing function of the stacking software will reject substandard frames that are slightly defocused. In fact, I have had very good results in seeing conditions where the image is slightly rolling and undulating. Such conditions can actually improve the final stacked image because of the fact that the moving cells of air that create the mediocre seeing are defocusing some parts of the image while at the same time sharpening the focus in other parts. By taking many frames and selecting editing and stacking parameters that reject all but the very best of the resulting video frames, the final stacked image will be quite good.

The majority of moving air cells that blur seeing are fairly small, being less than a foot in diameter. Small telescopes in the 6- to 8-inch range will usually be affected by no more than one moving cell of air at any one time. But instruments whose apertures are larger than the individual moving cells of air will look through multiple cells at the same time. Each cell has its own effect on seeing and it is unlikely that all cells that a telescope is viewing through will be stable at the same time. Therefore, the larger the aperture, the less likely it is for the air to be fully stable across the full aperture of the instrument.

Ron Dantowitz of the Clay Center Observatory in Boston has been a pioneer in the use of video imaging in astronomy. His experience with several telescopes using video techniques at resolutions and fields of view similar to that achieved with webcam imaging has enabled him to determine how often a "good" video frame will be captured using various size instruments. His results are summarized in Table 11.1 and emphasize the need for steady seeing to maximize success. The larger the instrument, the higher the number of individual air cells it views through, and the greater the chance of image blurring.

However, Ron's data should not discourage users of larger telescopes from trying their hand at imaging (even those using 100-inch apertures!). While a big telescope will achieve fewer images that exhibit good seeing, the ones taken at those favorable times will display the increased resolution possible with the large instruments. The key with big apertures is to

Table 11.1
Good Video Frames on an "Average" Night

Aperture (inches)	Good Frames
8	1 in 10
12	1 in 30
24	1 in 300
60	1 in 10,000
100	1 in 1,000,000

take many video frames to increase the chances of capturing one during the increasingly elusive moments of best seeing.

11.4 Predicting Seeing and Transparency

Some types of astrophotography such as wide-field imaging only needs a clear sky, however, high-resolution lunar and planetary imaging requires steady seeing. Anyone with Internet access has the same weather data available to them that professional forecasters have. With some study of how the atmosphere in your region reacts to certain influences, you can learn how to interpret and improve on forecasts for your area. By monitoring a number of factors such as the jet stream, the passage of cold fronts, and the proximity of thick clouds, a reasonably good assumption can be made about local seeing conditions.

Any time the jet stream is within 225 miles of your location in the winter or 300 miles in the summer, the local high-resolution seeing conditions are likely to be poor. The current location of the jet stream can be tracked at http://squall.sfsu.edu/crws/jetstream.html. The passage of cold fronts will also create increased atmospheric turbulence. Anytime a cold front is within 250 miles in the winter or 300 miles in the summer, there is a probability of poor seeing. The movement of frontal systems across the United States can be tracked at http://www.wunderground.com/US/Region/US/Fronts.html. Thick clouds form when ascending air reaches its dew point at cooler altitudes. Thick clouds have colder cloud tops. By monitoring the infrared cloud images from http://www.ssec.wisc.edu/ you can detect the nearby presence of thick clouds even at night. Anytime there are thick clouds with upper temperatures lower than –50 degrees Fahrenheit within 150 miles of your location, related atmospheric turbulence will likely create poor seeing.

The National Oceanographic and Atmospheric Administration (NOAA) Aviation Digital Data Service (ADDS) web site[1] provides a wealth of meteorological data that allow the user to form a good idea of

[1] http://adds.aviationweather.noaa.gov

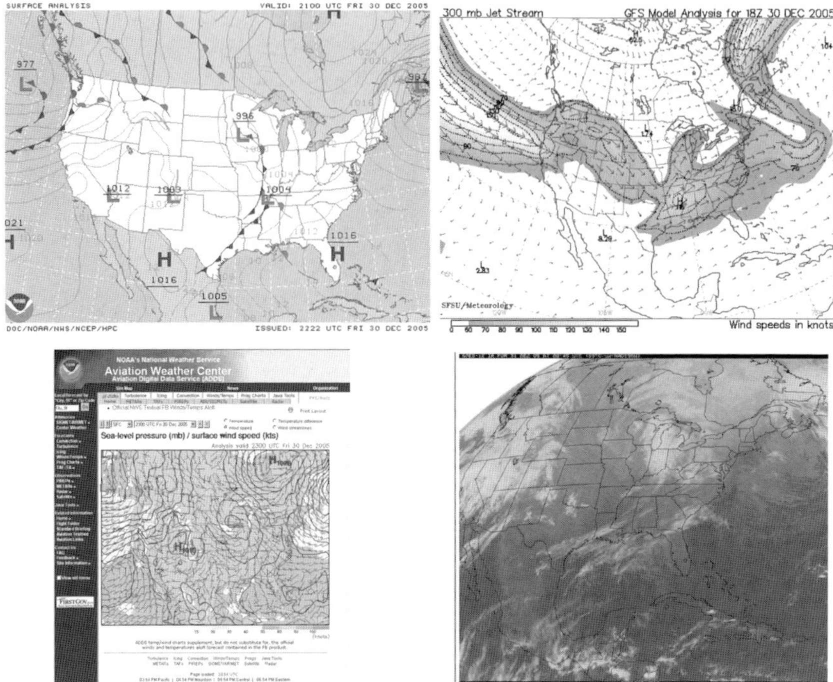

Fig. 11.3 *Internet sites provide anyone with all the climate data available to weather forecasting professionals. Critical data such as the location and passage of weather fronts (upper left), the location of the jet stream (upper right), surface winds (lower left), and daytime visible and night-time infrared satellite images (lower right) are just a mouse-click away. Screen shots by Robert Reeves.*

how weather conditions will affect seeing in a certain locale. Two areas of this government site are particularly useful in gauging potential astronomical seeing.[2]

Weather Underground at http://www.wunderground.com/ is another site that provides all the relevant weather data needed to make a personal "seeing analysis." On the main page, there are tabs above the map of the United States for Current Maps, Forecast Maps, and Aviation Maps. Under Current Maps, items such as temperature, dew point, wind, the jet stream, and a real-time visible satellite image can be accessed. Under the Forecast Maps, the user can find items such as dew point and sky cover forecasts for up to a week in advance. Under the Aviation Maps tab, winds at altitudes from 10,000 to 45,000 feet can be analyzed to determine the affect on local seeing. Additional seeing forecasts for extended areas of the U. S. and

[2] http://adds.aviationweather.noaa.gov/winds/ and http://adds.aviationweather.noaa.gov/turbulence/

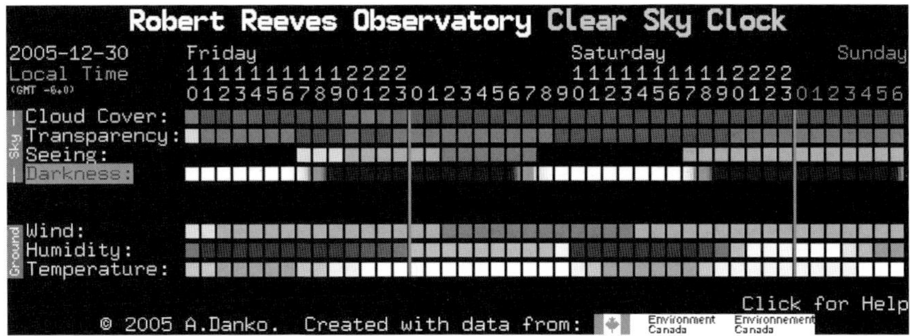

Fig. 11.4 *The Clear Sky Clocks are read from left to right. Each vertical column represents a different hour with the first column being the present time. The top row of blocks is the visible-light cloud cover forecast. Each block is color-coded from dark blue to light gray to white; blue represents ideal conditions, white represents overcast. This forecast may miss low clouds and afternoon thunderstorms. The second row is the transparency forecast. It is calculated from the total amount of water vapor in the air. Dark blue means excellent transparency, light blue is better than average, and pale blue is worse than average. White means that there is at least 20 percent cloud cover and transparency was not calculated. The third row is the astronomical seeing forecast and attempts to predict turbulence and temperature differences that affect seeing for all altitudes. The seeing scale is calibrated for instruments in the 11- to 14-inch range. There are gaps in the forecast because the prediction does not consider daytime heating, and thus only forecasts nighttime seeing. Screen shot by Robert Reeves.*

Canada are also available at http://www.cmc.ec.gc.ca/cmc/htmls/seeing_e.html.

For imagers, one of the most useful astronomy weather sites on the Internet is Atila Danko's Clear Sky Clock. Danko is a Canadian amateur who became tired of being frustrated by clouds when he wanted to do his astronomy. He set out to create an accurate method of forecasting the astronomically relevant weather in his area. This project evolved into the Clear Sky Clock that has become an indispensable tool for amateurs throughout North America. In conjunction with the Canadian Meteorological Center, Danko created a script that interprets weather data for a particular location in the U.S. or Canada and displays it in a graphical format representing a 48-hour forecast in three-hour blocks. An example of Danko's now famous Clear Sky Clock is shown in Figure 11.4. My observatory location is one of over 2000 with its own Clear Sky Clock. To check if there is an existing Clear Sky Clock location near your area, check http://cleardarksky.com/csk/ and check the state or province you live in. If there is no other Clear Sky Clock already in existence for a location within 15 miles, contact Danko and ask him to create a unique one for your location. User instructions are provided at Clear Sky Clock website.

I have also found the weather data and real time satellite images presented by the Space Science and Engineering center at the University of Wisconsin-Madison to be quite useful. These can be accessed at http://www.ssec.wisc.edu/. Another interesting Internet site is http://www.inquinamentoluminoso.it/worldatlas/pages/fig1.htm. This location maps the artificial sky brightness and apparent light pollution for any location on the planet. With the sky brightness data presented on this site, you can compare your observing site with nearby areas, and perhaps choose a darker sky location.

11.5 Light Pollution

It is a sad fact that astrophotography is affected by light pollution more than any other aspect of amateur astronomy. This is because, in addition to accumulating star light, photos accumulate light pollution, blotting out the target just as surely as a bright light blinds a visual observer. As urban areas grow larger and brighter, the imager is faced with the choice of traveling further away from cities and towns to escape light pollution, resorting to the use of specialized filters to block it.

There are two kinds of light pollution—natural and man-made. The two main forms of natural light pollution are scatter and airglow. The daytime sky is blue because air particles scatter sunlight. At night, the air scatters starlight in a similar way to make a photographically detectable sky fog. Fortunately, the digital astrophotographer does not need to worry about the effects of scatter or low-level airglow because the nature of webcam imaging sensors prevents lengthy continuous exposures that would run up against such phenomena. An extreme form of airglow is the aurora. This event occurs when Earth's magnetic field channels energetic solar particles into the upper atmosphere over the polar regions, where they interact with atmospheric atoms and air molecules. Even a faint aurora that is not visible to the eye can be bright enough to interfere with digital astrophotography. Under such circumstances, there is nothing to do except await for another clear night.

Man-made light pollution is primarily from nighttime outdoor lighting. High-pressure sodium lights radiate strongly in the yellow but appear pink to the eye because they also weakly emit from the ultraviolet to the green portion of the spectrum. Mercury vapor lights are blue to green.

If light pollution is a problem, our goal is to block the offending wavelengths with the proper kind of filter before they enter the camera, while passing the wavelengths unique to celestial sources. Fortunately, the wavelengths from light pollution sources are not spread uniformly over the spectrum, but are concentrated in a number of peaks that can be filtered out while allowing the desired astronomical wavelengths to get through.

Just as filters can be used to enhance terrestrial photographs, there is an arsenal of filters that can be used to combat light pollution. Light pollution restricts the length of exposures, but proper filters allows us to increase the exposure and contrast of a celestial photograph. Fortunately, light pollution filters are becoming extremely sophisticated and the advent of "narrow band" imaging, which concentrates on specific emission lines produced by deep-sky objects, has opened the door to taking stunning images from urban settings. Indeed, a skilled urban astrophotographer working with filters can often achieve images equal to unfiltered ones acquired an hour's drive away from the city lights.

11.6 Dewing

The formation of dew on a lens will slowly obscure starlight and eventually extinguish a star's image during a guided exposure. In severe dewing conditions, a camera lens or telescope can be completely covered in five minutes, which will end an astrophotography session as surely as if the shutter were closed. Thus, the morning dew, the poetic fairy's breath that makes a sunrise landscape beautiful, is not a pretty sight to astrophotographers.

The warmer the air, the more water vapor it can hold, because with increased temperature water vapor molecules move faster and remain suspended easier. By night, as the air cools, these molecules slow down. When the have slowed sufficiently, they begin to stick together to form liquid water. The temperature at which this happens is called the "dew point."

The dew point temperature is strictly a measure of the amount of water vapor in the air and is independent of the air temperature—except that the former can never be greater than the latter. When the dew point and the air temperature are the same, the air is saturated and water will begin to precipitate in the form of fog, dew, or, if the dew point is below 32 degrees Fahrenheit, frost. (A dew point below 32 degrees Fahrenheit can be called a "frost point.") "Relative humidity" is a measure of how much water vapor is in the air compared to how much it can potentially hold. When the air temperature has fallen to the dew point, the relative humidity is 100 percent. If, for example, the air temperature and the dew point are both 45 degrees Fahrenheit, the air is completely saturated, feels damp, and the relative humidity is 100 percent. But if the air temperature is 100 degrees Fahrenheit and the dew point is still 45 degrees, the relative humidity is very low and the air feels quite dry.

A telescope or camera will cool down to the dew point and begin to become moisture-covered before the surrounding night air itself has cooled down to the dew point. The reason is that metal and glass are more efficient radiators of heat than trees or grass, the cooling of which is what makes the air itself cool. (Air is transparent to heat—which is infrared radiation—and allows heat from solid objects like trees and telescopes to radiate into space. The air itself cools by contact with solid objects that are already cool.) A gentle night breeze is beneficial because it will keep the telescope and camera from cooling and collecting dew.

11.7 Dealing With Dew

When Schmidt-Cassegrain telescopes with their thin, exposed corrector plates became popular in the 1970s, they changed dew from an inconvenience to a major headache. An aggressive approach to the problem was clearly necessary; namely, to prevent dew from forming on the optics in the first place rather than to remove it after it had formed.

The first real line of defense against dew is to shield the optics from the cold and damp of the night sky. But this is difficult to accomplish with a camera lens or corrector plate, which of necessity must be aimed upwards. The oldest solution is to trap a column of warm air over the optics with a cylindrical extension of the scope's tube called a dew cap, or dew shield. At best, this will only delay the onset of dewing, but in moderate humidity this is usually long enough to complete the observing session.

All brands of SCTs have commercial dew caps available. Alternatively, a very effective one can be made out of the hard ⅞-inch foam-rubber camping mats that are available from sporting goods stores. Roll the foam into a cylinder and glue the overlapping ends together. Make it fit snugly so it will stay put when slipped over the end of the tube. The length from the corrector plate to the far end of the sheild should be 1.5 times the telescope's aperture. To prevent possible vignetting of the field of view, flare the far end of the shield out slightly.

Although the telescope may be affected by dew, the natural heat from a webcam's electronics will usually keep it dew-free. In times of heavy dew, or dusty conditions, a simple way to protect the webcam electronics is to slip a plastic bag over the camera and seal it shut as far as possible.

The most effective long-term defense against dew is continuous low heat to keep the optics just above the temperature of the night air. An excellent commercial device is the Kendrick Dew Removal System (www.kendrick-ai.com) that has many options for heating different types of optical equipment. Its single 12-volt DC controller powers as many heaters as needed as long as their total power drain does not exceed three amps. Because it uses a relatively benign 12-volts DC instead of potentially lethal

120-volts AC, the Kendrick is a safe system. For permanent installations, an inexpensive 120-volt AC to 12-volt DC transformer can be obtained.

Convenient 12-volt DC hair-dryer-style dew removers can be obtained from several sources. Auto supply stores sell them as portable rear window defoggers. A 12-volt unit can also be obtained from Orion Telescopes and Binoculars that plugs into a cigarette lighter receptacle. The dew blower can be powered directly from a 12-volt battery by using an "alligator clip to cigarette receptacle" adapter. Those who also run their telescope directly off receptacle can use an adapter that permits two cigarette lighter-style accessories to run off the same power supply receptacle. Twelve-volt dew blowers do not work as well as their 120-volt cousins, but in the field they can make the difference between saving the observing session and going home.

11.8 Mosquito Control

Anyone who spends a significant amount of time outdoors in the evening knows that mosquitoes are a problem. The outbreak of mosquito-borne diseases in the U.S. like St. Louis encephalitis, dengue fever, and the West-Nile virus raises mosquitoes from a mere nuisance to a genuine health threat.

There are more than 150 different species of mosquito in the United States alone and 2500 species worldwide. It is only the female that bites humans, doing so in her search for protein for her developing eggs. Males feed on nectar and other sugar sources, so they aren't responsible for the welts and itching that result when a female mosquito inserts her blood-sucking probe into our skin.

Mosquitoes live in tall grass where they can keep cool by day. Since they are most active around sundown, just when astronomers are setting up their telescopes, it is best to set up earlier in the day, if possible, then return after sundown.

11.9 Mosquito Repellents and Traps

Much discussion in astronomy circles centers on the non-celestial topic of which mosquito repellent works best. Indeed, a can of OFF repellent is a standard item in most telescope accessory kits. Urban folklore has suggested that various items such as garlic or a certain brand of fabric softener have value as mosquito repellent, but their effectiveness is variable depending on body chemistry and local environmental circumstances. Sonic devices and smoldering mosquito coils also have varying degrees of effectiveness depending on conditions. Natural repellents include essential oil of citronella, which is available from health food stores. Just rub a few drops on exposed skin. This should not be confused with the citronella oil sold for use in yard torches. Essential oil of geranium also works well. Mix a few drops with water and spray it onto your skin. A decidedly low-tech alternative is a simple 15-inch oscillating electric fan to keep the air moving, which makes it more difficult for a mosquito to alight on you.

Jonathan Day, a medical entomologist at the University of Florida, and Mark Fradin, a dermatologist at the University of North Carolina at Chapel Hill, studied commercial mosquito repellents and concluded that those containing the chemical DEET, listed on a product label as "N, N-diethylmeta-toluamide," are best for maximum protection. Introduced commercially in 1957, DEET is the "gold standard" of mosquito repellents. Products containing 24 percent DEET have been shown to prevent bites for up to five hours; lower concentrations are effective for shorter periods, sometimes less than a half-hour with products containing only 6.6 percent DEET. Be aware, though, that DEET is a solvent that reacts with certain plastics and synthetic fibers, so if you use repellents containing this chemical, wear old clothes and be careful not to have any on your hands when handling telescope or camera gear.

In recent years the safety of DEET has been questioned, leading the Environmental Protection Agency to review the product. Their conclusion was that "normal use of DEET does not present a health concern to the general U.S. population." Adverse effects are usually traced to gross overuse of the product. Still, many people think DEET has been classified as less toxic than it is, and the wisdom of slathering yourself with a chemical that is toxic enough to be offensive to other living beings is debated with some vigor. Logic suggests that if the chemical is that potent, it can't be good for the person wearing it.

Chemical safety concerns have led to the development of a wide range of products that are designed to bait and trap mosquitoes. New products on the market include the Mosquito Deleto that captures them with a sticky cartridge, the Dragonfly that zaps them with an electrical grid, and the propane-powered Mosquito Magnet that uses carbon dioxide emissions as bait and then vacuums victims into a net where they dehydrate and die. These products are available at most large hardware and home improvement chain stores.

I've had personal experience with the Mosquito Magnet and can vouch for its effectiveness. Within a week of using the device, my back yard went from "forbidden territory" to a mosquito-free area my family can enjoy, and in which I can set up my telescope without suffering constant bites and itching. (I have no personal association with the manufacturer of the Mosquito Magnet or any of its retailers. I am merely a very satisfied user of the product and highly recommend it.) It requires a 110V power supply and operates for three weeks on a standard 20-pound propane cylinder. The device would be perfect for mosquito control at an astronomy club observing site that is visited weekly.

A word of caution to those who do purchase a Mosquito Magnet: follow the maker's advice about refilling the propane tank to the letter. I found from experience that when they say do not use a pre-filled exchange propane tank, it is for an important reason. Such tanks have a certain percentage of air in them to allow for expansion when the tanks are stored in outdoor racks at the retailer. This air will not affect its use with a barbecue grill, but will cause the Mosquito Magnet to shut down. Have the tank refilled at a propane charging station that will purge air from the tank during refilling.

11.10 Other Nighttime Pests and Critters

Wasps and Yellow Jackets are other flying pests to watch out for in the evening. They can nest inside the roof of an observatory or other structures and their sting can quickly end an observing session. You should also take care not to set up your telescope near an ant bed.

Animals are another nighttime hazard. Raccoons are especially nosey. They will rummage through your telescope goodies while looking for food and cause all sorts of mischief. Skunks will forage within several feet of your telescope if you are perfectly still. Once one waddled under my tripod as I guided on a celestial target. As long as you don't startle them, skunks are oblivious to people at close range. But we are all aware of what can happen if they are disturbed. In wilderness areas, bobcats, mountain lions, and packs of coyotes can be a hazard. Music from a radio or CD player can be an effective large animal repellent. Keep your car unlocked in case it is needed as a temporary shelter from an aggressive animal.

11.11 Keeping Warm at the Telescope

Astronomy and astrophotography can be chilly activities even if it is not winter. For one thing, when we are at the telescope we are nearly stationary so our muscles are generating as little as 100 watts of heat. By contrast, a continuous activity like walking creates up to 2000 watts of heat. Moreover, during the day our body absorbs heat from the Sun, whereas at the eyepiece it radiates what little heat it produces into space. Consequently, unless we are observing on a warm night, we need to prevent loss of body heat.

Most people have misconceptions about how to keep warm at the eyepiece. A simple coat, which suffices during the day, will not do under the stars. Two pairs of socks sounds like a good way to keep feet warm, but often the extra thickness makes shoes too tight and therefore restricts blood flow, making feet colder. Ski ware might keep you warm on the slopes, but it will not work at the eyepiece; it is not designed for insulation, but to look good, fit snugly, and allow freedom of movement. A waterproof jacket may actually make you feel colder because it will not "breathe" allowing perspiration escape, and the water trapped next to the skin will make you feel colder.

The most effective way to stay warm at the telescope is to dress in layers of clothing. This insulates us with the "dead space" trapped between the multiple layers: the trapped air is isolated from the outside cold and makes us feel warm after our body heats it. In very cold weather, use a foundation of long underwear. A turtleneck shirt is necessary to stop heat loss around the neck. Corduroy pants are warmer than denim jeans. "Warm-up" pants over regular trousers will provide the air space vital to comfort. Dress as if it were going to be 20 degrees cooler than expected.

Extremities like the head, hands and feet need special attention. Because one-fourth of the body's heat can be lost from the head, a warm hat is essential. A wool stretch cap will keep your ears warm. Mittens are best for your hands, but gloves will be more practical for operating telescope and camera controls. Feet need loose shoes or boots to allow for an insulating air space between the layers of heavy socks (cotton socks under wool socks). A double boot designed for mountaineering can be obtained from sporting goods stores. A simple blanket draped over your chair and then wrapped over your legs will work wonders when you are sitting for long periods. Also, stand on a foam camping pad to insulate your feet from the cold ground. In very cold weather, a down sleeping bag that unzips at both ends is useful. Zip it up as far as will permit work at the telescope, but leave it open at your feet so you can move around.

Hooded snowmobile suits recently have gained popularity for astronomy. Their one-piece construction simplifies putting them on, and provides a warm environment for the whole body. Soft fur-lined boots and split mittens that snap to the sleeves of the suit complete the ensemble. Split mittens are useful because the fingers can be slipped out of them to operate controls without having to completely remove the mitten.

Appendix A
Using K3CCD Tools

Most users agree that K3CCD Tools (K3), created by Peter Katreniak, is the best program available for operating an astronomical webcam. There are both basic freeware and advanced shareware versions of it. The freeware version, K3v1, is very adequate for controlling either a standard or a long-exposure capable webcam and capturing its AVI output. K3v3, released in December 2005, offers a host of additional features that allow the software to continue serving the user as their webcam expertise grows and they begin utilizing more advanced cameras and techniques. K3v3 is trialware that must be registered to allow continued use beyond 30 days.

One of the compelling reasons to use either version of K3 instead of the webcam maker's own camera control software is that the former will automatically rename sequential AVI files so that none are overwritten by the next capture sequence. The Philips VRecord software available with their ToUcam, for instance, does not have this feature and will overwrite the previous video without any warning. The user must manually rename each succeeding video sequence to prevent the Philips software from assigning the same file name as the previous video. K3 offers a user-edited AVI file name prefix so a series of videos can be identified by a specific name followed by numbers in ascending order. The default file name is "K3CCD" followed by 0001, 0002, etc. The program also has a date-and-time option for naming files so no two AVIs will ever have the same file name.

K3 has no direct means of controlling the camera exposure. Instead it calls up the camera maker's default control software. If a ToUcam is being used for instance, the basic Philips VRecord software is called to allow setting the exposure parameters. An annoying side effect of this is the Philips camera control panel is overlaid on the K3 window and often blocks the image preview window. Fortunately, an additional freeware program called WcCtrl (Webcam Control) created by Martin Burri is designed to interface with K3 (see Figure A.1), and will allow setting the camera control parameters without need for the camera manufacturer's control panel being visible. K3 and WcCrtl should be considered companion programs

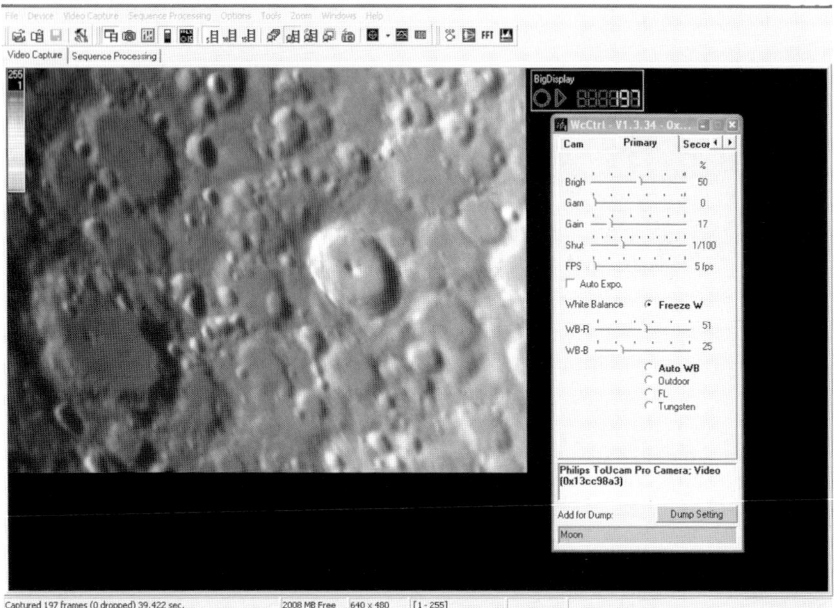

Fig. A.1 *The main screen for K3 displays a live image from the camera. The exposure meter, placed at the upper left, shows the values of the brightest and dimmest pixels in the image and graphically shows the average mean pixel values for the entire scene. During image capture, the <Big Display> box, placed at the upper right, shows the current frame number in the capture sequence. The camera control program WcCtrl can run concurrently with K3 allowing exposure settings to be controlled without accessing a separate window within K3. The status bar at the bottom displays the number of frames captured (or dropped), the elapsed capture time, remaining hard drive free space, and highest and lowest pixel values in the image. Screen shot by Robert Reeves.*

because K3 alone cannot control the camera while WcCtrl alone cannot display or record the camera's output. With K3v1, WcCtrl is started separately and run at the same time as K3. K3v3 has the option of calling WcCtrl from within K3, and WcCtrl that does not have to be started separately. A valuable feature of WcCtrl is that the camera control parameters are also displayed numerically instead of just being unnamed positions along a slider.

Both versions of K3 display a level meter to allow accurate adjustment of the camera's exposure setting. Most users have been surprised by the poor results achieved when they trust their eyes to manually adjust the shutter speed and gain at night while viewing a laptop screen. The K3 level meter graphically shows the highest and lowest pixel values, and if the target is a full-frame scene like the Moon the level meter will show the values for the main concentration of pixels in the scene. The meter provides

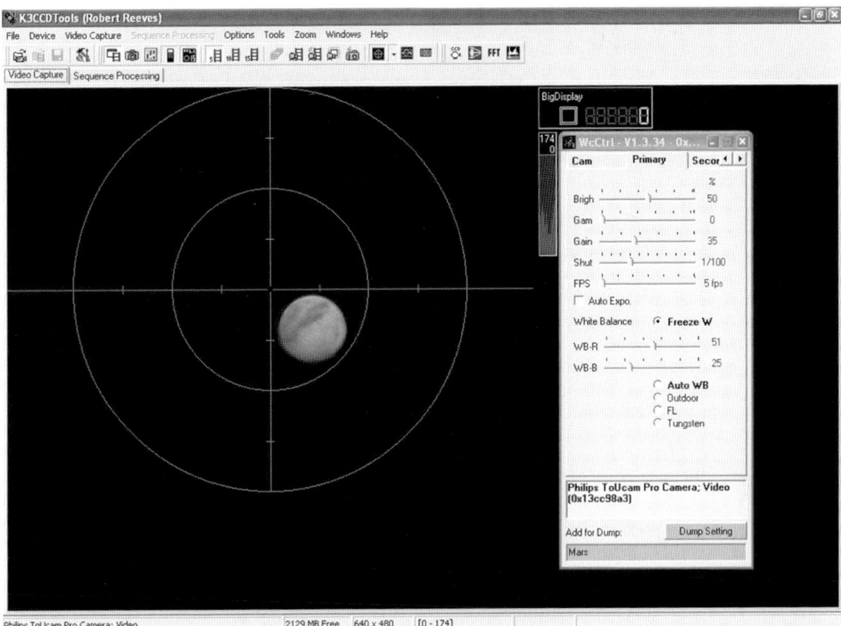

Fig. A.2 *The crosshair option helps show the location of a planet or deep-sky object relative to the image frame's dark boundaries. It is also helpful for adjusting the telescope's polar alignment using the star-drift method. Screen shot by Robert Reeves.*

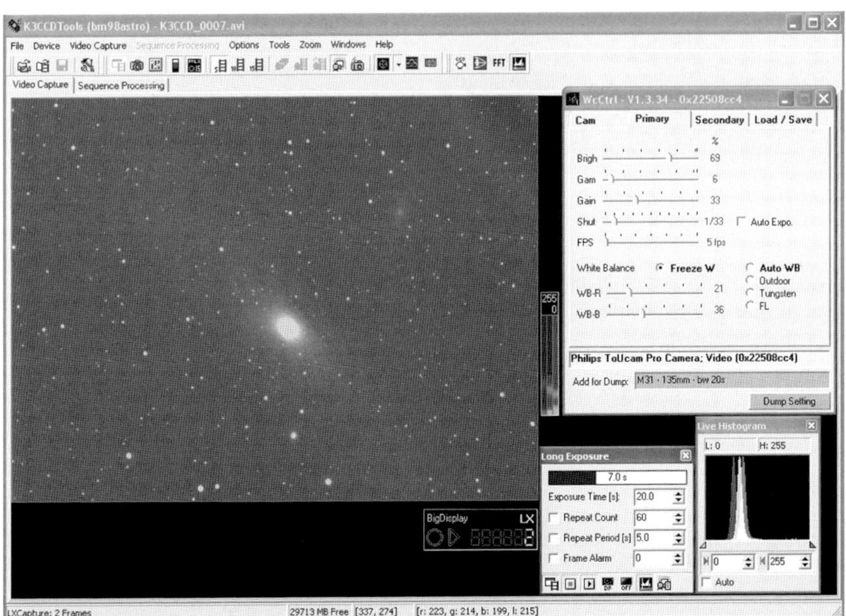

Fig. A.3 *The <Long Exposure> window in K3 allows automated exposure control of long-exposure-capable webcams and webcam-style cameras. The <Live Histogram> window allows adjustment of the exposure parameters using WcCtrl. Screen shot by Martin Burri.*

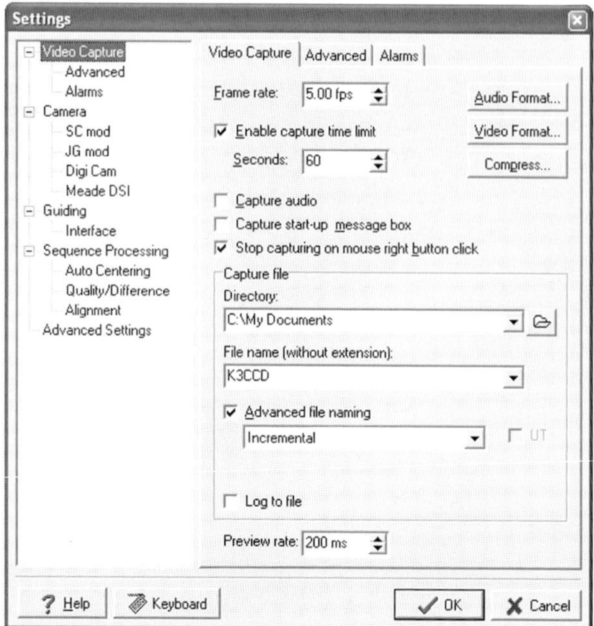

Fig. A.4 *The <Settings> window allows extensive configuration and control of image-capture parameters within K3. Screen shot by Robert Reeves.*

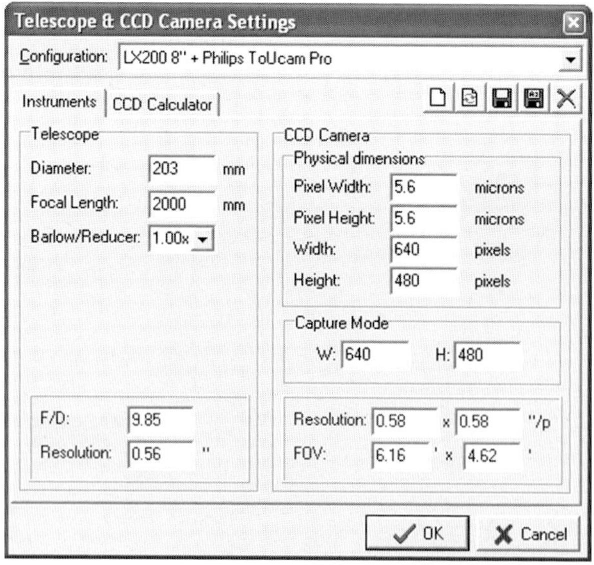

Fig. A.5 *The performance parameters of various telescope and image sensor configurations can be evaluated using the <Telescope & CCD Camera Settings> calculator window. Screen shot by Robert Reeves.*

objective information to remove errors in human vision from the exposure- and gain-setting process.

The level meter is very useful in preventing an artifact that plagues many planetary images, the "onion-ring effect." This appears as a concentric ring, or series of rings, inside the ball of a planet. It produces disturbing artifacts on Jupiter that disrupt the smooth flow of the planet's belt structure, and can produce some distinctly "canal-like" artifacts on Mars images. The onion-ring effect is caused by gain settings that are too low when capturing the raw video. The effect is predominantly visible with Jupiter and Saturn. To control it, expose using gain settings of at least 50 percent and adjust exposure-time settings accordingly.

Both versions of K3 have an on-screen crosshair option. In addition to keeping a target centered in the field of view, the crosshairs provide a reference for the camera's field. When imaging a planet, which normally appears against a black background, it is difficult to gauge the edge of the field because the K3 background is also black. The crosshairs help define the camera field border so a planet will not wander out of view.

When using K3 to control long exposures, make sure the camera frame rate is set to five per second. Even though long exposures are being taken, K3 needs to accurately time the frames in order to capture the completed frame from the camera. In some instances, using 10 frames per second with an "SC"-modified camera will result in black frames interspersed with good ones.

Appendix B
Using RegiStax

It has long been known by film astrophotographers that registering two negatives together so they perfectly overlap suppresses some of the apparent film grain. This is the film world's way of suppressing image noise. This technique was carried over to the digital realm. But until recently, the process of sorting through dozens, or even hundreds, of lunar or planetary images in order to find the best ones to combine was a painstaking, tedious manual process. In the webcam astrophotography era, this is no longer the case.

RegiStax is a powerful program that relieves the photographer of the labor-intensive task of manually sorting and stacking hundreds of video frames to create a single combined image. RegiStax will quickly and automatically sort through all the video frames in an AVI and select the best, then align them into perfect registration before combining them into a sharper, smoother, less noisy final image. The program also contains powerful image processing options that make it invaluable to the webcam astrophotographer. Even to an old hand at astrophotography like myself, RegiStax produces seemingly miraculous results that constantly amaze me. Indeed, it is so powerful and full-featured that webcam imagers would be willing to pay a lot for it. But you can't buy RegiStax anywhere. The reason it is not for sale is because the program's author, Cor Berrevoets, gives it away for free! The program can be downloaded from http://registax.astronomy.net/.

To begin processing an AVI (including proprietary file format movies produced by the advanced Luminera cameras), click on the <select> button and choose the desired AVI. The latest version of RegiStax has the ability to register multiple AVI clips in order to work around the two-gigabyte AVI file size barrier.

Next select the best frame within the AVI sequence for RegiStax to use as a reference frame. RegiStax will use the reference frame as a quality comparison against the other frames in the video clip. After the comparison, the program decides whether to add each frame to the image stack or reject it. Selecting the best frame can be accomplished by using the left and right keyboard arrow keys to scroll through the series of video frames. Pressing the spacebar will manually deselect frames with poor detail. When the frame being displayed is deselected, the green status button at the left of the frame selection slider will turn red.

The RegiStax default image-quality setting is 80 percent. That is, an image must be at least 80 percent as good as the reference frame in order to be added to the stack. It may be tempting to raise the quality setting to 90 or 95 percent and use only the very best frames, but under less than ideal seeing conditions this may actually reduce the quality of the final image. The reason is that if a majority of images are rejected, each individual frame may be more detailed, but noise will be more prominent in the stacked image and degrade the final result. Lowering the quality setting to 80 percent to allows more images to be stacked and often results in a smoother, better looking final image even if it possesses less fine detail.

Using the cursor, select and click on an alignment detail within the reference image. This will allow RegiStax to use this same feature in all frames to align and register them. The alignment feature can be a bright lunar feature or the entire ball of a planet. The size of the alignment feature box can be adjusted from 32 pixels to 512 pixels in the <Alignment Box> option on the RegiStax main page. For lunar images, it is best to use as small an alignment box as possible. With planetary images, it may be necessary to scale the alignment box to fit the entire planet. Once the alignment feature is defined, click on the <Align> button and RegiStax begins the frame sorting, alignment, and stacking process.

The default RegiStax settings perform well enough that outstanding results will be produced on the first pass through an AVI. Once the user understands the flow of the program, they are encouraged to try different user-selectable processing parameters to see if they work better with a particular AVI. Other program options allow such operations as de-rotating sequential video frames, applying a flat frame or dark frame to each video frame being processed to further reduce image noise, applying post-stacking corrections to contrast, brightness, and gamma, working with the histogram, shifting any of the color channels to align colors on a planet, and the creation of custom wavelet filters.

If RegiStax has a fault, it is the inability to reset the program parameters once the processing cycle has started. The only way to change any parameter is to start all over again, a time-consuming operation if a slow computer has to restart a lengthy video processing sequence.

The key to the marvelous image processing functions in RegiStax is

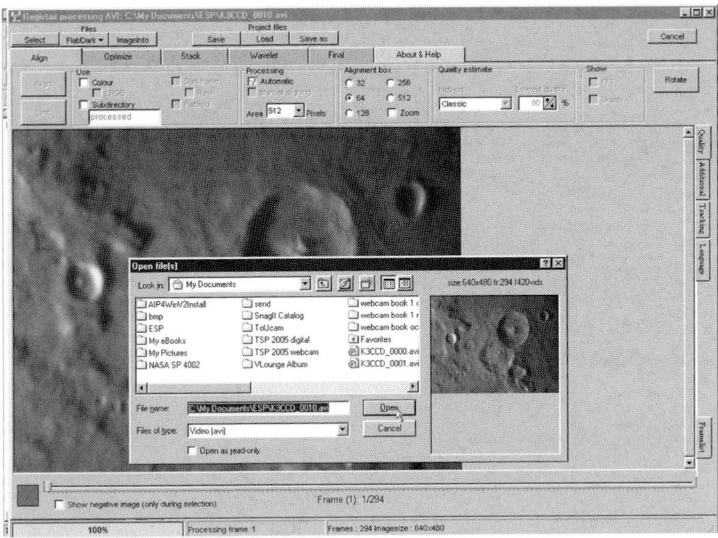

Fig. B.1 *Choose the AVI to be processed by clicking on the RegiStax <Select> button to bring up the Windows file-selection box. Screen shot by Robert Reeves.*

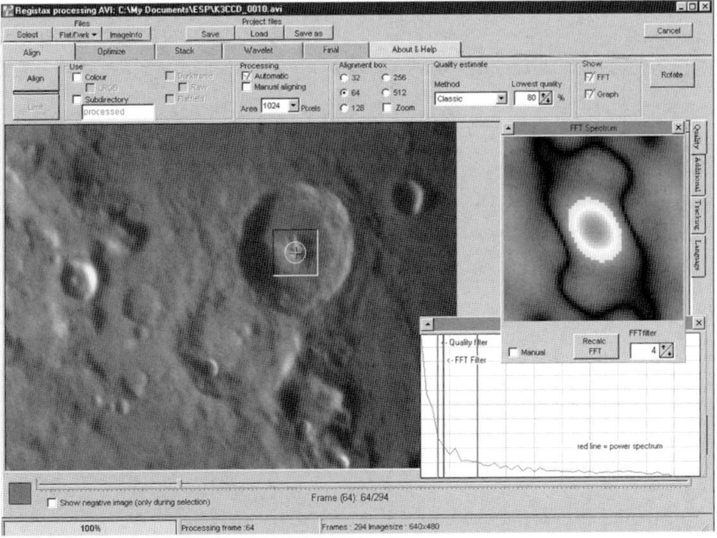

Fig. B.2 *Scroll through the AVI frames using either the slider under the preview window or the right and left arrow keys, and select the best frame as a master image for comparison against other frames. Select 1024 in the <Area Pixel> window to allow processing the entire video frame area. Use the mouse to click on a bright alignment feature. Select the appropriate size alignment box in the <Alignment box> window. When the FFT Spectrum window for the selected alignment feature appears, click the <Align> button to begin processing the video frames. Screen shot by Robert Reeves.*

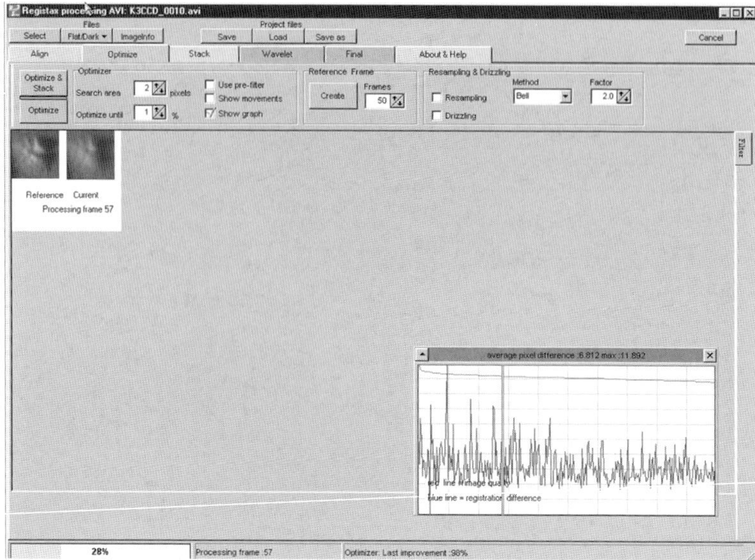

Fig. B.3 *RegiStax compares all AVI frames to the master image and rejects those that are below a given quality threshold. Screen shot by Robert Reeves.*

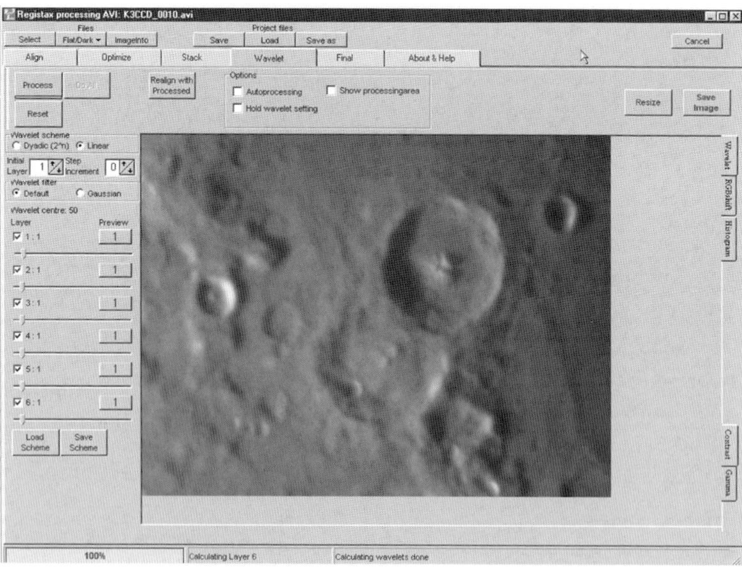

Fig. B.4 *After image stacking is complete, RegiStax offers six layers of wavelet processing, allowing sharpening control of both large and small features. Screen shot by Robert Reeves.*

a form of unsharp masking called wavelet sharpening. Unsharp masking is a procedure that increases the perceived contrast along borders showing different brightness levels. To accomplish this the original is blurred by a user-scalable amount. The blurred image is then subtracted from the original (the difference is called the "mask") and the mask is added to the original. The amount of blurring on the original image controls the point where the unsharp mask creates a perceived contrast change, thus making the image appear sharper. Large blurs change the edges over a larger distance but omit changes over a smaller area. Small blurs will increase sharpness over smaller areas, but miss the larger edges. The user has to define the blur size to get the maximum benefit.

The problem with most images is that one sharpening setting will not work for all areas of the image. This is where wavelet sharpening like that found in RegiStax is a very useful tool. Wavelet sharpening, sometimes known as Laplacian sharpening, is basically the application of many unsharp masks at the same time. The image is separated into its high-, medium-, and low-frequency components and a scalable mask is applied for each frequency "layer" in the image. For instance, RegiStax divides the image into six different layers. By increasing the amount of blur masking to the high-frequency components, the small features are sharpened. By applying increased blur masking to the low-frequency components, large features are sharpened. Each layer has its own sharpening controls. Experience is the best teacher in learning how to control each layer as it interacts with the results generated by the other layers. Additionally, there is a choice of default or Gaussian wavelet systems, each producing different results. Which system to choose depends on the circumstances of the image being processed. I often switch between the two.

The advantage of wavelet sharpening is that you can control the degree of sharpening to all features in an image at the same time. More control can produce better results, but wavelet sharpening is a very powerful tool so care must be exercised to prevent applying too much enhancement to an image or it will do more harm than good. The processing of lunar images illustrates the need for compromise. Higher wavelet settings may bring out finer details, but at the expense of creating donuts around bright crater rims.

Unfortunately, no set of wavelet settings works universally with all images. The settings for each RegiStax layer are unique for each object and will change as seeing, magnification, exposure, and the telescope change. Fortunately, multiple videos of an object taken through the same camera and telescope with the same exposure settings and similar seeing conditions will use the same wavelet settings. RegiStax has a wonderful option

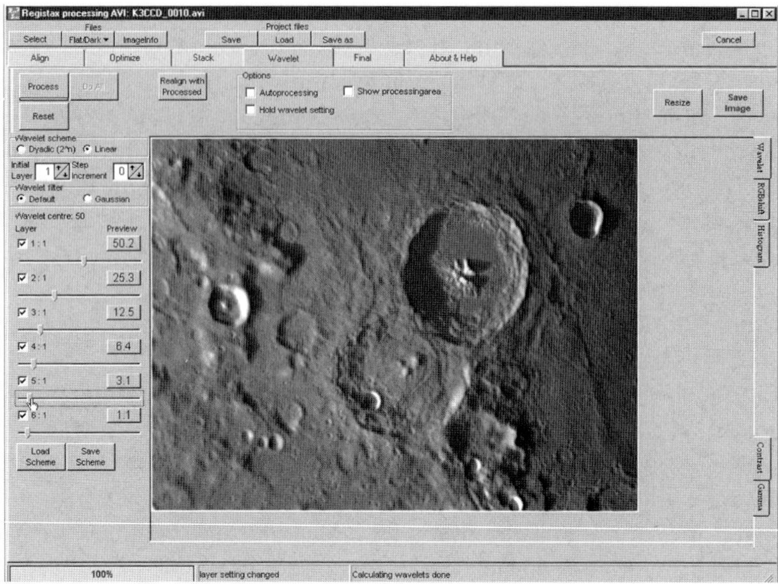

Fig. B.5 *Care must be exercised not to over-apply wavelet processing or fine detail may be obscured. Select settings that allow maximum detail while retaining natural appearance. With the author's setup, using successive wavelet layer settings that are half that of each previous layer gives good results. Screen shot by Robert Reeves.*

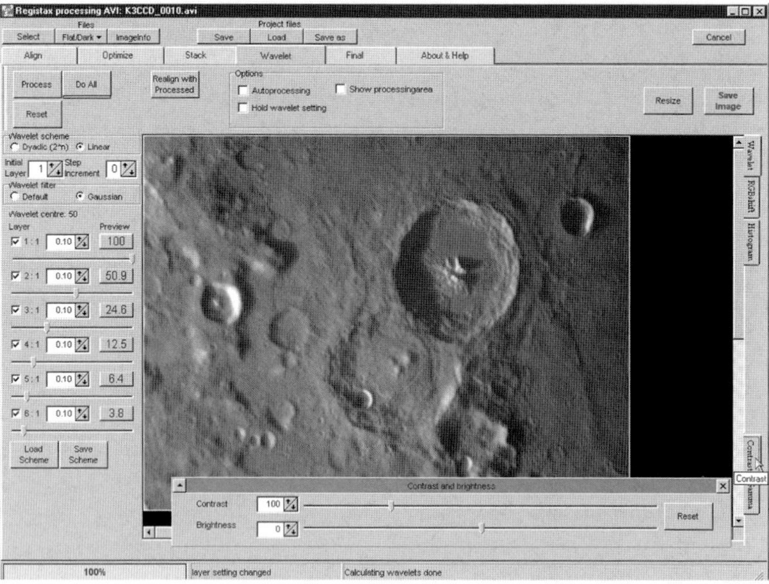

Fig. B.6 *Image contrast and brightness can be adjusted by clicking on the <Contrast> tab. Screen shot by Robert Reeves.*

that enables the program to remember wavelet settings and automatically apply them to all videos processed in a series. This greatly simplifies the processing of many segments of a lunar mosaic, for example.

When photographing planets at low elevations above the horizon, all telescopes are notorious for presenting "rainbow" images, that is, a planet with a red fringe on one side and a blue fringe on the other. This effect is caused by atmospheric dispersion of the planet's image where the low-elevation dense layer of air basically acts as a prism, resulting in a chromatically smeared planet. Fortunately, this effect can be partially corrected by allowing RegiStax to separate the red and blue color channels and then realign them. A helpful hint when manually correcting chromatic blur caused by atmospheric dispersion is to align the camera so either the vertical or horizontal axis is aligned with the direction of dispersion. This way only the red channel or the blue channel needs to be moved in one direction to align it with the other color.

After RegiStax has gone through the trouble of aligning all the images in a video, it would be a shame to lose that alignment if the video is being archived for future use. Fortunately, this program offers the option of saving the images as an aligned AVI. After the alignment and stacking steps have been performed, the program presents the completed stack in the wavelet window so the user can perform wavelet image processing. At this point, the aligned video can be saved by clicking on the <save> or <save as> buttons under the "project files" header at the top center of the RegiStax main page.

RegiStax will save the completed stacked image in BMP, JPG, FIT, 16-bit TIF, or 16-bit PNG format. To retain the most image information, it is best to use the16-bit TIF format if the processing program to next import the image will support 16-bit images.

B.1 Tips and Tricks

RegiStax is a computationally intense program. Working through all the operations performed on a two-minute AVI can be time consuming with slow computers. If the laptop used to acquire images is considerably slower than an available desktop computer, consider transferring the AVI files to the desktop for processing. A wireless LAN connection or USB 2.0 jump drive can quickly transfer files. If your laptop only has a USB 1.1 connection, you can buy a PCMCIA-USB 2.0 card to provide the proper interface. Upgrading to USB 2.0 is a good idea anyway since all future cameras will be USB 2.0.

An alternative to a dedicated USB jump drive is the use of a digital

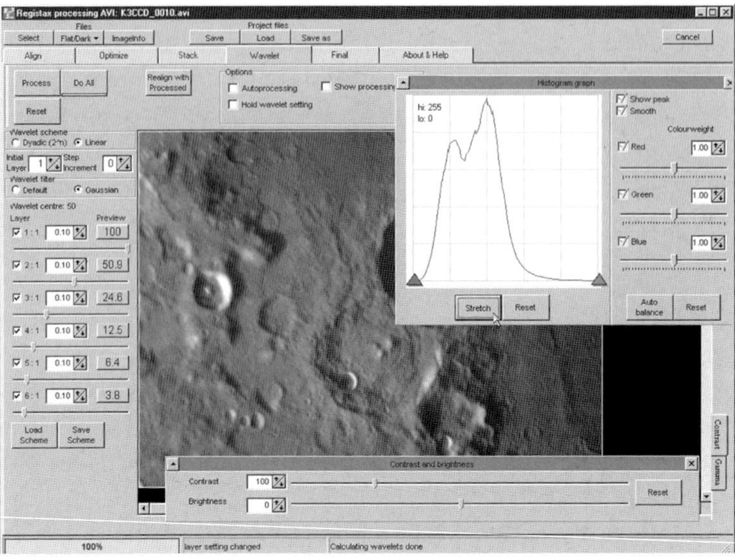

Fig. B.7 *Clicking on the <Histogram> tab allows the option of histogram stretching to enhance faint detail. Screen shot by Robert Reeves.*

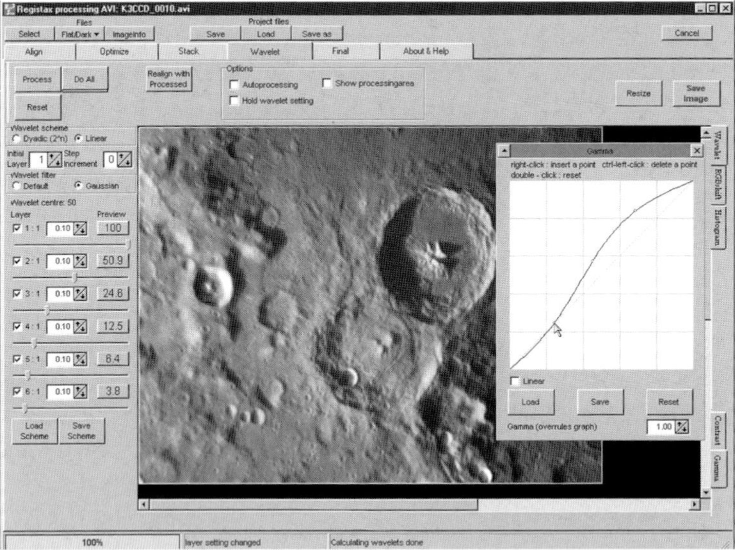

Fig. B.8 *Clicking on the RegiStax <Gamma> tab allows using the Curves tool to enhance image brightness and contrast. When the image is satisfactory, click on the <Save Image> button to save the processed image as a BMP, FIT, JPG, PNG, or TIF file. Screen shot by Robert Reeves.*

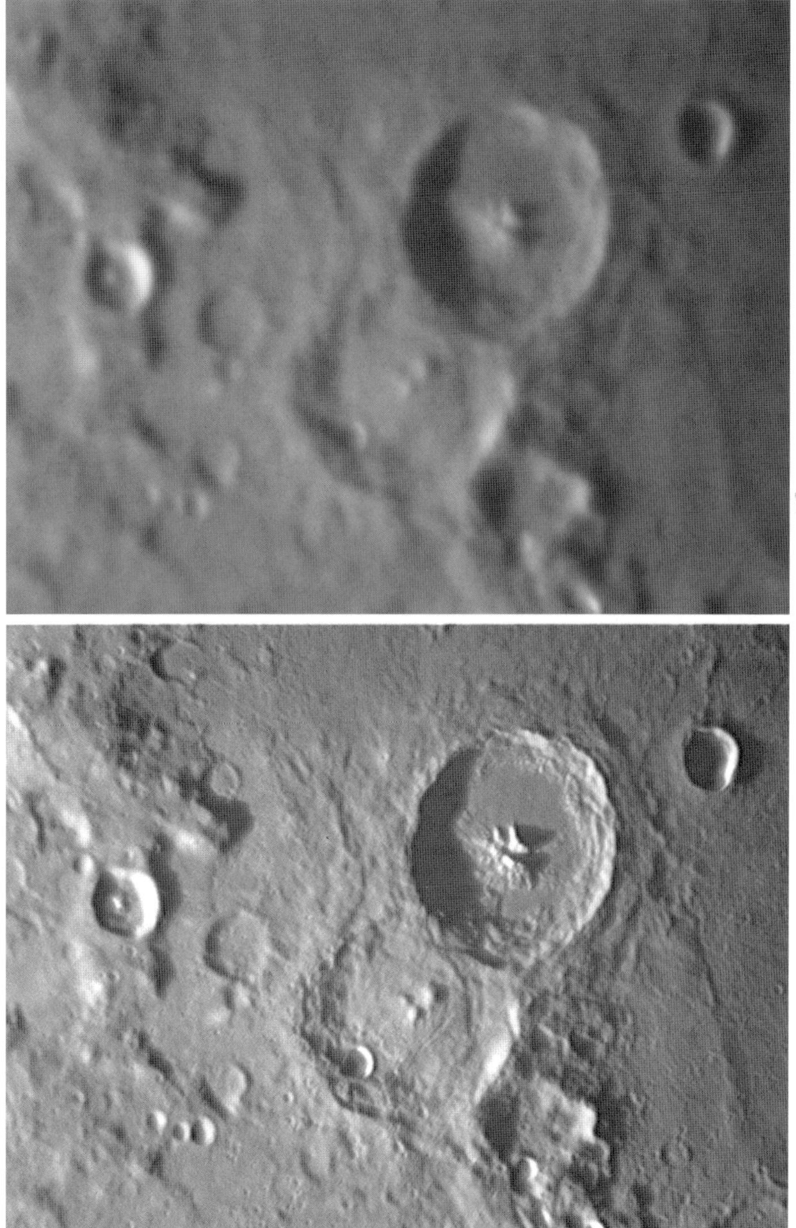

Fig. B.9 *The improvement in image quality gained by stacking and wavelet processing in RegiStax can be seen by comparing an average single frame (top) showing the craters Cyrillus and Theophilus, to a stacked and processed composite of all images (bottom) from that AVI. Photos by Robert Reeves.*

camera memory card and reader. Data can be written to a memory card through the card reader just as easily as images can be read out. A 256 Mb Compact Flash card can transfer a two-minute long 5-frame-per-second video.

One way for finding which series of AVIs of the same object has the greatest image detail is to ZIP compress the AVI files and see which one has the largest file size. The video sequence with the greatest amount of detail will compress to a larger file size than those with less.

When processing planetary images, it is tempting to overdo the colors. Although image details may be more visible with high-contrast, vibrant colors, this deviates from the true appearance of a planetary surface. One must remember that most images of planets presented on the Internet are color enhanced. Even NASA presents images that are unrealistically processed to allow the untrained observer to easily see planetary detail. Jupiter, the most detailed planet available to astrophotographers, is a world of pastel colored clouds, not brilliant colors as often presented. It is best to try to record planetary colors as accurately as possible at the time the image is taken instead of trying to restore them during processing. Take the eyepiece view of the planet as a guide and use the webcam's control software to set the image color using the red and blue slider bars. Raise the saturation level to high and adjust the color so white cloud bands appear as a neutral white on the monitor.

Most image processing programs have the capability to rotate images 90 or 180 degrees and some programs, like RegiStax, can rotate an image to an arbitrarily defined angle. This capability can be a great help for properly reframing an object so it appears in a more conventional north-south orientation. Users should beware of using too many non-90 degree rotations. Arbitrary angle rotations use an algorithm that works by interpolation and thus any non-90 degree rotation will create some degree of degradation that is most apparent along the edge of the image. Perform image rotation last in the chain of processing events. Some operations, such as aligning multiple RGB images may make this impractical, but if possible, rotation should be the final step so further processing does not accentuate any artifacts created by rotating the image early in the process.

Appendix C
Using GuideDog Autoguiding Software

A wonderful feature of webcam and webcam-style cameras is their ability to be used as an autoguider. Typically, another long-exposure capable webcam, or a film or conventional digital camera—is used to image a deep-sky target while the webcam uses a software program to interpret the guide-star movement on the CCD sensor, and issues telescope commands to keep it on the same pixel in the guiding camera. A number of commercial software packages, including K3CCD Tools 3, MaximDL, and the Meade Autostar Suite allow autoguiding with a webcam.

One of the most popular stand-alone software packages is GuideDog, a program created by Steve Barkes. Like many useful webcam programs, GuideDog is available both as a basic free version and an advanced GuideDog Pro version requiring a registration fee. The basic version works on older laptops that have parallel ports and interface with a telescope control system using the Shoestring Astronomy GPINT-PT parallel interface. Both standard and long-exposure capable webcams can be used as a guide camera. GuideDog is compatible with nearly all mounts that can accept either CCD autoguider inputs through a 6-pin phone-jack style connector or input from a serial cable connection to a computer.

Typically, a basic unmodified webcam like a ToUcam or SPC900NC on an 8-inch *f*/10 SCT can detect guide stars as dim as 6th magnitude with exposure times limited to a fraction of a second. This provides enough sensitivity to allow autoguiding a piggy-back wide-field 35mm or digital camera. Camera gain settings can be used at maximum, and although the image visually looks poor because of noise, GuideDog will discriminate the star against the background and track it. It can accept input from long-exposure SC-modified webcams, Atik, SAC, and Meade cameras and allows exposures as long as 30 seconds. With my own Atik ATK-2HS, exposures in the five-second range allow accurate guiding with enough sensitivity to detect guide stars in almost any telescopic field. GuideDog has flawlessly guided exposures up to three hours long on my Losmandy GM-8 mount.

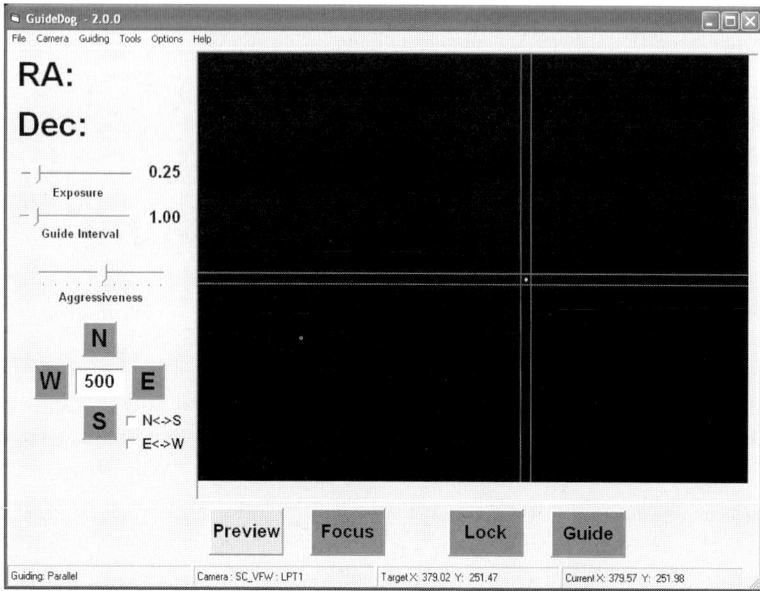

Fig. C.1 *While in GuideDog's preview mode, right-mouse click on a chosen star to place a dual crosshair over it. This will allow checking the orientation of the camera. For proper autoguiding, north must be at the top and west must be to the left. The crosshairs can also aid in refining telescope polar alignment. Click on the N, E, S, and W buttons to verify that guiding commands move the telescope in the proper direction. Use the N-S or E-W reverse button as necessary. The length in milliseconds of telescope guiding commands can be set in the window at the center of the N, E, S, and W buttons. Screen shot by Steve Barkes.*

It uses a centroid algorithm to determine the exact position of the center of the guide star on the CCD. This allows the star's location to be determined with sub-pixel accuracy and the program will, if atmospheric seeing is steady enough, keep the telescope aimed with sub-pixel tracking accuracy. Because of the centroid algorithm, it is better to guide on "fat" stars instead of very small stars. Choosing a star that covers a single pixel on the CCD will actually result in less accurate tracking because GuideDog will jump back and forth between pixels. Choosing a star that covers multiple pixels allows the program to calculate the actual center of the combined multi-pixel star image.

The GuideDog controls are simple to use. Press the <Focus> button to receive live streaming input from the camera to allow focusing on a bright star. Then, align the camera on the telescope so east-west right ascension movement will move the star left and right on the preview screen. Next, right-mouse click on the chosen guide star to place a dual crosshair over it. This will allow the star's movement on the CCD to be

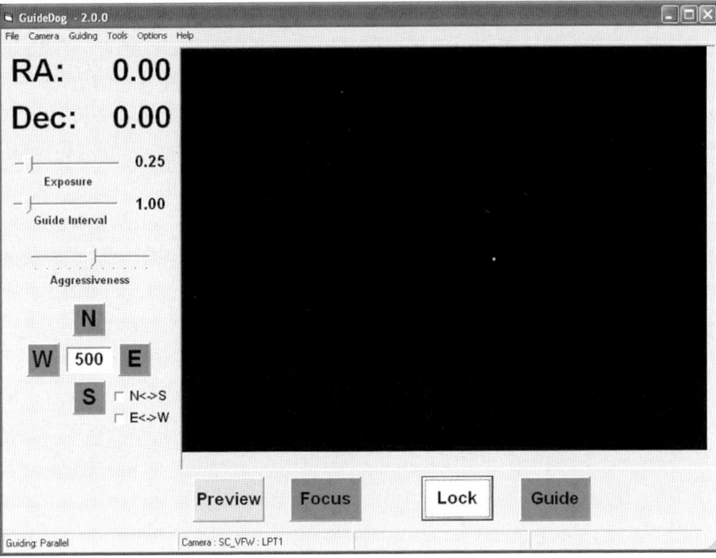

Fig. C.2 *Click on the chosen star and then click the <Lock> button to designate that star as the guide star to be used for autoguiding control. Screen shot by Steve Barkes.*

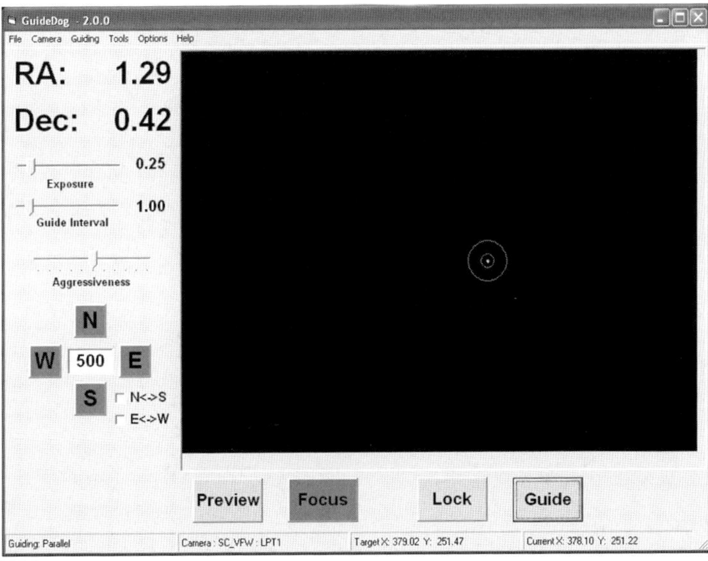

Fig. C.3 *After the guide star is locked, click on the <Guide> button to begin autoguiding. Screen shot by Steve Barkes.*

perceived more easily. Now command the guide star to move using the on-screen N, E, S, and W buttons. GuideDog expects the star to move up for north and left for west. If needed, click on the reverse N-S and E-W buttons to switch star movement to the proper direction. Press the <Focus> button again to disable the focus mode, and then press <Preview> to view the star as it will appear while autoguiding with longer exposure times. Adjust the <Exposure> slider to set the exposure time length with long-exposure capable cameras. Adjust the <Guide Interval> slider to issue telescope-guiding commands at given intervals. Click on the desired guide star, and then press the <Lock> to assign guiding functions to that star. Press <Guide> to allow the software to track the star and issue telescope control commands. The length in milliseconds of each autoguiding command sent to the mount is adjusted in the window at the middle of the N, E, S, and W buttons. For best results with telescope drives having adjustable tracking and guiding rates, set the guide correction rate at 2x the normal tracking rate.

Appendix D
Assembling Wide-Field Lunar Mosaics

A typical webcam CCD has a 4.5 millimeter diagonal measure, thus its field of view through a telescope is very small, essentially the same as visually using a 5mm eyepiece. But the magic of webcam planetary imaging is that a well-processed webcam image makes up in resolution what it lacks in field of view. Indeed, during periods of good seeing, sub-kilometer resolution on the lunar surface is possible with a webcam on a well-collimated large amateur telescope.

In spite of the high-resolution results, the fact remains that most of us are more familiar with the wider fields of view available with conventional film and digital cameras. The keyhole-sized views taken with a webcam leave something to be desired, even if they are highly detailed. But there is a solution to this dilemma. Assembling many overlapping narrow-field webcam lunar images into a high-resolution wide-field mosaic is an intriguing project that produces images very reminiscent of those returned by the exciting NASA lunar probes in the 1960s. Lets take a look at how you can use an inexpensive off-the-shelf webcam and backyard telescope to create seamless high-resolution lunar mosaics using only free software and easily-acquired imaging skills.

D.1 Basic Equipment Used

My webcam experiences are with the Philips ToUcam 840 and SPC900NC, and an Atik ATK-2HS. The two popular Philips cameras use the same ¼-inch Sony ICX098BQ color CCD with 5.6-micron pixels. The Atik ATK-2HS is a commercially-modified and repackaged ToUcam that is equipped with a ⅓-inch Sony ICX424AL CCD with 7.4-micron monochrome pixels. The larger CCD produces less electronic noise, and since it does not have a Bayer color mosaic filter with which to create color images, each photosite on the sensor receives all colors of incoming light, not just red, green, or blue. The larger monochrome pixels give the Atik cam-

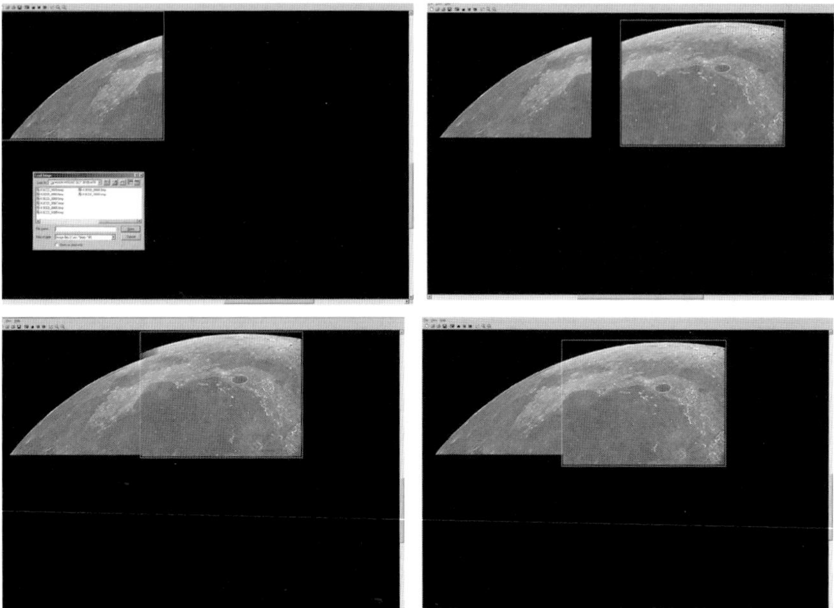

Fig. D.1 *The operation of iMerge is so intuitive that documentation is not needed. Simply open the first image using the <file> and <import image> selections. Use the same commands to open the next image to be added to the mosaic and drag the new image over the first one. Unmerged images are semi-transparent so detail in each can be aligned to the nearest pixel. Screen shots by Robert Reeves.*

era more sensitivity, allowing higher shutter speeds to freeze atmospheric turbulence that can degrade a lunar image. This is especially helpful when imaging at high *f*-ratios through a Barlow. The result is smoother, richer images with higher dynamic range. Although the Atik's larger pixels cover a greater area of the Moon with the same 640 x 480 array as the standard Philips webcams, and so theoretically have less resolution, I feel the Atik images are of better quality and thus I primarily use the Atik ATK-2HS when imaging the Moon. The lunar imaging techniques discussed here can also be applied to the Meade, Celestron, SAC, and Orion webcam-style cameras.

Regardless of which camera is used, it is important to install an infrared filter. First, it keeps dirt off the CCD sensor. As small as the CCD chip is, even a tiny dust mote will make a noticeable splotch on the resulting images. Secondly, webcams are sensitive to infrared wavelengths. Most refractor telescope optics do not focus infrared wavelengths on the same focal plane as the "visible light" image, and thus the infrared will degrade the focus. This is not a problem with reflector telescopes, but the filter will

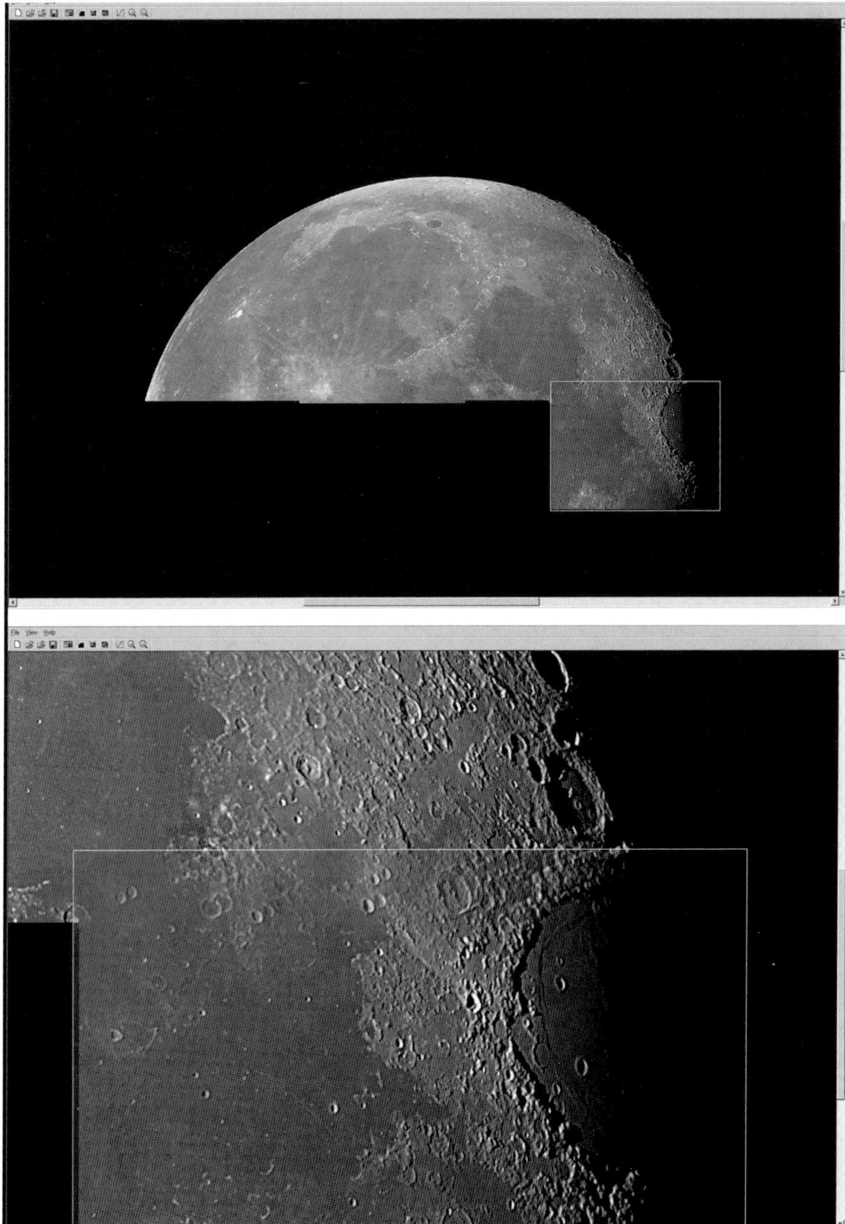

Fig. D.2 *As new segments are added to the mosaic, the image canvas expands as necessary. The + and - magnifying glass icons on the toolbar control overall image zoom. The view can be zoomed out to see the entire mosaic or zoomed in to the sub-segment level to examine alignment of the segments. Screen shots by Robert Reeves.*

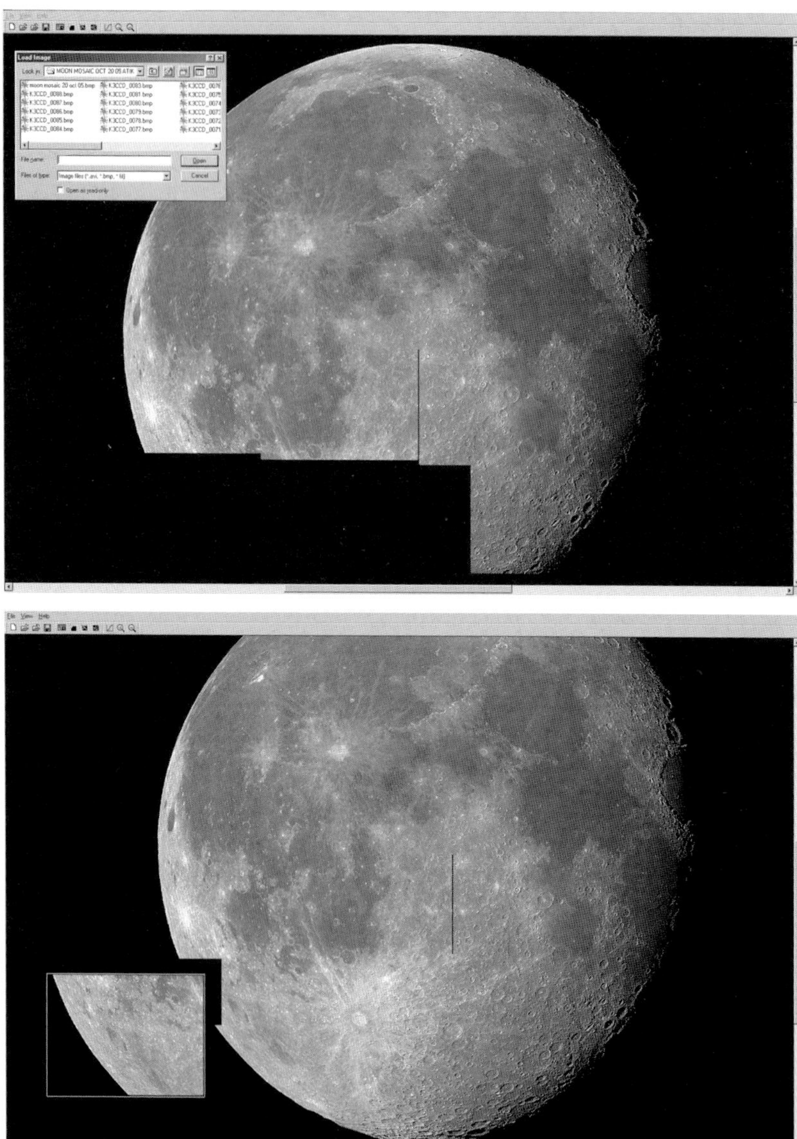

Fig. D.3 *Continue to add mosaic segments until the complete image is assembled. iMerge has no problem handling dozens of mosaic segments. Once the mosaic is complete, be sure it is displayed at 1:1 zoom and save it as a single merged image using the <file> and <save image as> commands. Often there is a gap between mosaic segments (seen here near center) due to telescope aiming error. Small gaps can be filled in with matching tone levels using an image editing program like Photoshop. Large gaps require reshooting the mosaic or careful substitution of additional image data acquired during a similar lunar phase. Screen shots by Robert Reeves.*

still protect the CCD from dust.

The main requirement for the computer used to capture webcam imagery is sufficient hard drive space to record the huge amount of video data. It is not unusual for a webcam to produce a 200-megabyte AVI file for each video sequence that will be processed into a single 900-kilobyte BMP image. With a ToUcam on an 8-inch *f*/10 telescope, it may take as a many as 40 to 50 overlapping images to create a mosaic of the full Moon. A rule of thumb for computers used with webcam imaging is to have at least 20 Gb of hard drive space.

D.2 Acquiring the Images

Before imaging the Moon for a mosaic, the camera must be properly oriented and focused. This is more of a challenge than it sounds because of the webcam's small field of view. Before critical focusing, aim at the Moon's northern limb and slew the telescope back and forth in right ascension. Orient the camera so the lunar limb passes parallel with the camera's left-right view. Properly orienting the camera will simplify moving left and right or up and down from one overlapping mosaic field to the next.

Because the field of view of a webcam's tiny sensor is so limited, just touching the telescope's focus knob will cause image shake, complicating finding exact focus. Ideally, an electric focuser should be used. If one is not available, one way to reduce focusing shake is to clip a wooden clothespin to the focuser so it can be turned with a gentle finger push instead of having to grasp it. Another focusing way is to view along the lunar terminator and then raise the camera's shutter speed high enough that the mountain peaks protruding out of the darkness are barely visible. Rack the focus in and out. The faint mountain peaks will vanish completely when out of focus, then pop back into view when they are well focused. Once good focus is achieved, reset the shutter speed to normal.

For lunar imaging, set the webcam to its highest non-interpolated resolution, 640 x 480. Do not use any video compression in order to minimize image artifacts. For a smooth, seamless mosaic with uniform density between segments, the exposures for all frames must be standardized to the same setting. While it is tempting to adjust the camera to fully expose the dim areas along the terminator, this will grossly overexpose the fully illuminated lunar limb resulting in a garish and unpleasing image. I find the best mosaic results are achieved by setting the exposure to properly record the brightest areas of the lunar limb and letting the dark terminator fade out naturally. A balance must be achieved between shutter speed and gain to allow normal contrast across the entire mosaic.

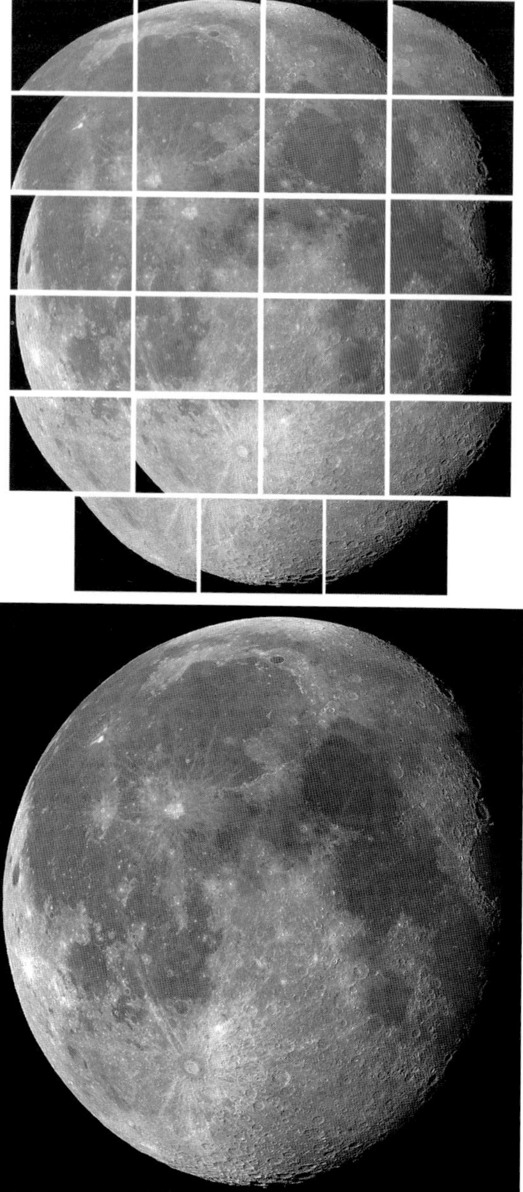

Fig. D.4 *The October 20, 2005 waxing gibbous Moon was captured using an Atik ATK-2HS camera on a Celestron-8 telescope at f/10. A total of 23 individual images, each stacked from a 600-frame AVI video, were combined using the iMerge mosaic-making program. The completed mosaic measures 2238 x 2409 pixels, and because it benefits from image stacking and wavelet processing in RegiStax, it displays greater detail than possible with single exposures using a larger format, conventional digital camera. Screen shots by Robert Reeves.*

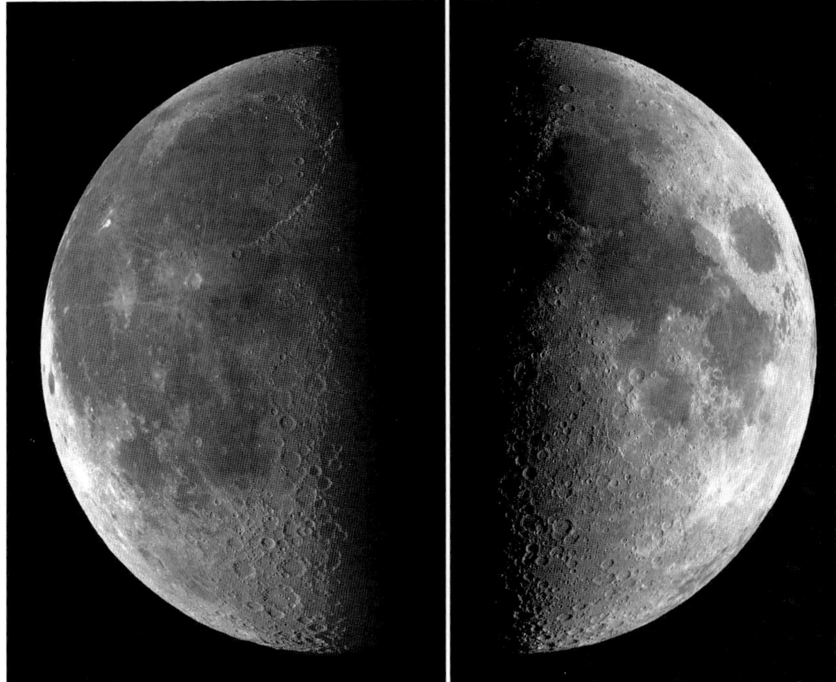

Fig. D.5 *The third- and first-quarter Moons were both captured using a Celestron-8 telescope at f/10. The August 26, 2005 third-quarter Moon was assembled from 16 images taken with an Atik ATK-2HS camera while the March 17, 2005 first-quarter Moon was composed of 24 images taken with a Philips ToUcam 840. Photos by Robert Reeves*

I use 600 frames of video for each segment of the mosaic. My cameras are all USB 1.1 and thus are used at five frames per second. This results in a two-minute exposure time, a practical limit that allows capturing an adequate number of frames without taxing the telescope's tracking ability. Once the basic exposure for all the segments is set, start the mosaic on one side of the Moon's northern limb and begin crisscrossing in an overlapping checkerboard pattern. Use a generous amount of overlap, 20 to 25 percent, to allow both navigation across the Moon and common ground between images to align the mosaic segments.

D.3 Creating the Mosaic

Regardless of how you assemble the mosaic, the conversion of the AVI files into still pictures begins with RegiStax (see Appendix B). Once the first video is processed, RegiStax can be set to "remember" the chosen parameters allowing the remaining videos to processed into still images using those same processing parameters with as few as three mouse clicks per video.

After the video frames are stacked there may be a narrow gray or white border around the edge of the completed image because small tracking errors or unsteady seeing caused the target to shift slightly during the video capture period. A means is needed to crop this border from each segment or it may appear on the mosaic. For this task, I recommend Irfanview, a freeware available over the web.[1] I have used it for years and consider this jewel created by Irfan Skiljan to be one of the most underrated Windows programs available. Not only is it a BMP, GIF, JPG, PNG, and TIF image-viewing program that is superior to the native Windows viewing program (assign it as the default image viewing utility in the Windows folder options), it will also play many movie formats, including the AVIs created by a webcam. Irfanview is also an excellent image-processing program that can crop, rotate, resize, adjust brightness and contrast, sharpen, and perform many other basic operations.

There are a number of processing options to turn a series of lunar images into a single high-resolution mosaic using a freeware program. One of the most elegant methods is to use iMerge, a freeware program created by Jon Grove. In the modern era of rampant bloatware, it seems a little odd to download a Windows program that is only 36 Kb in size and expect it to actually do something. But Grove applies his skills well and this small program produces big results. The program is simplicity itself. Simply drag and drop mosaic segments into place then save the completed mosaic as a single image.

So what is the bottom line? After purchasing the camera, telescope adapter, and infrared filter, what does it cost to create NASA-like high-resolution lunar mosaics? Nothing! As demonstrated here, the camera control, video processing, image processing, and mosaic-making software are all free. All that remains is to invest time to learn these easy programs and get out under the night sky and capture some video. Give it a try!

---

[1] www.irfanview.com

Appendix E Webcam Astrophotography Resources

Webcams and Webcam-Style Astrophotography Cameras

Atik Cameras (ATK-1C, ATK-1HS II, ATK-2C, ATK-2HS) http://www.atik-instruments.com

Celestron NexImage http://celestron.com

Logitech Webcams (Quickcam 3000, Quickcam 4000) http://www.logitech.com/

Meade LPI and DSI Cameras http://www.meade.com

Orion Starshoot Cameras http://www.telescope.com

Philips Webcams (Vesta, ToUcam, SPC900NC) http://www.Philips.com

SAC Cameras (SAC-4, SAC8.5) http://www.sac-imaging.com/

Essential Freeware for Webcam Astrophotography

Aberrator http://aberrator.astronomy.net/ A program that generates graphics showing the effects of aberrations and distortions on the appearance of images formed by a telescope undergoing a star test.

AVI2BMP http://www.daansystems.com/freestuff/ A program that allows the manual selection of the very best frames in a video sequence and writes them to a separate folder so they can be stacked by another program.

FITS Liberator www.spacetelescope.org/projects/fits_liberator/ APhotoshop plugin that allows working with FITS format images with Photoshop.

iMerge http://www.geocities.com/jgroveuk/iMerge.html A "donationware" program that allows creating seamless mosaics from multiple images by "dragging and dropping" images into place.

Irfanview http://www.irfanview.com A popular image-viewing program with image-capture and basic image-processing capabilities.

K3CCD Tools http://www.pk3.org/Astro/index.htm?k3ccdtools.htm A popular free-to-try-register-to-use webcam control and video image-processing program that works with most webcams and webcam-style astrocameras.

PoleFinder http://www.polarfinder.co.uk/ A freeware program for determining the location of the true celestial pole for instruments having polar scope equipped with Polaris Circle.

QCFocus http://www.astrosurf.org/astropc/qcam/programme.html A freeware program that measures a star image's full-width half-maximum to provide graphical and numeric data about the telescope's focus.

Registax http://registax.astronomy.net/ A very popular freeware image-stacking and processing program that allows stacking video frames and wavelet processing of the combined single image.

Virtual Dub http://www.virtualdub.org A powerful freeware AVI video repair, editing, and modification program.

Visual Spec http://astrosurf.com/vdesnoux/ A freeware program for processing astronomical spectra.

Wcctrl http://www.burri-web.org/bm98/ An independent camera settings control program that works in conjunction with K3CCD Tools. Allows camera adjustments without calling the default camera manufacturer's program that will overlay the K3CCD Tools screen.

Macintosh Webcam Video Capture and Processing Programs

http://www.unm.edu/~keithw/software/keithsAstroImager.html

http://www.unm.edu/~keithw/software/keithsImageStacker.html

PixInsight http://pleiades-astrophoto.com A popular powerful astronomical image-processing program that includes the SGBNR noise reduction program.

Windows Drivers for Philips cameras

http://www.driverscollection.com/?V=Philips&S=41

http://www.pc-cameras.philips.com/index_drivers.html

Macintosh Drivers for Philips cameras

Macintosh users will find no support for their computers on the Philips installation disk. However, Apple drivers for Philips Vesta and ToUcam models can be found at IOXperts (http://www.ioxperts.com/) and Bensoftware (http://www.bensoftware.com/help/toucam.html).

Commercial Software for Webcam Astrophotography

AstroVideo for COAA http://www.coaa.co.uk/astrovideo.htm An alternative free-to-try-register-to-use webcam control propgram.

AstroSnap http://www.astrosnap.com/ An alternative free-to-try-register-to-use webcam control program.

AstroStack http://www.astrostack.com/ An alternative free-to-try-register-to-use video image processing program.

Neat Image http://www.neatimage.com/ A commercial free-to-try-register-to-use image noise reduction program.

Internet Resources For Webcam Astrophotography

Ashley Roecklein's website contains an extensive collection of webcam-related software. http://astro.ai-software.com/

Dave Molyneaux's list of webcams and webcam properties http://homepage.ntlworld.com/molyned/images_of_web-cams.htm

Heavens Above, satellite pass prediction website http://heavens-above.com

Jan Timmermans' amazingly detailed website is filled with advice, tips, and demonstrations of webcam astrophotography topics. http://www.madpc.net/~firmament/

JPL Solar System Simulator allows viewing any object in the solar system for any given time and from any perspective. http://maps.jpl.nasa.gov/

QuickCam and Unconventional Astronomical Imaging Group mail list http://groups.yahoo.com/group/QCUIAG/

Steve Chambers' web site for "SC" modifications http://www.pmdo.com

ToUcam mail list http://groups.yahoo.com/group/ToUcam/

Astronomical Accessory Dealers

Adirondack Astronomy, one-stop source for all things related to digital imaging http://www.astrovid.com

Astronomik, filters and astronomical accessories http://www.astronomik.com/english/enghome.html

Baader Planetarium, filters and astronomical accessories http://www.baader-planetarium.de

Bob's Knobs, SCT collimation tools http://www.bobsknobs.com/

Camera lens adapters in the UK http://www.gtrain.freeserve.co.uk/Cphoto.htm

Hutech, telescope and accessory dealer known for IDAS light pollution filters http://www.science-center.net/hutech/

JMI, telescope and accessory dealer known for Motofocus electric focusers http://www.jimsmobile.com/

Kendrick, dew control products http://www.kendrick-ai.com/astro/

Losmandy Astronomical Products, excellent aftermarket telescope mounts http://www.losmandy.com

Lumicon, filters and astrophotography accessories http://www.lumicon.com

Mogg, webcam adapters and focal reducers http://webcaddy.com.au

Mosquito Magnet, on-site mosquito control http://www.mosquitomagnet.com

Outback Cooler, thermoelectric cooler for Meade DSI cameras. http://webcaddy.com.au

Paton Hawksley Education Ltd., source for Star Analyser webcam spectrograph http://www.paton-hawksley.co.uk/staranalyser.html

Rigil Systems, Puls-Guide accessory for illuminated reticle eyepieces www.flashlightsunlimited.com

ScopeStuff, hundreds of hard to find but essential items for telescopes and astrophotography http://www.scopestuff.com/

Scopetronix, a pioneer in the field of digital astrophotography http://www.scopetronix.com

Shoestring Astronomy, parallel port and USB autoguider adapters http://www.store.shoestringastronomy.com/products.htm

SpeckGRABER, optics cleaning device http://www.kinetronics.com/

Telescope Warehouse, new and used telescope and astrophotography gear http://www.telescope-warehouse.com/

Tele Vue Optics, outstanding eyepieces and Powermates http://www.televue.com

True Technology, flip-mirror finders http://www.trutek-uk.com/

WearGuard, cold weather outwear www.wearguard.com

Xantrex, maker of 12-volt to 120-volt inverters http://www.xantrex.com

Weather Information

http://adds.aviationweather.noaa.gov

http://adds.aviationweather.noaa.gov/turbulence/

http://adds.aviationweather.noaa.gov/winds/

http://cleardarksky.com/csk/

http://squall.sfsu.edu/crws/jetstream.html

http://www.cmc.ec.gc.ca/cmc/htmls/seeing_e.html

http://www.inquinamentoluminoso.it/worldatlas/pages/fig1.htm

http://www.ssec.wisc.edu/

http://www.wunderground.com/US/Region/US/Fronts.html

Appendix F
Personal Websites of Photo Contributors to this Book

Barkes, Steve: http://barkosoftware.com/GuideDog/

Burri, Martin: http://www.burri-web.org/bm98/index.htm

Croman, Russell: http://www.rc-astro.com/

Ferreira, Jim: http://www.lafterhall.com/astro.html

Foster, Steve: http://www.rmastro.com/astrophotos-nebulae.php

Hinson, Frank: http://www.rmastro.com/astrophotos-nebulae.php

Leadbeater, Robin: http://www.leadbeaterhome.fsnet.co.uk/astro.htm

Maxson, Paul: http://glory.gc.maricopa.edu/~pmaxson/

Matthias Meijer: http://www.home.zonnet.nl/m.m.j.meijer/astronomy.htm

Peach, Damian: http://www.damianpeach.com

Pujic, Zac: http://users.bigpond.net.au/metaplace/home.html

Reeves, Robert: www.robertreeves.com

Roecklein, Ashley: http://astro.ai-software.com/

Schaeffer, Glenn: http://home.houston.rr.com/gschaeffer/astropix/astropix.html

Stark, Craig: http://home.earthlink.net/~celstark/

Street, Dave: http://home.satx.rr.com/despage/

Tasto, Tim: http://www.cloudnet.com/~dsastro/astroimage.htm

Taylor, Matt: http://www.pbase.com/mataylor

Thommes, Jim: http://www.jthommes.com/Astro/

Ware, Jason: http://www.galaxyphoto.com/

Bibliography

Unk. "Advances in Long Exposure Video", http://www.cometdust.demon.co.uk/QCUIAG/LONGEX-POSURE/index.htm

Unk. MTD/PS-0237 "Application Note: Cover Glass Cleaning For Imaging Sensors", Eastman Kodak Company, Rochester, N. Y. 2001

Unk. "Barlows", Tele Vue Optics Article Page, http://www.televue.com

Unk. ""FITS Introduction", http://fits.gsfc.nasa.gov/fits_overview.html

Unk. "FITS Overview", http://heasarc.gsfc.nasa.gov/docs/heasarc/fits_overview.html

Unk. "Imaging Methods", Tele Vue Optics Article Page, http://www.televue.com

Unk. "Input-Output Standards", http://www.pctechguide.com/26interface3.htm

Unk. "Powermates", Tele Vue Optics Article Page, http://www.televue.com

Unk. "Resolution", http://www.buytelescopes.com/container.asp?dest=/promo/Resolution.htm

Unk. Telescope Optics & Pixel Size, http://www.ccd.com.xohost.com/ccd113.html

Unk. "Windows BMP Bitmap File Format", http://web.uccs.edu/wbahn/ECE1021/STATIC/REFER-ENCES/bmpfileformat.htm

Anderson, Dave, ed. "Recordable Formats", http://www.pctechguide.com/09cdr-rw.htm

Anderson, Dave, ed. "Recordable Formats", http://www.pctechguide.com/10dvd.htm

Anderson, Dave, ed. "Recordable Formats", http://www.pctechguide.com/10dvd3.htm

Arnholm, Carsten. "Polar Alignment", http://arnholm.org/astro/polar_alignment/index.html

Ashby, Peter. "Slo-Mo Focuser", http://www.geocities.com/grendel1960a/slo-mo/Slo-mo-focus-er.htm

Berry, Richard, and Burnell, James. *The Handbook of Astronomical Image Processing*, Willmann-Bell, Inc., Richmond, VA, 2000

Botha, Alwyn. "Why do I do Webcam Astrophotography?", http://www.webcam-astrophotogra-phy.com/why-do-webcam-astrophotography.html

Campbell, Peter. "Peter's Astronomy Page", http://home.austin.rr.com/campbelp/astro/

Carlin, Nils Olof. "More Power to your Barlow", http://www.atmsite.org/contrib/Carlin/barlow/

DeVercelly, William. "USB 2.0 – High-Speed USB – FAQ", http://www.everythingusb.com/usb2/faq.htm

Dixon, Douglas. "AVI Video File Formats", http://www.manifest-tech.com/media_pc/avi_formats.htm

Emerson, Richard. "Web Cam Astro Imaging Tutorial", http://hometown.aol.com/astrofx/WebCamTutorial.html

Emerson, Richard. "Web Cam Astro Imaging Tutorial Page 2", http://hometown.aol.com/astrofx/WebCamTutorial2.html

Evans, Tony. "Calibration", http://www.myastrostuff.com/CCD-Imaging/calibration.htm

Evans, Tony. "CCD Basics", http://www.myastrostuff.com/CCD-Imaging/ccdbasic.htm

Evans, Tony. "Noise Reduction", http://www.myastrostuff.com /CCD-Imaging/noise%20reduction.htm

Evans, Tony. "Projection", http://www.myastrostuff.com/CCD-Imaging/projection.htm

Evans, Tony. "Resolving Power", http://www.myastrostuff.com/CCD-Imaging/resolving.htm

Ireland, R. Scott. *Photoshop Astronomy*, Willmann-Bell, Inc., Richmond, VA, 2005

Leadbeater, Robin. "Using An IR Blocking Filter for Astro Imaging With a Webcam or Video Camera" http://www.leadbeaterhome.fsnet.co.uk/IR_FAQ.htm

Legault, Thierry. The Collimation, http://legault.club.fr/collim/html

Lloyd, Peter. "Resolution", http://homepage.ntlworld.com/peter.lloyd3/Saturn/Resolution.html

Meijer, M. M. J. "How to Identify Your CCD Chip", http://www.home.zonnet.nl/m.m.j.meijer/D_I_Y/CCD_type.htm

Meijer, M. M. J. "Pixel Size and Field of View", http://www.home.zonnet.nl/m.m.j.meijer/D_I_Y/pixel_size.htm

Meijer, M. M. J. "ToUcam SPC900NC", http://www.home.zonnet.nl/m.m.j.meijer/D_I_Y/spc900nc.htm

Nash, Dave. "Techniques: Capturing Images", http://www.davesastro.co.uk/techniques/capture2.html

Nash, Dave. "Techniques: Focussing", http://www.davesastro.co.uk/techniques/focussing.html

Peach, Damian. "An Examination of the Effects of Optical Aberrations, Obstruction and Seeing on Real Planetary Images", http://www.damianpeach.com/simulation.htm

Pujic, Zac. "The Galilean Moons in 2005", http://users.bigpond.net.au/metaplace/galilean.html

Reeves, Robert. *Introduction to Digital Astrophotography*, Willmann-Bell, Inc., Richmond, VA, 2005

Reeves, Robert. *Wide-Field Astrophotography*, Willmann-Bell, Inc., Richmond, VA, 2000

Roecklein, Ashley. Simple Hartmann Mask, http://astro.ai-software.com

Stark, Craig. "SAC/MOGG Focal Reducer", http://home.earthlink.net/~celstark/id7.html

Thorkildson, Ron. "Astronomical Seeing", http://www.rca-omsi.org/seeing.htm

Timmermans, Jan. "CCD Cleaning", http://www.madpc.net/~firmament/astro/ccd_cleaning.html

Timmermans, Jan. "Collimation of a Newtonian Telescope (or how I overcame my fear to collimate)", http://www.madpc.net/~firmament/astro/collimation.html

Timmermans, Jan. "Determining Field-Of-View (FOV)", http://www.madpc.net/~firmament/astro/hints_and_tips_fov.html

Timmermans, Jan, editor. "Imaging Cookbook for Beginners", http://www.madpc.net/~firmament

Timmermans, Jan. "Mastering Your Camera and Imaging Software", http://www.madpc.net/~firmament/astro/hints_and_tips_camera_handling.html

Timmermanns, Jan. "Polar Alignment", http://www.madpc.net~firmament/astro/polar_alignment.html

Walters, Steve. "Collimation Guide for Schmidt Cassegrain Telescopes", http://www.asterism.org/tutorials/tut14-1.htm

Webb, Colin. "Colin Webb's FAQ Page," http://www.cometdust.demon.co.uk/QCUIAG/FAQ/colinfaq.htm

Index